Natural Polymers and Biopolymers III

Natural Polymers and Biopolymers III

Editor
Sylvain Caillol

MDPI • Basel • Beijing • Wuhan • Barcelona • Belgrade • Manchester • Tokyo • Cluj • Tianjin

Editor
Sylvain Caillol
University of Montpellier
France

Editorial Office
MDPI
St. Alban-Anlage 66
4052 Basel, Switzerland

This is a reprint of articles from the Special Issue published online in the open access journal *Molecules* (ISSN 1420-3049) (available at: https://www.mdpi.com/journal/molecules/special_issues/biopolymer_III).

For citation purposes, cite each article independently as indicated on the article page online and as indicated below:

LastName, A.A.; LastName, B.B.; LastName, C.C. Article Title. *Journal Name* **Year**, *Volume Number*, Page Range.

ISBN 978-3-0365-7244-4 (Hbk)
ISBN 978-3-0365-7245-1 (PDF)

© 2023 by the authors. Articles in this book are Open Access and distributed under the Creative Commons Attribution (CC BY) license, which allows users to download, copy and build upon published articles, as long as the author and publisher are properly credited, which ensures maximum dissemination and a wider impact of our publications.

The book as a whole is distributed by MDPI under the terms and conditions of the Creative Commons license CC BY-NC-ND.

Contents

About the Editor vii

Preface to "Natural Polymers and Biopolymers III" ix

Sylvain Caillol
A Blooming Season for Natural Polymers and Biopolymers
Reprinted from: *Molecules* **2023**, *28*, 3207, doi:10.3390/molecules28073207 1

Dattatray K. Bedade, Cody B. Edson and Richard A. Gross
Emergent Approaches to Efficient and Sustainable Polyhydroxyalkanoate Production
Reprinted from: *Molecules* **2021**, *26*, 3463, doi:10.3390/molecules26113463 5

Wan Hazman Danial, Nur Fathanah Md Bahri and Zaiton Abdul Majid
Preparation, Marriage Chemistry and Applications of Graphene Quantum Dots–Nanocellulose Composite: A Brief Review
Reprinted from: *Molecules* **2021**, *26*, 6158, doi:10.3390/molecules26206158 61

Max Petitjean and José Ramón Isasi
Locust Bean Gum, a Vegetable Hydrocolloid with Industrial and Biopharmaceutical Applications
Reprinted from: *Molecules* **2022**, *27*, 8265, doi:10.3390/molecules27238265 81

Elizabeth Diederichs, Maisyn Picard, Boon Peng Chang, Manjusri Misra and Amar Mohanty
Extrusion Based 3D Printing of Sustainable Biocomposites from Biocarbon and Poly(trimethylene terephthalate)
Reprinted from: *Molecules* **2021**, *26*, 4164, doi:10.3390/molecules26144164 99

Aleksandra A. Wróblewska, H. Y. Vincent Ching, Jurrie Noordijk, Stefaan M. A. De Wildeman and Katrien V. Bernaerts
Radical Formation in Sugar-Derived Acetals under Solvent-Free Conditions
Reprinted from: *Molecules* **2021**, *26*, 5897, doi:10.3390/molecules26195897 113

Lucie Quinquet, Pierre Delliere and Nathanael Guigo
Conditions to Control Furan Ring Opening during Furfuryl Alcohol Polymerization
Reprinted from: *Molecules* **2022**, *27*, 3212, doi:10.3390/molecules27103212 121

Vasylyna Kirianchuk, Bohdan Domnich, Zoriana Demchuk, Iryna Bon, Svitlana Trotsenko, Oleh Shevchuk, Ghasideh Pourhashem and Andriy Voronov
Plant Oil-Based Acrylic Latexes towards Multisubstrate Bonding Adhesives Applications
Reprinted from: *Molecules* **2022**, *27*, 5170, doi:10.3390/molecules27165170 135

Muriel Józó, Róbert Várdai, András Bartos, János Móczó and Béla Pukánszky
Preparation of Biocomposites with Natural Reinforcements: The Effect of Native Starch and Sugarcane Bagasse Fibers
Reprinted from: *Molecules* **2022**, *27*, 6423, doi:10.3390/molecules27196423 149

Simona Rizzo, Tiziana Benincori, Francesca Fontana, Dario Pasini and Roberto Cirilli
HPLC Enantioseparation of Rigid Chiral Probes with Central, Axial, Helical, and Planar Stereogenicity on an Amylose (3,5-Dimethylphenylcarbamate) Chiral Stationary Phase
Reprinted from: *Molecules* **2022**, *27*, 8527, doi:10.3390/molecules27238527 165

Jorge Iván Castro, Diana Paola Navia-Porras, Jaime Andrés Arbeláez Cortés, José Herminsul Mina Hernández and Carlos David Grande-Tovar
Synthesis, Characterization, and Optimization Studies of Starch/Chicken Gelatin Composites for Food- Packaging Applications
Reprinted from: *Molecules* **2022**, *27*, 2264, doi:10.3390/molecules27072264 . . . . **175**

Dwi Hudiyanti, Muhammad Fuad Al Khafiz, Khairul Anam, Parsaoran Siahaan and Sherllyn Meida Christa
In Vitro Evaluation of Curcumin Encapsulation in Gum Arabic Dispersions under Different Environments
Reprinted from: *Molecules* **2022**, *27*, 3855, doi:10.3390/molecules27123855 . . . . **191**

Alexander V. Levdansky, Natalya Yu. Vasilyeva, Yuriy N. Malyar, Alexander A. Kondrasenko, Olga Yu. Fetisova, Aleksandr S. Kazachenko, et al.
An Efficient Method of Birch Ethanol Lignin Sulfation with a Sulfaic Acid-Urea Mixture
Reprinted from: *Molecules* **2022**, *27*, 6356, doi:10.3390/molecules27196356 . . . . **205**

Ali Fazli and Denis Rodrigue
Biosourced Poly(lactic acid)/polyamide-11 Blends: Effect of an Elastomer on the Morphology and Mechanical Properties
Reprinted from: *Molecules* **2022**, *27*, 6819, doi:10.3390/molecules27206819 . . . . **225**

Maria Teresa Ferrandez-Garcia, Antonio Ferrandez-Garcia, Teresa Garcia-Ortuño and Manuel Ferrandez-Villena
Assessment of the Properties of Giant Reed Particleboards Agglomerated with Gypsum Plaster and Starch
Reprinted from: *Molecules* **2022**, *27*, 7305, doi:10.3390/molecules27217305 . . . . **241**

About the Editor

Sylvain Caillol

Sylvain Caillol is a Research Director at CNRS. He graduated as an engineer from the National Graduate School of Chemistry of Montpellier in 1998 and then received his M. Sc. Degree in Chemistry from the University of Montpellier. He received his PhD degree in 2001 from the University of Bordeaux. Subsequently, he joined Rhodia Company, where he headed the Polymer Research Department in the Research Center of Paris. In 2007, he joined the CNRS at the Institute Charles Gerhardt of the University of Montpellier, where he started a new research topic dedicated to Green Chemistry and Biobased Polymers. He is co-author of more than 250 articles, patents and book chapters. He won the Green Materials Prize in 2018 and 2020 and he entered the list of World Top Scientists (Stanford) in 2021.

Preface to "Natural Polymers and Biopolymers III"

There is a blooming interest for natural polymers and biopolymers. The past 20 years have seen a booming number of articles and reviews describing the use of bio-resources as a starting point for original polymer chemistry. Indeed, the use of renewable resources could help the chemical industry to answer to some of the current challenges of our society: development facing global warming and limited fossil resources. Hence, the latest developments have not only created a library of polymeric materials exhibiting a wide range of properties to fulfill the requirements of various industrial applications, but have also improved our knowledge and understanding of the structure and reactivity of complex biomasses. Additionally, these biopolymers can address unmet needs and obtain new properties that cannot be achieved with petrobased chemicals. They could also help to avoid the use of harmful substances, thus contributing to restoring the chemical industry's sustainability.

This Special Issue on "Natural Polymers and Biopolymers" is prompted by the increasing attention that the field of "green polymers" is receiving. It presents cutting-edge research works focusing on the use of bio-resources for polymeric materials and shows how natural polymers and biopolymers, with their interesting and original properties, are destined to replace and outperform oil-based polymers. This themed issue can be considered as a collection of highlights within the field of Natural Polymers and Biobased Polymers which clearly demonstrate the increased interest in this field. We hope that this will inspire researchers to further develop this area and thus contribute to future more sustainable societies.

Sylvain Caillol
Editor

Editorial

A Blooming Season for Natural Polymers and Biopolymers

Sylvain Caillol

ICGM, University of Montpellier, CNRS, ENSCM, 34000 Montpellier, France; sylvain.caillol@enscm.fr

The year 2023 is particularly remarkable because we are celebrating the 25th anniversary of the 12 principles of Green Chemistry described in the groundbreaking book *Green Chemistry: Theory and Practice* co-authored by Paul Anastas and John C. Warner and published in 1998. The use of renewable resources is one of the most important of these principles. This approach is in particular very important for polymers which find very varied industrial applications. Hence, the past 25 years have seen a booming number of articles and reviews describing the use of bio-resources as a starting point for original polymer chemistry. Indeed, the use of renewable resources could help the chemical industry to answer to some of the current challenges of our society, including development facing global warming and limited fossil resources. Hence, the latest developments not only have created a library of polymeric materials exhibiting a wide range of properties to fulfill the requirements of various industrial applications, but also have improved our knowledge and understanding of the structure and reactivity of the complex biomass. Additionally, these biopolymers could allow addressing unmet needs and obtaining new properties that cannot be achieved with petroleum-based chemicals. They could also help avoiding the use of harmful substances, thus contributing to restoring the chemical industry's sustainability.

This Special Issue on "Natural Polymers and Biopolymers" is prompted by the increasing attention that the field of "green polymers" is receiving. It presents cutting-edge research works focusing on the use of bio-resources for polymeric materials and shows how natural polymers and biopolymers, with their interesting and original properties, are destined to replace and outperform oil-based polymers.

The first article is a perspective paper dedicated to the evaluation of production of sustainable production of polyhydroxyalkanoates (PHAs), which remain promising candidates for commodity bioplastic production [1]. This article focused on defining obstacles and solutions to overcome cost performance metrics that are not sufficiently competitive with current commodity thermoplastics. To that end, this review described various process innovations that build on fed-batch and semi-continuous modes of operation as well as methods that lead to high-cell-density cultivations. Finally, future directions for efficient PHA production and relevant structural variations are discussed.

The second paper is a perspective paper on the synthesis and applications of a graphene quantum dots–nanocellulose Composite. Graphene quantum dots (GQDs) are zero-dimensional carbon-based materials, while nanocellulose is a nanomaterial that can be derived from naturally occurring cellulose polymers or renewable biomass resources [2]. The unique geometrical, biocompatible, and biodegradable properties of both these remarkable nanomaterials have caught the attention of the scientific community in terms of fundamental research aimed at advancing technology. This study reviewed the preparation, marriage chemistry, and applications of GQDs–nanocellulose composites and unlocks windows of research opportunities for GQDs–nanocellulose composites.

The purpose of third paper, a review on locust bean gum, a vegetable hydrocolloid, was to report the structural characteristics of locust bean gum, its biosynthetic origin and its chemical isolation, and its applications and derivatives either by functionalization or cross-linking [3].

Citation: Caillol, S. A Blooming Season for Natural Polymers and Biopolymers. *Molecules* **2023**, *28*, 3207. https://doi.org/10.3390/molecules28073207

Received: 31 March 2023
Accepted: 31 March 2023
Published: 4 April 2023

Copyright: © 2023 by the author. Licensee MDPI, Basel, Switzerland. This article is an open access article distributed under the terms and conditions of the Creative Commons Attribution (CC BY) license (https:// creativecommons.org/licenses/by/ 4.0/).

The fourth paper considered new bio-based printable materials [4]. In this work, bio-based poly(trimethylene terephthalate) (PTT) blends were combined with pyrolyzed miscanthus biocarbon to create sustainable and novel filaments for extrusion 3D printing.

The aim of the fifth paper was to carefully investigate the stability of acetal-containing monomers, mainly focusing on reaction conditions during melt polycondensation [5]. Hence, it allows shedding new light on the underlying mechanism governing the observed behavior, thus aiding the development of solvent-free experiments, and later, material design.

Furfuryl alcohol is a promising bio-based furan derivative. The objective of the sixth paper was to study the control of ring opening polymerization of furfuryl alcohol according to different parameters in order to open the way to various applications [6].

Adhesion onto polypropylene (PP) and poly(ethylene terephthalate) (PET) is a challenge. The seventh paper successfully addressed this challenge with new emulsion polymers based on vegetable oil monomers [7]. In this study, the best-performing latex adhesives containing up to 45 wt. % of high-oleic soybean-oil-based monomer fragments demonstrated promising efficiency in the testing of PET to PP and coated-to-uncoated paperboard substrate pairs, resulting in substrate failure during the adhesive testing.

The eighth paper interestingly compared the effect of native starch and sugarcane bagasse fibers for the preparation of biocomposites with natural reinforcements [8]. It demonstrated that although the environmental benefit of the prepared biocomposites is similar, the overall performance of the bagasse-fiber-reinforced polylactide (PLA) composites is better than that offered by the PLA/starch composites.

The ninth paper focused on the chiral resolving ability of the commercially available amylose (3,5-dimethylphenylcarbamate)-based chiral stationary phase (CSP) toward four chiral probes representative of four kinds of stereogenicity [9]. This study confirmed that the use of the amylose-based CSP in HPLC is an effective strategy for obtaining the resolution of chiral compounds containing any kind of stereogenic element

The indiscriminate use of plastic in food packaging contributes significantly to environmental pollution, promoting the search for more eco-friendly alternatives for the food industry. The tenth paper studied five formulations of biodegradable cassava starch/gelatin films [10]. In the outcome of this study, an optimal formulation was obtained to develop cassava starch/gelatin-based films in a 53/47 ratio, plasticized with glycerol using the casting method that would meet the expectations of the model polyethylene film for food-packaging applications

Biopolymers, especially polysaccharides (e.g., gum arabic), are widely applied as drug carriers in drug delivery systems due to their advantages. Curcumin, with high antioxidant ability but limited solubility and bioavailability in the body, can be encapsulated in gum arabic to improve its solubility and bioavailability. The eleventh paper studied the released of curcumin in various conditions [11].

The aim of the twelfth paper was to experimentally and numerically optimize the process of the sulfation of ethanol lignin birch wood with a mixture of sulfamic acid and urea in a 1,4-dioxane medium and to characterize the structure and thermochemical properties of the sulfated ethanol lignin [12]. Using experimental and computational methods, the optimal conditions for the process of birch ethanol lignin sulfation with a sulfamic acid–urea mixture to provide a high yield of sulfated product (more than 96 wt. %) with a sulfur content of 8.1 wt. % were established.

Despite a number of studies addressing the issues of low deformation and the poor impact resistance of PLA blends without sacrificing stiffness and strength, it is still not clear how the phase structure and phase interaction in multicomponent blends contribute to the impact modification of ternary blends. The objective of the thirteenth article was to report on the effect of adding an intermediate elastomer phase and the blend composition on the morphology development of fully bio-based PLA-Polyamide (PA) blends prepared by melt blending [13].

Starch is a macroconstituent of many foods and 40% of starch is used in nonfood industries as an additive in cement, paper, gypsum, adhesives, bioplastics, composites, etc.

In particleboards, they are used as a substitute for binders such as urea-formaldehyde, phenol–formaldehyde, and other petroleum derivatives. The aim of the fourteenth article was to study the manufacturability of particleboards made from giant reed with gypsum plaster and starch following a method based on the wood industry dry process but with variations so that it can be produced in the particleboard industry [14].

Taken together, the articles in this issue reveal not only the growing attention paid to the use of renewable resources and the substitution of petroleum-based substances, following the trends in society, but also all the interest and promises of this chemistry for the improvement of current materials and the design of new sustainable materials.

Funding: This research received no external funding.

Conflicts of Interest: The author declares no conflict of interest.

References

1. Bedade, D.K.; Edson, C.B.; Gross, R.A. Emergent Approaches to Efficient and Sustainable Polyhydroxyalkanoate Production. *Molecules* **2021**, *26*, 3463. [CrossRef] [PubMed]
2. Danial, W.H.; Md Bahri, N.F.; Abdul Majid, Z. Preparation, Marriage Chemistry and Applications of Graphene Quantum Dots–Nanocellulose Composite: A Brief Review. *Molecules* **2021**, *26*, 6158. [CrossRef] [PubMed]
3. Petitjean, M.; Isasi, J.R. Locust Bean Gum, a Vegetable Hydrocolloid with Industrial and Biopharmaceutical Applications. *Molecules* **2022**, *27*, 8265. [CrossRef] [PubMed]
4. Diederichs, E.; Picard, M.; Chang, B.P.; Misra, M.; Mohanty, A. Extrusion Based 3D Printing of Sustainable Biocomposites from Biocarbon and Poly(trimethylene terephthalate). *Molecules* **2021**, *26*, 4164. [CrossRef] [PubMed]
5. Wróblewska, A.A.; Ching, H.Y.V.; Noordijk, J.; De Wildeman, S.M.A.; Bernaerts, K.V. Radical Formation in Sugar-Derived Acetals under Solvent-Free Conditions. *Molecules* **2021**, *26*, 5897. [CrossRef] [PubMed]
6. Quinquet, L.; Delliere, P.; Guigo, N. Conditions to Control Furan Ring Opening during Furfuryl Alcohol Polymerization. *Molecules* **2022**, *27*, 3212. [CrossRef] [PubMed]
7. Kirianchuk, V.; Domnich, B.; Demchuk, Z.; Bon, I.; Trotsenko, S.; Shevchuk, O.; Pourhashem, G.; Voronov, A. Plant Oil-Based Acrylic Latexes towards Multisubstrate Bonding Adhesives Applications. *Molecules* **2022**, *27*, 5170. [CrossRef] [PubMed]
8. Józó, M.; Várdai, R.; Bartos, A.; Móczó, J.; Pukánszky, B. Preparation of Biocomposites with Natural Reinforcements: The Effect of Native Starch and Sugarcane Bagasse Fibers. *Molecules* **2022**, *27*, 6423. [CrossRef] [PubMed]
9. Rizzo, S.; Benincori, T.; Fontana, F.; Pasini, D.; Cirilli, R. HPLC Enantioseparation of Rigid Chiral Probes with Central, Axial, Helical, and Planar Stereogenicity on an Amylose (3,5-Dimethylphenylcarbamate) Chiral Stationary Phase. *Molecules* **2022**, *27*, 8527. [CrossRef] [PubMed]
10. Castro, J.I.; Navia-Porras, D.P.; Arbeláez Cortés, J.A.; Mina Hernández, J.H.; Grande-Tovar, C.D. Synthesis, Characterization, and Optimization Studies of Starch/Chicken Gelatin Composites for Food-Packaging Applications. *Molecules* **2022**, *27*, 2264. [CrossRef] [PubMed]
11. Hudiyanti, D.; Al Khafiz, M.F.; Anam, K.; Siahaan, P.; Christa, S.M. In Vitro Evaluation of Curcumin Encapsulation in Gum Arabic Dispersions under Different Environments. *Molecules* **2022**, *27*, 3855. [CrossRef] [PubMed]
12. Levdansky, A.V.; Vasilyeva, N.Y.; Malyar, Y.N.; Kondrasenko, A.A.; Fetisova, O.Y.; Kazachenko, A.S.; Levdansky, V.A.; Kuznetsov, B.N. An Efficient Method of Birch Ethanol Lignin Sulfation with a Sulfaic Acid-Urea Mixture. *Molecules* **2022**, *27*, 6356. [CrossRef] [PubMed]
13. Fazli, A.; Rodrigue, D. Biosourced Poly(lactic acid)/polyamide-11 Blends: Effect of an Elastomer on the Morphology and Mechanical Properties. *Molecules* **2022**, *27*, 6819. [CrossRef] [PubMed]
14. Ferrandez-Garcia, M.T.; Ferrandez-Garcia, A.; Garcia-Ortuño, T.; Ferrandez-Villena, M. Assessment of the Properties of Giant Reed Particleboards Agglomerated with Gypsum Plaster and Starch. *Molecules* **2022**, *27*, 7305. [CrossRef] [PubMed]

Disclaimer/Publisher's Note: The statements, opinions and data contained in all publications are solely those of the individual author(s) and contributor(s) and not of MDPI and/or the editor(s). MDPI and/or the editor(s) disclaim responsibility for any injury to people or property resulting from any ideas, methods, instructions or products referred to in the content.

Review

Emergent Approaches to Efficient and Sustainable Polyhydroxyalkanoate Production

Dattatray K. Bedade [1], Cody B. Edson [2] and Richard A. Gross [1,2,*]

[1] Center for Biotechnology and Interdisciplinary Studies, Rensselaer Polytechnic Institute, 110 8th Street, Troy, NY 12180, USA; bedadedattu03@gmail.com
[2] New York State Center for Polymer Synthesis, Department of Chemistry and Chemical Biology, Rensselaer Polytechnic Institute, 110 8th Street, Troy, NY 12180, USA; edsonc@rpi.edu
* Correspondence: Grossr@rpi.edu

Abstract: Petroleum-derived plastics dominate currently used plastic materials. These plastics are derived from finite fossil carbon sources and were not designed for recycling or biodegradation. With the ever-increasing quantities of plastic wastes entering landfills and polluting our environment, there is an urgent need for fundamental change. One component to that change is developing cost-effective plastics derived from readily renewable resources that offer chemical or biological recycling and can be designed to have properties that not only allow the replacement of current plastics but also offer new application opportunities. Polyhydroxyalkanoates (PHAs) remain a promising candidate for commodity bioplastic production, despite the many decades of efforts by academicians and industrial scientists that have not yet achieved that goal. This article focuses on defining obstacles and solutions to overcome cost-performance metrics that are not sufficiently competitive with current commodity thermoplastics. To that end, this review describes various process innovations that build on fed-batch and semi-continuous modes of operation as well as methods that lead to high cell density cultivations. Also, we discuss work to move from costly to lower cost substrates such as lignocellulose-derived hydrolysates, metabolic engineering of organisms that provide higher substrate conversion rates, the potential of halophiles to provide low-cost platforms in non-sterile environments for PHA formation, and work that uses mixed culture strategies to overcome obstacles of using waste substrates. We also describe historical problems and potential solutions to downstream processing for PHA isolation that, along with feedstock costs, have been an Achilles heel towards the realization of cost-efficient processes. Finally, future directions for efficient PHA production and relevant structural variations are discussed.

Keywords: Polyhydroxyalkanoate; PHA; production; high cell density cultivations; productivity; downstream processing

1. Introduction

Polyhydroxyalkanoates (PHAs) are a class of biopolymers produced as intracellular energy/carbon storage materials that also possess versatile material properties. PHA was first discovered in *Bacillus megaterium* as granular inclusion bodies by the French scientist Lemoigne. Later these granular bodies were extracted and identified as poly (3-hydroxybutyrate) (PHB) [1]. PHAs remain of high interest as potential substitutes to conventional plastics in numerous fields of application due to their widespread applicability in various fields such as food packaging, agriculture, tissue-engineering scaffolds, bioresorbable implants and for drug delivery [2]. To date, various PHAs and their copolymers have been isolated from different bacterial species. More than 150 constituent repeat units have been reported to have been incorporated as PHA units along chains [3,4]. The monomer composition of PHAs can be altered so that the polymers have tailored physicochemical and mechanical properties [5].

PHAs are classified by the chain length of 3-hydroxyalkanoate (3HA) repeat units. PHAs with short chain-length repeat units (scl-PHA) contain primarily 4 and 5 carbon atoms in repeat units (e.g., 3-hydroxybutyrate [3HB] and valerate units [3HV]). In contrast, PHAs that consist of medium-chain-length PHA (mcl-PHA) have repeat units with chain lengths of 6-14 carbon atoms (e.g., poly [3-hydroxynonanoate]) [6]. The structure of the scl- and mcl-PHA are presented in Figure 1.

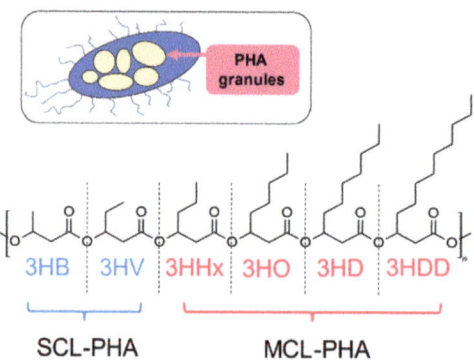

Figure 1. Polyhydroxyalkanoates with short chain-length repeat units (Scl-PHA) monomers include 3-hydroxybutyrate (3HB) and 3-hydroxyvalerate (3HV). Mcl-PHA monomers include 3-hydroxyhexanoate (3HHx), 3-hydroxyoctanoate (3HO), 3-hydroxydecanoate (3HD), and 3-hydroxydodecanoate (3-HDD). PHA granules accumulate within the cytoplasm of the bacteria cell.

Principal microbial producers of mcl-PHAs include *Pseudomonas* sp. Characteristic features of mcl-PHAs are that they are soft, ductile materials due to their low glass transition temperatures and crystallinities [7,8]. P(4-hydroxybutyrate) (P4HB), that has an extended methylene group between carbonyl and oxygen moieties, is just one example of a comonomer in 3HB copolymers that has shown promising material properties. Increasing the content of 4HB units in poly(3HB-*co*-4HB) tends to increase copolymer ductility while decreasing its melting point and % crystallinity. Comprehensive reviews have been published that elaborate on the effects of PHA composition on its physico-mechanical properties and it is not our intent herein to recapitulate this information [2,9–11].

A major bottleneck in the commercialization of PHAs is the high cost of production where metrics such as carbon conversion yield (g/g), titer or volumetric yield (g/L) and productivity (g/L/h) are critically important [12]. In addition to production metrics, low cost downstream processing methodologies and PHA manufacturing that meets cost-performance requirements have remained challenging. This has inspired researchers to work towards increasing PHA fermentation and downstream processing efficiency to reduce the overall cost [13–16].

To date, more than 300 bacterial species were identified that produce PHA under aerobic/anaerobic conditions and extremophiles such as halophiles that provide genetic diversity and diverse production conditions [17,18].

PHA is accumulated in the form of granules (size range from 0.2–0.5 μm) in the cytoplasm of bacteria. Based on PHA production capability, bacteria are classified as to: (1) whether PHA predominantly accumulates during the stationary phase under nutrient (e.g., nitrogen, phosphorous, oxygen and magnesium) limiting conditions or (2) PHA is formed during the growth phase without nutrient limitation [2]. *Pseudomonas putida*, *Pseudomonas oleovorans* and *Ralstonia eutropha* require nutrient limiting conditions for PHA production whereas recombinant *Escherichia coli* and *Alcaligenes latus* do not [19].

PHA synthesis occurs by the consecutive action of a β-ketoacyl-CoA thiolase (*PhaA*), acetoacetyl-CoA reductase (*PhaB*), and P(3HB) polymerase (*PhaC*) (Figure 2).

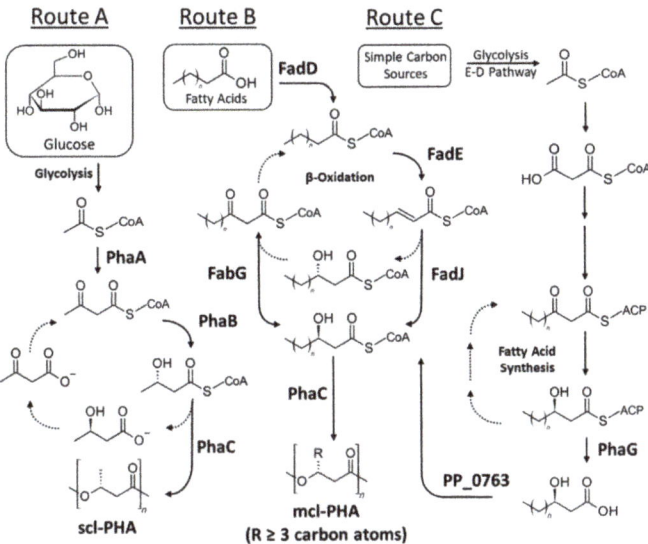

Figure 2. Metabolic routes for PHA biosynthesis.

The *phaCAB* operon encodes these enzymes and the upstream promoter of *phaC* transcribes the complete operon. The biosynthetic pathway involved in biosynthesis of mcl-PHA is displayed in Route B (Figure 2). Indeed, significant efforts have been made to engineer these enzymes with enhanced activity and vary their substrate specificity. Detailed discussions of metabolic and protein engineering of PHA biosynthetic enzymes are reviewed elsewhere [20].

For production, achieving high PHA cell contents during high cell-density cultivations (HCDC) is a key objective that leads to high product titers. HCDC was first established using yeasts that fabricate bioethanol and single-cell proteins [21]. Later, HCDC was explored for production in high titers of antibiotics and PHA using mesophilic strains [6,22]. HCDC processes are favored over low-density processes due to their advantages such as reduction of culture volume and residual liquids, reduced cost of production and lower capital investment [23,24]. Continuous and fed-batch cultivations are crucial operation modes used to attain HCDC of bacteria for PHA production. While cell dry weight (CDW) above 50 g/L are considered as high for production of recombinant proteins [21,25], cell densities and residual biomass above 100 g/L and 30–40 g/L, respectively, are considered as HCDC for PHA production [15,26–30].

Figure 3 illustrates the potential sustainable production of PHAs. This figure encompasses much of what will be discussed in this review. To meet sustainability metrics measured by life-cycle analysis, it is critical that carbon sources are derived from non-food sources such as lignocellulose; conversions of feedstocks to products occurs with high carbon conversion efficiencies; downstream processing is achieved with minimal process steps, inputs of chemicals (i.e., enzymes), and energy utilization; and, after use, the products are disposed of in bioactive environments such as composts or are degraded chemically or enzymatically to building blocks that can be re-used.

In this review article, we summarize HCDC methods for the biosynthesis of scl- and mcl-PHA and associated strategies that lead to increased productivity. We will also discuss approaches such as nutrient limitation, genetic and metabolic engineering, use of mixed culture and renewable carbon sources for enhancement of PHA production efficiency. Recent developments on cost-effective downstream processing are also discussed herein.

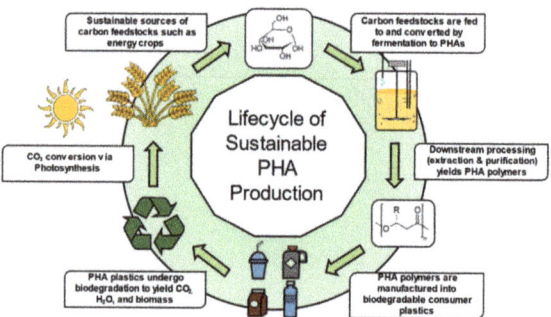

Figure 3. Using renewable, low-cost carbon sources such as energy crops or waste streams in combination with optimized fermentation strategies supports high yielding PHA production processes. The PHA polymers are processed and manufactured into consumer plastics that will biodegrade when disposed. The cycle begins again when biodegradation products of PHA plastics are consumed during photosynthesis or are recovered in waste streams from composting facilities.

2. Modes of Operations for Production of Polyhydroxyalkanoates (PHAs) in High Cell Density Cultivations

The productivity of PHAs is dependent on many factors such as the bacterial strain, carbon/nitrogen ratio, pH, temperature, cultivation time, and presence of micro and macro nutrients [31]. For optimization of PHA yield, different fermentation strategies such as batch, fed-batch and continuous processes have been used. Figure 4 provides a diagram of the overall PHA production pathway with options of alternative processes. Fermentation processes can be separated into two categories: continuous or discontinuous processes. The upper half of the figure illustrates discontinuous processes: batch fermentations, repeated fed-batch and fed-batch coupled with cell recycling. The bottom half of the figure illustrates continuous processes: single-stage chemostat, two-stage chemostat and a multi-stage bioreactor cascade. Unlike batch fermentations, continuous processes maintain static fermentation parameters. The use of multi-reactor fermentation strategies, especially multi-stage cultivation systems, are recommended to obtain high yields of PHA. R1–R5: Five continuous stirred tank reactors. F1, F3, F5, F7 and F9: Feed streams for supply of nutritional medium to the bioreactors R1, R2, R3, R4 and R5, respectively. F2, F4, F6 and F8: Continuous transfer of fermentation broth to the subsequent reactors. F10: Outlet stream containing final product. Microbial cell growth occurs in R1, whereas PHA accumulation takes place in R2–R5.

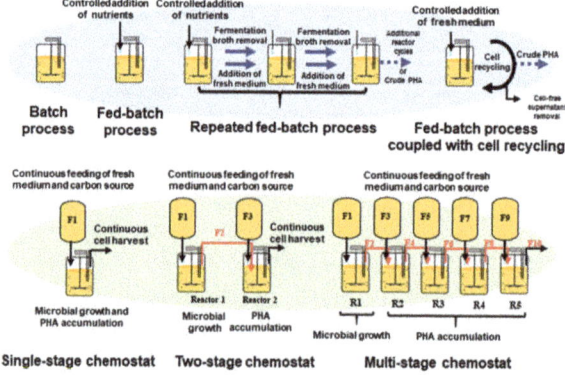

Figure 4. Process regimes for PHA fermentative synthesis.

2.1. Batch Cultivations

In batch cultivations, the carbon/nitrogen sources are added to the system at time zero of the cultivation. That is, during batch cultivations, additional nutrients are not added during the course of the fermentation [32]. Generally, batch fermentation processes result in low PHA yields which is partly attributed to PHA degradation that occurs at the later during cultivations [18]. Singh et al., investigated P3HB production from industrial sugar waste using *Bacillus subtilis* NG220. Cultivations resulted in 10.22 g/L of biomass that contained 51.8% PHA [33]. Rai et al., investigated the batch production of P(3-hydroxyoctanoate) from *Pseudomonas mendocina* using sodium octanoate. This resulted in very low biomass yield of 0.8 g/L with 31% of homopolymer production using sodium octanoate as sole carbon source [34].

2.2. Fed-Batch Fermentations

The fed-batch fermentation method was originally developed in the early 1900s by yeast producers such as cultivation of *Saccharomyces cerevisiae* [35]. Later, this concept was applied for production of antibiotics, amino acids, microbial cells, enzymes, growth hormones, vitamins, organic acids and PHA [36]. Fed-batch processes are extensively used for industrial fermentations due to its distinct advantages over other modes of operation of a bioreactor. In a fed-batch process, cells are grown under continuous feeding of carbon sources and essential nutrients at a certain rate until the desired cell density is attained. The feed solution containing carbon sources and essential nutrients maintains a specific growth rate that reduces by-product formation.

There are two types of fed-batch cultivations: product formation that is either growth-associated or that occurs under non-growth-associated conditions. PHA production is usually carried out in two phases, first, the growth phase is conducted such that cells reach the desired biomass. Polymer production occurs in the second phase in which all essential nutrients required for production are fed to the bioreactor [37]. In most cases, the second phase is conducted where an essential nutrient, such as phosphorus, nitrogen, oxygen and sulfur, is at a limiting concentration such that metabolic pathways supporting growth are suppressed and the cells respond by focusing their efforts on PHA accumulation to store carbon and energy. Fed-batch cultivations must have suitable strategies to supply carbon sources and other nutrients. Under circumstances where a carbon source for PHA becomes limiting, the rate of PHA degradation via a PHA depolymerase increases.

The use of fed-batch cultivations has proved valuable to achieve PHA production under HCDC conditions [29,38,39]. Cultivations of *Cupriavidus necator* under fed-batch HCDC has been a target of interest by numerous research groups and the results of this work are displayed in Table 1.

Mcl-PHAs have also been produced by HCDCs and Table 1 summarizes the results of published work. The pH stat and DO (dissolved oxygen) stat involves maintaining the pH and DO at certain levels during the fermentation process. Lee et al., reported a strategy for achieving HCDCs using *P. putida* as the microbial catalyst and oleic acid as the carbon source [15]. Oleic acid feeding was controlled by a pH-stat during the growth phase and a DO-stat in the polymer production phase. This results in a total biomass, %-PHA in the CDW and overall productivity were 141 g/L, 51% and 1.9 g/L/h, respectively [15].

A two-stage fed-batch HCDC of *P. putida* KT2440 using glucose and nonanoic acid was reported by Davis et al. Cells were grown on glucose in biomass accumulation phase, and nonanoic acid was fed in the PHA production phase. Using this two-stage feeding strategy, *P. putida* KT2440 accumulated a total biomass, %-PHA CDW content and overall productivity of 102 g/L, 32% and 0.98 g/L/h, respectively [26]. The authors claimed that, this two-stage feeding strategy resulted in the highest ever reported value of biomass for a *P. putida* strain.

While cell concentration increases during a fed-batch fermentation, one can impose a slowly decaying specific growth to attain high cell densities while preventing other perturbations that could result from rapid decreases in the specific growth rate. Maclean et al., reported a decaying exponential feeding of nonanoic acid during a *P. putida* KT2440 cultivation to form an mcl-PHA. A linear and quadratic decaying exponential feeding rate were used to control biomass accumulation and, subsequently, to control the oxygen uptake rate of the cells. The latter strategy resulted in in a total biomass, CDW, % PHA content and overall productivity of 109 g/L, 63% (i.e., 69 g/L PHA) and 2.3 g/L/h, respectively [27]. The larger final biomass concentration and mcl-PHA content is explained by the directly correlation between the highest rates of cell formation and oxygen uptake.

PHA-producing organisms (e.g., *R. eutropha*) can use acetic, propionic and butyric acids as primary substrates for both biomass accumulation and PHA formation. Propionate/propionic acid introduced in culture media serves as a precursor for the formation of 3HV units in P(3HB-*co*-3HV). The incorporation of 3HV units increases the ductility while decreasing the copolymer melting point relative to P3HB. Huschner et al., reported fed-batch cultivations of *R. eutropha* that functioned to decrease the toxicity of organic acid substrates. The rate of organic acid feeding was pO_2-dependent. This approach resulted in highly reproducible cultivations that reached a total biomass, %-P(3HB-*co*-5.6 mol%3HV) content and overall productivity of 112 g/L, 83% and >2 g/L/h, respectively [28].

Yamane et al., reported a fed-batch cultivation of *A. latus* where sucrose, inorganic elements and an ammonia solution were fed into a bioreactor by the pH-stat method. This work highlighted the importance of carbon and nitrogen feed concentrations in obtaining high PHA productivity. Consumption rates were used to inform when to supply carbon/nutrient sources that enabled nutrient concentrations to be maintained at nearly constant levels throughout cultivations. The feeding solutions were supplied based on their consumption rates, thereby maintaining the nutrient concentrations at nearly constant levels during the fermentation. This approach resulted in a total biomass, PHA content in the CDW and overall productivity of 143 g/L, 50% and 3.97 g/L/h, respectively [40].

PHB production by *C. necator* DSM 545 from glucose by a two-phase fed batch cultivation was reported by Mozumder et al. The first phase was dedicated to biomass production after which, in the second phase, a specific PHB production rate and nutrient-limiting conditions induced PHB formation. Process optimization led to a total biomass, PHB content in the CDW and overall productivity of 164 g/L, 76.2% and 2.03 g/L/h, respectively [41].

Burkholderia sacchari was identified as an efficient micro-organism for P3HB production. Sucrose from sugarcane was the primary carbon source while γ-butyrolactone (GBL) was used as a co-substrate for the formation of 4HB units. P3HB formed in the absence of GBL reached a PHB concentration and overall productivity of 36.5 g/L and 1.29 g/L/h, respectively [42]. However, addition of GBL results in P(3HB-*co*-1.6 mol%4HB) at a concentration and volumetric productivity of 54 g/L and 1.87 g/L/h, respectively [42].

Haas et al., reported the use of a membrane bioreactor and HCDC for P3HB formation. PHB was formed by continuously feeding of a synthetic medium with 50 g/L glucose to *C. necator*. This strategy resulted in a total biomass, P3HB content in the CDW and overall productivity of 148 g/L, 76% and 3.10 g/L/h, respectively [43]. The bacterium rapidly consumed fed sugar resulting in low contents in the final medium.

Cell recycle has been successfully implemented in fed-batch and continuous cultures resulting in HCDC and efficient PHA formation [44,45]. Ienczak et al., demonstrated that by coupling repeated fed-batch cultivations with cell recycle, HCDC of *C. necator* DSM 545 was achieved from glucose and fructose (90 g/L) [45]. Culture media depleted of nutrients was removed from the bioreactor without loss of cells during recirculation. The results showed that total biomass, P3HB yield, P3HB CDW content and overall productivity reached 61.6 g/L, 42.4 g/L, 68.8% and 1 g/L/h, respectively [45]. It was noteworthy that, by this cultivation strategy, carbon source concentration about 7-fold below that used in other studies was effectively converted to product. This is particularly meaningful for waste feedstocks that contain low sugar concentrations. Schmidt et al., used *C. necator* DSM 545 for P(3HB-*co*-3HV) by external cell recycling under a production phase where nitrogen limitation was imposed. The glucose concentration fed to the culture medium simulated that often found in agro-industrial wastes (90 g/L). To induce 3HV formation, propionic acid was used as a co-substrate. The results showed that the total biomass, PHB content in the CDW and overall productivity reached 80 g/L, 73% and 1.24 g/L/h, respectively [46].

Rodríguez-Contreras et al., substituted glucose for glycerol during the production of P3HB by *C. necator* and *B. sacchari*. The maximum biomass, P3HB CDW content and productivity for these strains were 68.6 g/L, 64.6%, 0.76 g/L/h and 43.8 g/L, 10.2%, 0.08 g/L/h, respectively [47]. The isolated bacterium *Zobellella denitrificans* MW1 that possess a high capability to accumulate PHA from glucose, was assessed in a pilot scale reactor (42 L) for P3HB formation from glycerol. Using fed-batch cultivation, the optimized feeding strategy of glycerol and sodium chloride results in 81.2 g/L total biomass, 67% PHA content and 1.09 g/L/h productivity [48].

Chanprateep et al., reported the formation of P(3HB-*co*-4HB) during fed-batch HCDC of *C. necator* A-04. Fructose served as the carbon sources for biomass accumulation while 1,4 butanediol functioned as a 4HB precursor. The authors varied the molar ratios of fructose to 1,4-butanediol and, consequently, altered the composition and productivity of P(3HB-*co*-4HB) formation. The culture in which P(3HB-*co*-38% mol 4HB) was produced reached a total biomass, PHA CDW content and overall productivity of 112 g/L, 65% and 0.76 g/L/h [49].

Le Meur et al., reported increased scl-PHA productivity by recombinant *E. coli*. Glycerol served as the carbon source for biomass accumulation while 4HB functioned as a precursor for 4HB units. Pulse, linear and exponential feeding strategies were evaluated, the exponential feeding of glycerol and butyric acid was found to be highly reproducible and results in biomass, PHA CDW and overall productivity were 43.2 g/L, 33% and 0.207 g/L/h [50].

Stanley et al., reported pH-based and pulse feeding strategies to improve PHA yields during *Halomonas venusta* cultivations. Usually, the fermentation broth pH moves toward lower values during microbial growth due to the production of organic acids. In contrast, the medium pH increases under carbon source limiting conditions due to the production of ammonium ions during protein catabolism. Consequently, the feed pump was altered such that, in the event that the pH increased above a set value (7.05 in the current study), this cues the automated feeding of the carbon source. Using this pH-based strategy, the authors reported a total biomass and PHA CDW content of 66.4 g/L and 39%, respectively [51]. Also, they found that, when the maximum utilization of glucose was reached, a single pulse (100 g/L in this study) was used to increase the available glucose in the bioreactor. The single pulse feeding approach resulted in an accumulated biomass and %-PHA content of 37.9 g/L and 88%, respectively [51]. The increase in PHA content in pulse feeding could be due to increased glucose flux towards PHA synthesis. *B. sacchari* IPT 189 was cultivated for P(3HB-*co*-3HV) production using sucrose and propionic acid and a two-stage bioreactor process [52]. During the first stage, a balanced culture medium was used for growth up to sucrose exhaustion. The second stage constituted feeding sucrose/propionic acid solution to the bioreactor. The sucrose/propionic acid ratio was varied while the feed flow rate was kept constant. The results showed that, by increasing the ratio of sucrose to propionic acid to 30:1, the %-PHA [P(3HB-*co*-10 mol%3HV)] cell content and productivity reached 60% and 1.04 g/L/h, respectively [52].

In pursuit of HCDC for P(3HB-*co*-3HV) production by *Aeromonas hydrophila* 4AK4, Chen et al., employed the cofeeding of glucose/lauric acid in a two-stage fermentation. Lauric acid pulsed feeding results in 20 g/L residual carbon source concentration and a total biomass, %-PHA CDW content and productivity of 50 g/L, 50% and 0.54 g/L/h, respectively [53].

Blunt et al., reported an oxygen-limited fed-batch cultivation process for enhanced productivity of mcl-PHAs using *P. putida* LS46. They used octanoic acid as the carbon source and a bench-scale (7 L) bioreactor. The resulting total biomass, %-PHA CDW content and productivity reached 29 g/L, 61% and 0.66 g/L/h, respectively [54]. The relatively low biomass accumulation may be due to the toxicity of octanoic acid to *P. putida* LS46 cells. Gao et al., conducted a fed-batch cultivation of *P. putida* KT2440 mcl-PHA production with a co-feed mixture of decanoic and acetic acids [55]. Acetic acid functioned to prevent decanoic acid crystallization. To identify co-feed ratios that would result in higher mcl-PHA yields, different ratios of decanoic acid/acetic acid/glucose was used. With the optimized ratio (5:1:4), the total biomass, %-PHA CDW content and overall productivity reached 75 g/L, 74% and 1.16 g/L/h, respectively [55].

Sun et al., reported the formation of mcl-PHA by *P. putida* KT2440 by co-feeding glucose and nonanoic acid during a carbon-limited fed-batch cultivation. Exponential and, thereafter, linear feeding of 1:1 (w/w) nonanoic acid: glucose resulted in a total biomass, %-PHA CDW content and overall productivity of 71 g/L, 56% and 1.44 g/L/h, respectively [56]. Cerrone et al., reported the HCDC of *P. putida* CA-3 by a two-stage fermentation of co-substrates decanoic and butyric acid [57]. To enhance the mcl-PHA volumetric productivity, the cells were initially grown on butyric acid (biomass growth phase) and, subsequently, during the PHA production stage, the carbon source used was a mixture of butyric and decanoic acid (20:80 v/v ratio). This strategy resulted in a total biomass, %-PHA CDW content and overall productivity reached 71.3 g/L, 65% and 1.63 g/L/h respectively [57].

Sun et al., reported the HCDC of *P. putida* KT2440 for mcl PHA formation from nonanoic acid. An exponential growth rate ($\mu = 0.15$ h^{-1}) under nonanoic acid-limited conditions resulted a total biomass, %-PHA CDW content and overall productivity of 70 g/L, 75% and 1.11 g/L/h, respectively [58]. However, by increasing the exponential feed rate to $\mu = 0.25$ h^{-1}, the overall productivity increased (1.44 g/L/h), however, the biomass (56 g/L) and mcl-PHA content (67%) decreased due to the higher oxygen demand [58]. Diniz et al., studied different feeding strategies such as pulse feed followed by constant feed, and exponential feed to produce mcl-PHA using *P. putida* IPT 046 [59]. The exponential feeding strategy results in total biomass of 40 g/L with 21% mcl-PHA content. However, under phosphate limitation, biomass accumulation, the CDW content of mcl-PHA and overall productivity reached 50 g/L, 63% and 0.8 g/L/h, respectively [59].

Cultivation of *P. oleovorans* ATCC 29347 was conducted under pH-stat fed-batch conditions using octanoic acid as the feedstock. The resulting total biomass, %-PHA in the CDW and overall productivity were 63 g/L, 62% and 1 g/L/h [60]. Kim et al., reported *P. putida* BM01 cultivation by a two-stage fed-batch process. These workers co-fed glucose and octanoate during both biomass growth and PHA production. This strategy resulted in a total biomass, %-PHA in the CDW and overall productivity of 55 g/L, 66% and 0.90 g/L/h, respectively [61].

With the objective of improving the distribution of both carbon and energy, Andin et al., coupled *P. putida* KT2440 growth and mcl-PHA production from fatty acids [62]. Experimental data validated a model that describes the energy flux distribution and carbon utilization in *P. putida* KT2440 during the simultaneous processes of growth and PHA formation. This approach explored the possibility of shifting available carbon and energy to PHA formation during the production phase. The resulting fed-batch culture had a total biomass, %-PHA in the CDW and overall productivity of 125.6 g/L, 54.4% and 1.01 g/L/h, respectively [62]. Thus, one can couple PHA formation and growth when substrate catabolism occurs via β-oxidation.

Dey and Rangarajan reported a HCDC of *C. necator* (MTCC 1472) on sucrose for P3HB formation by a fed batch fermentation [63]. Under nitrogen limited fed-batch cultivation, the concentration during feeding of sucrose was varied from 100–200 g/L. The total biomass, %-P3HB in the CDW and overall productivity reached 38 g/L, 62% and 0.58 g/L/h, respectively, at a dilution rate of 0.046 h^{-1} and by feeding a 200 g/L sucrose solution [63]. The authors claimed their approach provided an economically attractive route to PHB production.

To maximize the P3HB accumulation rate of *Azohydromonas lata* DSM 1123, Penloglou et al., adopted an intensified fed-batch process based on a model. The models were validated to determine optimal feeding and operating conditions that optimize P3HB productivity. By a continuous feeding strategy under non-limiting nitrogen conditions, a maximum PHB CDW content of 94% overall productivity of 4.2 g/L/h, was reported [64]. However, the authors did not provide the value of the total biomass concentration accumulation.

Table 1. PHA production by high cell-density cultivation (HCDC) fed-batch operations.

Microorganisms	Stages of Fermentation	PHA Type	Carbon Source	Biomass (g/L)	PHA Content(%)	Overall Productivity (g/L/h)	Carbon Conversion Efficiency ($Y_{p/s}$) g/g	Reference
P. putida KT2440ATCC 47054	One stage	PHO/PHD	Oleic acid	141	51	1.91	NA	[15]
		PHN	Glucose/nonanoic acid	102	32	0.97	0.56	[26]
		PHN	Nonanoic acid	109	63	2.3	NA	[27]
		PHD	Decanoic acid/glucose	75	74	1.16	0.86	[55]
		PHN	Glucose/nonanoic acid	71	56	1.44	0.66	[56]
		PHN	Glucose/nonanoic acid	56	67	1.44	0.60	[58]
	Two stage	mcl-PHA	Oleic acid/linoleic acid/palmitic acid/stearic acid	125	54.4	1.01	0.70	[62]
R. eutropha H16	Two stage	P(3HB-co-3HV)	Acetic acid/propionic acid/butyric acid	112	83	2	NA	[28]
Alcaligenes latus DSM 1122	One stage	PHB	Sucrose	143	50	3.97	NA	[40]
Cupriavidus necator DSM 545	Two stage	PHB	Glucose	164	76.2	2.03	NA	[41]
	One stage	PHB	Glucose	148	76	3.1	0.33	[43]
		PHB	Glucose/Fructose	61.6	68.8	1.0	NA	[45]
		P(3HB-co-3HV)	Glucose/propionic acid	80	73	1.24	NA	[46]
		PHB	Glycerol	68.56	64.55	0.76	0.34	[47]
Burkholderia sacchari DSM 17165	One stage	P(3HB-co-4HB)	Saccharose/γ-butyrolactone	77	72.6	1.87	0.275	[42]
		PHB	Glycerol	43.79	10.22	0.08	0.41	[47]
Zobellella denitrificans MW1	NA	PHB	Glycerol	82.2	67	1.09	NA	[48]
C. necator A-04	One stage	P(3HB-co-4HB)	Fructose/1,4 butanediol	112	65	0.76	NA	[49]
Halomonas venusta KT832796	One stage	PHB	Glucose	38	88	0.25	0.22	[51]
Burkholderia sacchari IPT 189	One stage	P(3HB-co-3HV)	Sucrose/propionic acid	NA	60	1.04	0.25	[52]
Aeromonas hydrophila 4AK4	One stage	P(3HB-co-3HHx)	Glucose/lauric acid	50	50	0.54	NA	[53]
P. putida LS46	One stage	PHO	Octanoic acid	29	61	0.66	0.62	[54]
P. putida CA-3	Two stage	PHD	Butyric acid/decanoic acid	71.3	65	1.63	0.55	[57]
P. putida IPT046	One stage	mcl-PHA	Glucose/fructose	50	63	0.80	0.19	[59]
P. oleovorans ATCC 29347	One stage	PHO	Octanoic acid	63	62	1	NA	[60]
P. putida BM01	Two stage	PHO	Octanoic acid/glucose	55	65	0.90	0.40	[61]
Cupriavidus necator MTCC 1472	One stage	PHB	Sucrose	37.56	61.82	0.58	0.20	[63]
Azohydromonas lata DSM 1123	One stage	PHB	Sucrose	-	94	4.2	0.15	[64]

$Y_{p/s}$: PHB yield to substrate (g PHB produced per g substrate consumed); %-PHA: Final intracellular PHA content (g/g) of DCW; NA: Data not available, PHB: Polyhydroxybutyrate, P(3HB-co-3HV): Poly (3-hydroxybutyrate-co-3-hydroxyvalerate), PHO: Polyhydroxyoctanoate, PHD: Polyhydroxydecanoate, PHN: Polyhydroxynonanoate, P(3HB-co-4HB): Poly (3-hydroxybutyrate-co-4-hydroxybutyrate), P(3HB-co-3HHx): Poly(3-hydroxybutyrate-co-3-hydroxyhexanoate), P(3HB -co-4HB): Poly (3-hydroxybutyrate-co-4-hydroxybutyrate)

2.3. Continuous Culture

By this technique, the rate of microbial growth is constant under steady-state conditions. A continuous cultivation process that runs at high specific growth rates can provide high productivities. Furthermore, continuous cultivations are desirable since they substantially decrease the frequency of bioreactor shutdown and cleaning operations. Also, continuous cultivation processes circumvent wash-out even at high dilution rates. This can lead to high productivity and concentrations of the product. To minimize the disruption of normal microbial cellular behavior, continuous or semi-continuous processes can be implemented in place of batch process. Continuous cultivation processes are characterized by the continuous addition at a constant flow rate of fresh media to the bioreactor which provides the cells with fresh nutrients. To keep the bioreactor working volume constant, products and effluents are continuously removed. Representative outcomes of PHA production under continuous HCDC conditions are displayed in Table 2.

Jung et al., reported a continuous two-stage process by which *P. oleovorans* converts *n*-octane to mcl-PHA. Two stage fermentations offer the opportunity to focus on biomass accumulation in the first bioreactor and PHA accumulation in a second bioreactor. In the first (D1) and second (D2) bioreactors, the dilution rate were $0.21\ h^{-1}$ and $0.16\ h^{-1}$, respectively. These conditions resulted in a total biomass, %-mcl-PHA content in the CDW and overall productivity of 18 g/L, 63% and 1.06 g/L/h, respectively [65].

Atlic et al., reported a continuous cultivation of *C. necator* for P3HB production from glucose [66]. The multistage reaction system consisted of five bioreactors in series. The first bioreactor functioned for biomass accumulation; thereafter, the fermentation broth was continuously fed into subsequent reactors for P3HB production under nitrogen limiting conditions. The dilution rate ($0.139\ h^{-1}$) for the cascade experiment was substantially higher when compared to the corresponding 2-stage process ($0.075\ h^{-1}$) since the authors assumed that the five-reactor series would have a relatively higher product throughput. Upon reaching steady state conditions, the total biomass, %-P3HB CDW content and overall productivity reached 81 g/L, 77% and 1.85 g/L/h, respectively [66]. This work highlights how, by adopting a continuous process using a series of bioreactors, high product titers and productivity can be achieved.

As above, Horvat et al., reported a continuous cultivation of *C. necator* for P3HB production from glucose [67]. The multistage reaction system consisted of five bioreactors in series. For the first bioreactor in the cascade, modelling was based on maintaining a nutrient balanced system with continuous biomass production. The second bioreactor adopted a model for process control using two substrates. Control of the next three bioreactors aimed to achieve high P3HB formation under nitrogen limitation with continuous glucose feeding. They reached a total biomass, %-P3HB in the CDW and overall productivity of 80 g/L, 77% and 2.14 g/L/h, respectively [67]. Du et al., adopted a continuous two-stage cultivation where the first and second bioreactors were optimized for biomass and P3HB accumulation, respectively [68]. P3HB formation in the second bioreactor was under nitrogen limiting conditions. After optimization of the dilution rates ($0.075\ h^{-1}$) and carbon source (50 g/L in first stage and 500 g/L in second stage), they reported a total biomass, %-P3HB content in the CDW and an overall productivity of 50 g/L, 73% and 1.23 g/L/h, respectively [68].

To study the kinetics of P3HB synthesis, Du et al., performed a continuous cultivation of *R. eutropha* containing two bioreactors in series [69]. In the first bioreactor, *R. eutropha* cells were cultivated under limiting glucose (feeding solution concentration 50 g/L) conditions. In the second bioreactor, P3HB accumulation occurs with excess carbon source (feeding solution concentration 500 g/L) and limiting nitrogen conditions. The specific P3HB production rate was dependent on the C/N molar ratio such that, the C/N ratio of 30 in PHB production phase gave optimal results: biomass accumulation reached 32.6 g/L and %-PHB content in dried cells was 75% [69]. Khanna and Srivastava explored the formation of P3HB formation by *Wautersia eutropha* NRRL B-14690 under continuous cultivation conditions [70]. P3HB formation was induced by imposing nutrient limiting conditions. Minimal P3HB formed during the exponential growth phase. P3HB formation in the second stage was increased by low dilution rates. Under these conditions the authors reported a total biomass, %-P3HB in the CDW and overall productivity of 49 g/L, 51% and 0.42 g/L/h, respectively [70].

Egli et al., used chemostat culture conditions to investigate PHA formation by *P. putida* GPo1. This work revealed that, when both carbon and nitrogen simultaneously limit growth, PHA formation can occur. Under these conditions, studies were conducted to determine how the C/N ratio, substrate type and the cell growth rate affected product formation. A correlation was found between increased *P. putida* GPo1 PHA formation and prolonged carbon and nitrogen limiting cultivation conditions [71]. Similarly, Zinn et al., reported on a cultivation of *R. eutropha* DSM 428 under both nitrogen and carbon limiting conditions. The carbon sources used were butyric and/or valeric acid while ammonium served as the source of nitrogen. This strategy results in a cellular PHA content of 40% [72]. Unfortunately, the authors did not provide sufficient information to calculate the cell concentration and productivity.

Yu et al., reported the continuous production by *R. eutropha* of P(3HB-*co*-3HV) using glucose and sodium propionate as co-substrates. Increased molar fractions of HV units in the final product resulted by increasing the relative concentration of sodium propionate in the feed. This resulted in a total biomass, P(3HB-*co*-60 mol%3HV) CDW content and an overall productivity of 8 g/L, 30% and 0.045 g/L/h, respectively [73]. While increase in the sodium propionate concentration correlated with higher copolymer 3HV content, 3HV can inhibit *R. eutropha* growth decreasing both the biomass and P(3HB-*co*-3HV) formation. A continuous cultivation of *P. putida* KT2442 for biosynthesis of mcl-PHA was studied by Huijberts and Eggink. Using oxygen limited continuous HCDC, the total biomass, mcl-PHA cell content and overall productivity reached 30 g/L, 23% and 0.69 g/L/h, respectively [74].

Halomonas sp. TD01 is highly tolerant to both high salt and pH conditions. This is advantageous since, such an environment is intolerant to other potential strains that pose contamination risks. Consequently, the rigors of processes normally used to maintain sterile conditions can be relaxed such that continuous and open fermentation processes can be used without concerns of contamination. In one example, an unsterile two-stage continuous cultivation for P3HB production by halophilic bacteria *Halomonas* TD01 was reported by Tan et al. [75]. *Halomonas* TD01 cells were cultivated on glucose in the first bioreactor for 2 weeks and, thereafter, the cells were transferred into the second bioreactor under nitrogen limiting conditions. While the continuous transfer of cultures from the first to the second bioreactor diluted the cells, the %-P3HB in the CDW remained at between 65–70% [75]. At 24 h, the first fermenter had a biomass of 40 g/L that contained 60% P3HB. These values were maintained during the entire cell growth period. In the second stage, the total biomass, %-mcl-PHA in the CDW and overall productivity reached 20 g/L, 65% and 0.26 g/L/h, respectively [75]. The low cell biomass in reactor 2 is a consequence of culture dilution while maintaining high P3HB content results from nitrogen limitation. These results highlight that *Halomonas* TD01 is an attractive cell bio-factory for P3HB accumulation since the culture conditions are highly amenable for commercial processes. However, further development of the organisms and process is needed to reach commercially viable PHA yields and productivities.

Another approach to conduct semi-continuous fermentation is by cyclic fed-batch fermentations (CFBF) at high cell densities. CFBF is performed by partially removing culture broth with subsequent refilling of fresh medium to the bioreactor [76]. This approach circumvents an accumulation of toxic concentrations of by-products and corresponding increased culture volumes that occur during fed-batch fermentations [77]. As a result, CFBF has proved useful in reducing the impact of media chemical changes enabling thermophiles to reach high final biomass and product concentrations. Ibrahim and Steinbuchel reported the application of CFBF in a stirred tank reactor for HCD cultivation of the PHB accumulating thermophile *Chelatococcus* sp. Strain MW10. The aim was to develop energy-saving PHB production processes. Using this strategy, total biomass reached 115 g/L but the %-PHB in the CDW was relatively low (12%) [78]. Nevertheless, CFBF is attractive for thermophilic strains for the reasons described above as well as the relatively simple fermenter set up and ability to monitor by the withdrawal/refilling process.

Karasavvas and Chatzidoukas, reported the modelling and dynamic optimization two continuous cascade bioreactors to optimize P3HB formation from sucrose by *Azohydromonas lata*. For the system at steady state they reported a total biomass and %-PHB CDW content of 20.52 g/L and 83.4%, respectively [79].

Table 2. PHA production by HCDC continuous mode operations without cell recycle.

Carbon Source	Microorganisms	Stages of Fermentation	Biomass (g/L)	%PHA	Overall Productivity (g/L/h)	Carbon Conversion Efficiency ($Y_{p/s}$) g/g	Reference
n-Octane	P. oleovorans ATCC 29347	Two stage	18	63	1.06	NA	[65]
Glucose	C. necator DSM 545	Five stage	81	77	1.85	NA	[66]
	C. necator DSM 545	Multi stage	80	NA	2.14	0.47	[67]
	R. eutropha WSH3	Two stage	50	73	1.23	0.36	[68]
	R. eutropha WSH3 (lab collection)	Two stage	32.6	75	NA	0.043	[69]
	Halomonas TD01	One stage	20	65	0.26	0.242	[75]
	Chelatococcus sp. Strain MW10	semi-continuous/cyclic fed-batch	115	12	NA	0.09	[78]
Fructose	Wautersia eutropha NRRL B-14690	Two stage	49	51	0.42	NA	[70]
Butyric acid and valeric acid	R. eutropha DSM 428	One stage	NA	40	NA	NA	[72]
Glucose/sodium propionate	R. eutropha (ATCC 17699, CCRC 13039)	One stage	7.96	30	0.045	NA	[73]
Oleic acid	P. putida KT2442	Two stage	30	23	0.69	0.15	[74]
Sucrose	Azohydromonas lata	Two stage	20.52	83.43	NA	NA	[79]

$Y_{p/s}$: PHB yield to substrate (g PHB produced per g substrate consumed); %-PHA: Final intracellular PHA content (g/g) of DCW; NA: Data not available.

3. Effect of Nutrient Limitations on Yield of PHA

Nutrient limitation is a key strategy for PHA production processes. Different nutrients have different effects on cell metabolism, growth, and PHA production. PHA production under nutrient limitations is generally conducted by a two-stage fed batch cultivation in which PHA accumulation occurs primarily during the nutrient-depleted stage [15,59,61]. Nutrient limiting conditions is imposed by continuous feeding of essential nutrients while reducing the concentration of the growth limiting nutrient (i.e., nitrogen) to reach a desired C/N ratio [60,80,81]. Increased PHA cell contents under conditions that are nutrient limiting is a direct result of imposing constraints such that, available carbon sources are not used for biomass accumulation but, instead, for PHA synthesis [82]. For example, the rate of scl-PHA synthesis by *R. eutropha* is significantly increased under nitrogen and phosphorus limitations [83]. The effect of nutrient limitations on PHA synthesis is summarized in Table 3.

Sun et al., (2007) reported simultaneous growth and accumulation of mcl-PHA using *P. putida* KT2440 where the rate of non-PHA biomass accumulation is below that of PHA biosynthesis [58]. Hence, PHA synthesis is not strictly associated with cell growth. Lee et al., reported mcl-PHA fed-batch production using *P. putida* KT2440 under phosphorous limitations that led to impressive mcl-PHA production. To increase PHA cell content, the initial phosphorus concentration in the feed was varied during fed-bath cultivations. By reducing the initial concentration of KH_2PO_4 from 7.5 to 4 g/L, the total concentration of total biomass and mcl-PHA cell content reached 141 g/L and 51.4% (i.e., mcl-PHA yield of 72.6 g/L), respectively, with an overall productivity of 1.91 g/L/h [15]. When the initial phosphate concentration was further reduced, the PHA content of the CDW remained unchanged but the overall productivity and concentration of PHA was reduced. These workers also provided a useful roadmap as to how the feeding strategy can be used to reach HCDC and productivity values. Furthermore, the time at which the nutrient limitation was applied significantly affected biomass accumulation, PHA cell concentration and productivity. The initial phosphorus concentration mainly affected the conversion efficiency of acetate to PHA. Ryu et al., reported *A. eutrophus* fed-batch cultivations under phosphorus limitation for P3HB production [29]. The dissolved oxygen (DO) concentration was used to control both the glucose feeding rate as well as to monitor its concentration. Variation in the glucose concentration was between 0-20 g/L. In addition, the influence of the initial phosphate concentration on P3HB formation was evaluated. At 5.5 g/L initial phosphorus concentration, the total biomass, mcl-PHA cell content and overall productivity reached 281 g/L, 82% and 3.14 g/L/h, respectively [29]. These results also stand out as highly impressive and provide valuable information on how high PHA yields can be attained.

Shang et al., investigated the effect of the glucose feeding rate on the formation of P3HB by *R. eutropha* under phosphate limiting and fed-batch cultivation conditions. By sustaining the glucose concentration in the medium at 2.5 g/L, P3HB formation and cell growth were restricted by the carbon source shortage. However, by sustaining the concentration of glucose in the culture at 9 g/L, the total biomass, %-P3HB in the CDW and overall productivity reached an impressive 208 g/L, 67% and 3.1 g/L/h, respectively [39]. However, further increase of the glucose concentration in culture media to 16 g/L resulted in significant decreases in P3HB productivity.

Tu et al., evaluated the effect of phosphorus limitation on accumulation of PHA from thermally hydrolyzed sludge. Decrease in the phosphorus concentration from 127.6 to 1.35 mg/L resulted in an increase in PHA cell contents from 23 to 51% [84].

Wen et al., evaluated how nitrogen and phosphorus limitation effected PHA formation from acetate. The microbes in this study were from activated sludge. Ratios of C:P and C:N were varied to investigate the effect of nitrogen and phosphorus limitation, respectively. The maximum %-PHA in the CDW reached 59% at the C:N 125 and 37% under phosphorus limitation experiments [85]. However, the authors did not provide information on how nutrient limitations affected the total biomass accumulated and overall productivity.

Portugal-Nunes et al., investigated the effect of nitrogen availability on PHB accumulation in two recombinant strains of *S. cerevisiae* using xylose as the carbon source. However, nitrogen deficiency did not enhance PHB accumulation in *S. cerevisiae*. Instead, the highest PHB contents (2.7-fold increase) were obtained under excess of nitrogen [86].

Grousseau et al., reported cultivation of *C. necator* on a butyric acid/propionic acid co-feed during fed batch fermentation conditions. They discussed how the distribution of 3HB and 3HV monomers was influenced by the ratio of propionic and butyric acids in the feed. Decreasing 3HB with sustained 3HV formation occurred under phosphorus limited conditions. In fact, under these conditions the PHA formed consisted of nearly 100% 3HV units [87]. By feeding phosphorus, which sustained cell growth, and using propionic acid as the carbon source, the maximum 3HV content in the P(3HB-*co*-3HV) copolymer was 33% [87]. This was explained by the fact that, by imposing a phosphorus limitation, the decarboxylation of propionic acid decreased thereby maximizing 3HV production. By cofeeding butyric and propionic acid (1:2 molar ratio), the total biomass, %-P3HB in the CDW and overall productivity reached 65.9 g/L, 88% and 0.65 g/L/h, respectively [87]. Furthermore, by moving from a feedstock that consists of only propionic acid to one with a butyric acid co-substrate, metabolism of propionic acid to 3HV occurs at higher efficiency.

Da Cruz Pradella et al., reported the HCDC of *B. sacchari* 189 on sucrose in an airlift bioreactor. In a two-phase fed-batch fermentation experiment, nitrogen limitation induced P3HB biosynthesis. In phase one, a limited sucrose feeding regime resulted in 60 g/L biomass and a low (13%) %-P3HB content in the CDW [88]. However, in phase two, P3HB accumulation was induced by nitrogen limitation leading to a total biomass, %-P3HB CDW content and overall productivity of 150 g/L, 42% and 1.7 g/L/h, respectively [88].

Grousseau et al., conducted cultivations of *C. necator* DSM 545 on butyric acid to determine how maintaining continued cell growth would influence P3HB formation kinetics. These authors showed that NADPH formation via the Entner-Doudoroff pathway was enabled by anabolic demand. The result was a high carbon conversion efficiency where 0.89 mol-carbon in P3HB resulted from one-mole of carbon in the feedstock [89]. Indeed, this is an extraordinarily high carbon utilization efficiency. The total biomass, %-P3HB in the CDW and the overall productivity reached 46.7 g/L, 82% and 0.57 g/L/h, respectively [89].

Kim et al., reported the fed batch cultivation of *Methylobacterium organophilum* for P3HB production under potassium limitation. The methanol concentration was maintained at 2–3 g/L to avoid cell growth inhibition. P3HB accumulation accelerated when the concentration of potassium in the culture broth was reduced to less than 25 mg/L. The total biomass, %-P3HB CDW content and overall productivity reached 250 g/L, 52% and 1.86 g/L/h, respectively [90]. In other words, the fermentation produced 130 g/L P3HB [90].

R. eutropha ATCC 17699 was cultivated using fructose in stage one, and fructose/γ-butyrolactone during stage two under nitrogen limitation. Since γ-butyrolactone is metabolized into 4HB, the resulting product was P(3HB-*co*-4HB) [91]. To improve copolymer yields, cultivations were performed by fed-batch with DO-stat control for controlled feeding. Using the DO-stat strategy and at a 1.5:1 molar ratio of fructose to γ-butyrolactone, the total biomass, %-P(3HB-*co*-1.64 mol%4HB) in the CDW and overall productivity reached 48.5 g/L, 50.2% and 0.55 g/L/h, respectively [91].

Table 3. Accumulation of PHA produced by different strains under nitrogen/phosphorus/oxygen limitation conditions.

Microorganisms	PHA Type	Limiting Nutrients	Carbon Source	Biomass (g/L)	%-PHA	Overall Productivity (g/L/h)	Carbon Conversion Efficiency ($Y_{p/s}$) g/g	Reference
P. putida KT 2440 ATCC 47054	mcl-PHA	Phosphorus (4 g/L)	Oleic acid	141	51.4	1.91	NA	[15]
		Phosphorus (22 g/L)		173	18.7	1.13		
Alcaligenes eutrophus NCIMB 11599	PHB	Phosphorus (5.5 g/L)	Glucose	281	82	3.14	0.38	[29]
R. eutropha NCIMB 11599	PHB	Phosphate	Glucose	208	67	3.1	NA	[39]
Activated sludge	PHB/PHV	Phosphorus	Acetate	NA	37	NA	NA	[85]
		Nitrogen		NA	59	NA		
C. necator DSM 545	P(3HB-*co*-3HV)	Phosphorus	Propionic acid/butyric acid	65.9	88	0.65	0.51	[87]
Burkholderia sacchari IPT 189	PHB	Nitrogen/oxygen	Sucrose	150	42	1.7	0.22	[88]
C. necator DSM545	PHB	Phosphate	Butyric acid	46.7	82	0.57	0.62	[89]
Methylobacterium organophilum NCIB 11278	PHB	Potassium	Methanol	250	52	1.86	0.19	[90]
Ralstonia eutropha ATCC 17699	P(3HB-*co*-4HB)	Nitrogen	Fructose + γ-butyrolactone	48.5	50.2	0.55	NA	[91]

$Y_{p/s}$: PHB yield to substrate (g PHB produced per g substrate consumed); %-PHA: Final intracellular PHA content (g/g) of DCW; NA: Data not available, PHB: Polyhydroxybutyrate, PHV: Polyhydroxyvalerate, P(3HB-*co*-3HV): Poly (3-hydroxybutyrate-co-3-hydroxyvalerate), P(3HB-*co*-4HB): Poly (3-hydroxy-butyrate-co-4-hydroxybutyrate).

4. PHA Production Using Genetically Modified Organisms

Non-PHA producing organisms can be genetically modified to biosynthesize PHAs (Figure 5). Furthermore, genetic modification of PHA and non-PHA producing strains provides a route to recombinant strains with improved kinetics for PHA production, a wider ability for substrate utilization (e.g., utilization of lignocellulose components), and changes in selectivity enabling the production of PHAs with unique structures. In addition, genetic engineering of strains has been used to improve substrate utilization from, for example, treated lignocellulose materials. In some cases, recombinant strain construction of, for example, *C. necator* and *E. coli* and are considered as strong candidate for commercial PHA production [92,93].

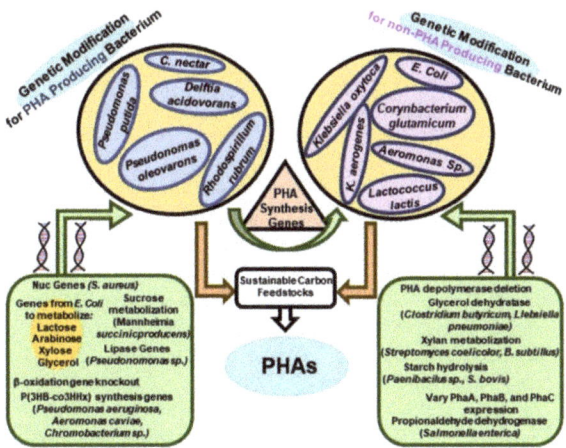

Figure 5. Genetic engineering strategies to improve bacterial strains for PHA production. PHA producing strains can be improved to convert various carbon substrates into PHAs via gene insertion or deletion. PHA synthase genes from PHA producing strains are often inserted in non-PHA producing strains. Using different strains of bacteria facilitates PHA production under conditions that are not suitable for naturally producing PHA strains.

Recent strain engineering studies have focused on manipulating the metabolic flux by, for example, gene deletion or reducing gene expression for competing pathways that would redirect carbon away from PHA monomer formation [94]. Furthermore, genetic modification of production strains has been used to inhibit β-oxidation so that the PHA composition will more closely resemble that of the carbon source-fed [95]. A wide range of wild-type strains such as *A. hydrophila* 4AK4 [96], *M. extorquens* [97], and *Pseudomonas* [98,99] have been modified for enhanced copolymer yield, modification of PHA structure and incorporation of scl-and mcl-PHA monomers, respectively.

E. coli has proved to be a valuable organism for genetic modification to attain highly productive PHA producing strains. Investigation of optimization of P3HB formation by *E. coli* has used tunable promotors to modulate expression levels of *phaA*, *phaB* and *phaC* [95]. Another powerful tool is the ability to construct ribosomal binding site (RBS) libraries where the copy number of plasmids can be systematically varied [100]. Omission of enzymes such as the PHA depolymerase is valuable as the PHA production organism has no mechanism to carry out PHA hydrolysis, the reverse of PHA synthesis [101,102]. Table 4 provides representative examples of PHA production by recombinant *E. coli* and *C. necator* using HCDC methodologies.

E. coli possesses a rich genetic background and multiple available tools making it an ideal host for PHA biosynthesis [12]. The design and generation of recombinant *E. coli* strains has enabled the synthesis of a variety of PHAs. Furthermore, one can construct E.

coli strains such that they can utilize a diverse set of feedstocks including carbon source mixtures derived from treatment of lignocellulose [103,104]. *E. coli* is capable of accumulating high contents of PHAs (80% to 90% of CDW) which enables large-scale PHA production. Also, Ren et al., claim there are economically viable method that can be applied to recover PHA from *E. coli* [105]. Due to the high accumulation of PHA, *E. coli* cells become fragile which enables efficient and easier product recovery [106]. Furthermore, produced PHA is not subjected to degradation during cultivations since it is normally constructed without an intracellular depolymerase enzyme [107].

Difficulties were encountered in identifying wild-type PHA producing strains that can use 4HB as the sole carbon source. To solve this problem, Le Meur et al., expressed a 4-HB-CoA transferase from *Clostridium kluyveri* orfZ in *E. coli*. In addition, the recombinant strain harbored the *PhaC* gene from *A. eutrophus*. By this approach, the developed strain gained the capability of producing P(4HB) during cultivations containing 4HB as the sole carbon source. That is, using glycerol for cell growth, 4HB for polymer formation, with exponential feeding to control the growth rate, and HCDC fed-batch operating conditions, P4HB formation reached a total biomass, %-P(4HB) in the CDW and overall productivity of 43.2 g/L, 33% and 0.21 g/L/h, respectively [50].

Ahn et al., expressed the *A. latus* genes *PhaA*, *PhaB*, and *PhaC* encoding in an *E. coli* strain. Interestingly, P3HB formation by this recombinant strain was more efficient than the corresponding recombinant *E. coli* strain was constructed using *PhaA*, *PhaB*, and *PhaC* from *R. eutropha*. The recombinant *E. coli* harboring the genes from *A. latus* was cultivated using a pH-state fed-batch culture and a 280 g/L sucrose-equivalent feed solution from concentrated whey. Whey is the liquid remaining after milk is curdled and strained. It is also a byproduct of cheese or casein manufacturing. The results are impressive as the total biomass, %-PHB in the CDW, overall productivity and carbon conversion efficiency reached 194 g/L, 87% (169 g/L P3HB), 4.6 g/L/h, and 0.45 g/g of P3HB per g of lactose [92]. These results are impressive and, if PHA copolymers could also be formed under similar conditions, this work provides guidance toward development of a commercially viable process.

The recombinant *R. eutropha* strain that expresses the *Rhodococcus aetherivorans* I24 PHA synthase gene and the hydratase gene (*phaJ*) from *P. aeruginosa* produced PHA containing HHx units [93]. The level of HHx in P(HB-*co*-HHx) was found to be a function of the acetoacetyl-CoA reductase activity. Cultivation of the recombinant *R. eutropha* strain on palm oil, under HCDC conditions, and by inducing nitrogen limiting conditions during PHA formation resulted in a total biomass, %-P(3HB-*co*-19 mol%3HHx) in the CDW and overall productivity of 139 g/L, 74% and 1.06 g/L/h, respectively.

Construction of the expression plasmids, pJRDTrcphaCABRe and pTrcphaCABRe was performed using the low and high copy number plasmid pJRDTrc1 and pTrc99a, respectively. Individually, these plasmids were expressed into *E. coli* XL1-Blue. The productivity of P3HB and biomass reached 2.8 g/L/h and 180 g/L, respectively, using glucose as the carbon source. The P3HB produced reached molecular weight values in the millions (3.5×10^6 to 5.0×10^6) and dispersity values remained low (~1.5) [108]. These ultrahigh molecular weight P3HB materials, like ultrahigh molecular weight polyethylene, provide advantaged mechanical properties.

The *E. coli* strain K24KL was constructed by deactivating the D-lactate synthesizing enzymes (ldhA) to produce P3HB from glycerol. This strain proved successful in increasing ethanol and P3HB production with a corresponding decrease in acetate formation. Analysis of the cofactor's NADPH/NADP$^+$ and NADH/NAD$^+$ showed that *E. coli* K24KL possesses a higher ratio of the former (NADPH/NADP$^+$). This led the authors to conclude that, the Idha mutation creates an intracellular environment with higher reducing capacity. By adopting fed-batch cultivation conditions, strain K24KL reached a total biomass, %-PHB in the CDW and overall productivity of 42.9 g/L, 63% and 0.45 g/L/h, respectively [109]. Insights from this work provide strategies for enhanced PHA formation from glycerol.

For improved plasmid stability, the kanamycin resistant gene was introduced into *E. coli* strain K1060. Cultivation of this strain was performed by a fed-batch operation in which the medium consisted of the agro-industrial by-products milk whey and corn-steep liquor. The total biomass, %-PHB in the CDW and overall productivity reached 70.1 g/L, 73% and 2.13 g/L/h, respectively [110].

Agus et al., expressed the *W. eutropha* PHA synthase (*PhaC*) in *E. coli* XL1-Blue cells. These workers than performed studies to assess how expression of PhaC effected P3HB production and molecular weight. IPTG (isopropyl-β-D-thiogalactopyranoside) functions as an inducer that controls the plasmid copies of PhaC. In other words, the concentration of IPTG in cultures provides a mechanism to control plasmid expression that results in PhaC formation. At 0.5 mM of IPTG that induces low PhaC expression, the recombinant *E. coli* strain reached a total biomass, %-PHB in the CDW and molecular weight of 178 g/L, 72% (128 g/L PHB), and 3 million g/mol, respectively [111]. In other words, the authors were successful in synthesizing PHB of ultra-high molecular weight.

The recombinant *E. coli* strain XL1-Blue which contained the *A. eutrophus* PHA synthase genes as well as the *E. coli* ftsZ gene that, in previous work, was found to increase P3HB formation efficiency by quelling filamentation, was used for P3HB formation. Cultivations were performed by fed-batch HCDC using glucose as the carbon source and thiamine for growth limitation. The total biomass, %-PHB in the CDW and overall productivity reached 156 g/L, 72% and 2.4 g/L/h, respectively [112].

The recombinant *E. coli* strain GCSC 6576 which contained the *R. eutropha* PHA synthesis genes as well as the *E. coli* ftsZ gene was studied to convert whey concentrate derived lactose to P3HB [113]. It consists of a 5% solution of lactose in water, with some minerals and lactalbumin (whey protein) [114]. Under fed-batch conditions, pH-stat control, and 210 g/L lactose equivalents from a concentrated whey solution, the total biomass, %-PHB in the CDW and overall productivity reached 87 g/L, 80% and 1.4 g/L/h, respectively [113]. These results highlight the potential to design of recombinant microorganisms for the efficient conversion of concentrated whey solution to PHB.

Subsequently, Riedel et al., reported that cultivation of *R. eutropha* on low quality waste animal fats results in 45 g/L of biomass with 60% of PHA and a productivity of 0.4 g/L/h [115]. Sato et al., reported that recombinant *C. necator* H16 is capable of synthesizing high levels of P(3HB-*co*-19 mol%3HHx) using palm kernel oil and butyrate as carbon sources. Moreover, the authors showed that butyrate increased the 3-HHx fraction in phaA-deactivated mutant strains of KNK005 (AS). This strategy results in high biomass (171 g/L), HHx copolymer content in cells (81%) and, corresponding, high PHA titers (139 g/L) [116].

Povolo et al., developed a *C. necator* recombinant strain capable of using inexpensive carbon sources such as lactose and hydrolyzed whey directly from whey permeate with an enhanced PHA production capability. A contributing factor to increasing PHA productivity was eliminating the metabolic pathway for polymer degradation. The recombinant *C. necator* utilized hydrolyzed whey permeate (composed of glucose and galactose) as the sole carbon source such that the cells contain 30% PHB [117]. However, the authors did not mention the total biomass concentration, PHB yield and overall productivity.

P. putida is a well-known producer of mcl-PHAs [118]. As discussed above, PHA production can be improved by deleting PHA depolymerase activity from the corresponding strain. Cai et al., constructed a recombinant *P. putida* KTMQ01 which accumulated 86% mcl-PHA of its CDW [119]. Le Meur et al., constructed a recombinant *P. putida* KT2440 strain to which the xylulokinase (XylB) and xylose isomerase (XylA) *E. coli* genes are inserted. The XylA and XylB genes help utilize the cost-effective substrate xylose, the main building block of the hemicellulose xylan, for mcl-PHA production. The resulting engineered *P. putida* KT2440 sequentially uptakes inexpensive carbon sources such as xylose and fatty acids (octanoic acid) for the cost-effective production of mcl-PHA. The cells reached 20% accumulation of mcl-PHA and authors did not determine final CDW and volumetric productivity [120].

Kahar et al., constructed a recombinant R. *eutropha* strain for high yield P(3HB-*co*-3HHx) production. R. *eutropha* PHA-negative mutant was built that harbored the phaC gene from *Aeromonas caviae*. Cultivations performed on soybean oil (20 g/L) by a fed batch process resulted in a 3HB copolymer consisting of 5 mol% 3HHx [121]. The total biomass, %-PHA in the CDW and overall productivity that reached 133 g/L, 72.5% and 1.0 g/L/h, respectively [121].

Aeromonas hydrophila 4AK4 produces P(3HB-*co*-3HHx) that contains 15 mol% HHx from dodecanoate [122]. To determine the factors that influence the incorporation of 3HHx in the copolymer, a recombinant *Aeromonas hydrophila* strain that expresses the genes *phaJ*, *phaC* and *phaP* from *Aeromonas punctate* were introduced individually or in combination. The authors discovered that expression of *phaC* alone enhanced the content of 3HHx in the copolymer from 14 to 22 mol% [122]. Co-expression of *phaC* with *phaP* and *phaJ* further increased the content of 3HHx in the copolymer to 34 mol%. The recombinant strain with *phaP* or *phaC* alone gave copolymer production in shake flask cultivations (48 h) that reached 4.4 g/L total biomass and 64% PHA in the CDW [122]. The authors concluded that by increasing the PHA synthase activity, higher contents of the 3HHx comonomer was incorporated whereas, by co-expressing *phaJ* with *phaP* or *phaC*, PHA production increased. Unfortunately, the authors did not assess whether high PHA production would be achieved in a fermenter using a HCDC protocol.

Towards the development of a strain that converts the unrelated carbon sources glucose and gluconates to P(3HB-*co*-3HHx), Qiu et al., built recombinant strains of *Pseudomonas putida* GPp104 and *Aeromonas hydrophila* 4AK4. This capability could eliminate the need for fatty acid substrates that can lead to foaming. The recombinant *A. hydrophila* 4AK4 expresses a cytosolic thioesterase-I, encoded by a truncated Tes A gene, to convert acyl-ACP into free fatty acids. Cultivation of the recombinant *A. hydrophila* 4AK4 strain on gluconate produced P(3HB-*co*-3HHx) containing 14 mol% HHx units [123]. Further genetic manipulations by overexpression of the P(3HB-*co*-3HHx) synthesis gene *phaPCJ* enlarged the copolymer content of 3HHx units to 19%. Moreover, these authors revealed that, recombinant *P. putida* GPp104, which harbors the *A. hydrophila phaC* gene that encodes the formation of 3HB/3HHx copolymers, *phaB* from *Wautersia eutropha* that encodes acetoacetyl-CoA reductase, and the *P. putida phaG* gene that encodes 3-hydroxyacyl-ACP-CoA transferase, resulted in PHA cell contents of 19% (*w*/*w*) with 5 mol% 3HHx units from glucose, a carbon source that is not related to HHx [123]. These results provide a roadmap to strategies that incorporate 3HHx units into PHA copolymers from unrelated carbon sources.

Ouyang et al., built a recombinant *A. hydrophila* 4AK4 strain that encoded the phbA and phbB from R. *eutropha* and *Vitreoscilla*, respectively. Cultivations of recombinant *A. hydrophila* 4AK4 were performed by a fed-batch process on the co-substrates dodecanoate and gluconate (1:1). Using dodecanoate only, the total biomass, %-P(3HB-*co*-12 mol%3HHx) content in the CDW and overall productivity reached 54 g/L, 52.7%, and 0.791 g/L/h, respectively [124]. In contrast, the wild-type strain produces total biomass, %-P(3HB-*co*-14.4 mol%3HHx) content and overall productivity of 40.4 g/L, 54.6% and 0.525 g/L/h, respectively [124].

The origin of sludge palm oil (SPO) is the palm oil milling industry. It is a solid that is generally considered difficult to use as a carbon source in cultivations. Budde et al., built a recombinant *C. necator* strain that encoded the phaC gene from R. *aetherivorans* I24 and phaJ gene from *P. aeruginosa* [125]. This engineered strain efficiently utilizes plant oils for P(HB-*co*-HHx) production. It was evaluated for its ability to utilize palm oil for P(3HB-*co*-3HHx) production [126]. To increase PHA productivity on SPO, fed-batch fermentations were conducted. The combination of the selected recombinant strain and cultivation conditions resulted in a total biomass, %-P(3HB-*co*-22 mol%3HHx) in the CDW and overall productivity of 88.3 g/L, 57% and 1.1 g/L/h, respectively [126].

A recombinant *E. coli* strain was built by encoding the *phaA*, *phaB* and *phaC* genes from R. *eutropha* PHA. This strain provided substantial benefits relative to the corresponding wild-type strain for PHA formation. Further improvements were realized by expressing the phaC gene from *A. latus* [127]. In 2002, Choi et al., used this strain for P(3HB-*co*-3HV) formation adopting a fed-batch feeding strategy and using glucose for biomass accumulation and the co-substrates propionic and oleic acids for PHA formation. They reported a total biomass, %-P(3HB-*co*-5.7 mol% 3HV) in the CDW and overall productivity of 42.2 g/L, 70% and 1.37 g/L/h, respectively [128].

Chen et al., constructed the recombinant *E. coli* strain by encoding the orfZ gene that expresses the *Clostridium kluyveri* 4HB-CoA transferase. The resulting strain, *Halomonas bluephagenesis* TD40, was evaluated for P(3HB-*co*-4HB) formation [129]. The use of this salt-tolerant strain allowed cultivations to be performed without taking precautions to maintain sterile conditions. HCDC of cultivations of *Halomonas bluephagenesis* TD40 were performed under fed-batch operating conditions, in 1 and 7 L fermenters for 48 h using glucose and γ-butyrolactone as carbon sources. The total biomass and %-PHA reached 70 g/L, and 63% of a P(3HB) copolymer containing 12 mol% 4HB [129]. Subsequently, this process was transferred to a 1000-L pilot scale fermenter which, by 48 h had a total biomass, %-PHA content and overall productivity of 83 g/L, 61% of the CDW that contained a P3HB copolymer with 16 mol-% 4HB and 1.04 g/L/h, respectively [129]. We conclude that *H. bluephagenesis* TD40 has excellent potential after further development to provide a platform for P(3HB-*co*-4HB) commercial production under open non-sterile conditions.

Poblete-Castro et al., constructed a recombinant *P. putida* KT2440 strain for PHA production on glucose by deletion of *gcd* (glucose dehydrogenase) and *gad* (gluconate dehydrogenase). The logic behind these deletions was to prevent gluconate and 2-ketogluconate formation. Fed-batch cultures were conducted under varying conditions and were used to assess mcl-PHA formation directly from glucose [130]. The first phase of biomass growth utilized exponential feeding with carbon limitation, whereas, for mcl-PHA formation in the second phase, substrate-pulse feeding, constant feeding and DO-stat feeding strategies were evaluated under nitrogen limiting conditions. The DO-stat feeding strategy gave the highest mcl-PHA formation such that the total biomass, %-mcl-PHA in the CDW and overall productivity reached 62 g/L, 67% and 0.83 g/L/h, respectively [130].

Yang et al., engineered *E. coli* strains to for PHAs containing aromatic repeat units from glucose, an unrelated carbon source [131]. For this purpose, the authors constructed a recombinant *E. coli* capable of producing D-phenyl lactate (PhLA). This involved the overexpression of isocaprenoyl-CoA:2-hydroxyisocaproate CoA-transferase as well as an engineered phaC from *Clostridium difficile*. The resulting recombinant *E. coli* was cultivated by a fed-batch process, using the co-substrates 3HB and glucose, such that the total biomass, %-P(38.1 mol% PhLA-*co*-61.9 mol% 3HB) in the CDW and overall productivity reached 25.27 g/L, 55% and 0.145 g/L/h, respectively [131]. Furthermore, the authors showed that other aromatic repeat units such as D-3-hydroxy-3-phenylpropionate and D-mandelate could be metabolized from glucose and incorporated into PHAs. This work highlights the ability to engineer PHA-producing strains that produce aromatic polyesters from unrelated renewable resources.

'PHAomics' highlights that a diverse range of PHAs, including those with block structures, can be formed by microbial PHA producers resulting in an expanded library of PHAs with unique properties [132]. The development of recombinant strains that enable the production of different types of random copolymer, homopolymers, block copolymers, and PHAs decorated with functional entities are displayed in Table 5.

Engineered *P. putida*, *P. entomophila*, *P. mendocina*, *P. oleovorans*, *H. bluephagenesis* and *E. coli*, have been reported to produce homopolymers [133–136], scl-PHA random copolymer [137], scl- and mcl-PHA random copolymers [138–141], mcl-PHA random copolymers [133,134,142–144], block copolymers [100,107,133,134,141,145–150], functional PHAs [151–154], and PHA monomers [155,156].

Pichia pastoris has proved highly useful for the high-level expression of heterologous proteins [157]. Furthermore, *P. pastoris* has proved amenable to genetic manipulation. Also, this organism naturally synthesizes mcl-PHAs [158]. Vijayasankaran et al., reported that by the introduction of *R. eutropha* PHA biosynthesis genes, the recombinant *P. pastoris* strain showed an enhanced ability to accumulate P3HB [159]. This provides the opportunity to co-express the formation of PHAs and high-value proteins, an approach that can improve PHA economics.

A recombinant *E. coli* strain was built by encoding a phaC that can convert lactic acid-CoA to PHA repeat units. Furthermore, the strain was developed for the in vivo formation of lactic acid [160–167]. The resulting strain formed P(3HB-*co*-LA) with 4 to 47 mol% LA units. Changing the copolymer composition was accomplished by regulating the anaerobic culture conditions. The resulting P(3HB-*co*-LA) films were found to be comparatively pliable, flexible, and semi-transparent compared to both rigid homopolymers.

Table 4. PHA production by recombinant starins under HCDC conditions.

Microorganisms	PHA Type	Carbon Source	Biomass (g/L)	%-PHA	Overall Productivity (g/L/h)	Carbon Conversion Efficiency ($Y_{p/s}$) g/g	Reference
Recombinant *E. coli*	P4HB	Glycerol/acetate/4-hydroxy-butyrate	43.2	33	0.207	NA	[50]
	PHB	Lactose	194	87	4.6	0.45	[92]
	PHB	Glycerol	42.9	63	0.45	NA	[109]
	PHB	Glucose/thiamine	156	72	2.4	NA	[112]
	PHB	Lactose	87	80	1.4	0.11	[113]
	P(3HB-co-3HV)	Glucose/oleic acid/propionic acid	42.2	70	1.37	0.5	[128]
Recombinant *R. eutropha*	P(3HB-co-3HHx)	Palm oil	139	74	1.06	0.52	[93]
	P(3HB-co-3HHx)	Waste animal fat	45	60	0.4	0.40	[117]
	P(3HB-co-3HHx)	Soybean oil	133	72.5	1	0.74	[121]
E. coli strain K1060	PHB	Corn steep liquor/milk whey	70.1	73	2.13	NA	[110]
Escherichia coli XL1-Blue	PHB	Glucose	178	72	NA	NA	[111]
Recombinant *C. necator* H16	P(3HB-co-3HHx)	Palm kernel oil/butyrate	171	78	-	NA	[116]
Recombinant *Aeromonas hydrophila* 4AK4	P(3HB-co-3HHx)	Dodecanoate	54	52.7	0.791	NA	
	P(3HB-co-3HHx)	Dodecanoate: sodium gluconate	38.4	52	0.475	NA	[124]
Recombinant *C. necator*	P(3HB-co-3HHx)	Sludge palm oil	88.3	57	1.1	0.7	[126]
Recombinant *Halomonas Bluephagenesis* TD01	P(3HB-co-4HB)	Glucose, γ-butyro-lactone	83	60.62	1.04	0.27	[129]
Recombinant *P. putida* KT2440	PHD	Glucose	62	67	0.83	NA	[130]
Recombinant *E. coli* XL1-Blue	P (PhLA-co-3HB)	Glucose	25.27	55	0.145	NA	[131]

$Y_{p/s}$: PHB yield to substrate (g PHB produced per g substrate consumed); %-PHA: Final intracellular PHA content (g/g) of DCW; NA: Data not available, P4HB: Poly(4-hydroxybutyrate), PHB: Polyhydroxybutyrate, P(3HB-co-3HV):Poly(3-hydroxybutyrate-co-3-hydroxyvalerate), P (PhLA-co-3HB): Poly(Phenyllactate-co-3-hydroxybutyrate), P(3HB-co-3HHx): Poly(3-hydroxybutyrate-co-3-hydroxyhexanoate), P(3HB-co-4HB): Poly(3-hydroxybutyrate-co-4-hydroxybutyrate).

Table 5. Production of different types of PHA polymers by recombinant strains.

Microorganisms	Polymer Type	PHAs Composition	Carbon Source	Reference
Recombinant *Escherichia coli*	Polylactic acid random copolymer	Poly(lactate-co-3-hydroxybutyrate)	Xylan	[17]
Recombinant *Pseudomonas putida*	Block copolymer	Poly(3-hydroxybutyrate-b-poly4-hydroxybutyrate)	sodium butyrate/γ-butyrolactone	[100]
Recombinant *Escherichia coli*	Block copolymer	poly(3-hydroxybutyrate)-b-poly(3-hydroxypropionate)	Glycerol	[107]
Recombinant *P. entomophila* LAC23	Homopolymer	Poly(3-hydroxyheptanoate) Poly(3-hydroxyoctanoate) Poly(3-hydroxynonanoate) Poly(3-hydroxydecanoate) Poly(3-hydroxyundecanoate) Poly(3-hydroxytetradecanoate) Poly(3-hydroxytridecanoate)	Sodium heptanoate/sodium octanoate/sodium nonanoate/decanoic acid/undecanoic acid/dodecanoic acid/tridecanoic acid/tetradecanoic acid	[133]
Recombinant *P. entomophila* LAC23	Random copolymer	P(3hydroxyoctanoate-co-3hydroxydodecanoate)	Sodium octanoate/dodecanoic acid	[133]
		P(3hydroxyoctanoate-co-3hydroxytetradecanoate)	Sodium octanate/tetradecanoic acid	
P. entomophila LAC32	Diblock copolymer	P(3-hydroxyoctanoate)-b-P(3-hydroxydodecanaote)	Sodium octanoate/dodecanoic acid	[133]
Recombinant *Pseudomonas entomophila* LAC23	Homopolymer	Poly(3-hydroxy-9-decenoate)	9-decenol	[134]
Recombinant *Pseudomonas entomophila* LAC23	Block copolymer	P(3-hydroxydodecanoate-b-3-hydroxy-9-decenoate)	dodecanoic acid/9-decenol	[134]
Recombinant *Pseudomonas putida*	Homopolymer	Poly(3-hydroxyhexanoate) Poly(3-hydroxyheptanoat) Poly(3-hydroxyoctanoate-co-2 mol% 3-hydroxyhexanoate) Poly(3-hydroxyvalerate) Poly(3-hydroxybutyrate) Poly(4-hydroxybutyrate)	Hexanoate/Heptanoate/Octanoate/Valerate/γ-butyrolactone	[135]
Recombinant *Pseudomonas putida*	Homopolymer	Poly(3-hydroxydecanoate) Poly(3-hydroxydecanoate-co-84 mol%3-hydroxydodecanoate)	Decanoic acid/Dodecanoic acid	[136]
Recombinant *E. coli*	Random copolymer	P(3-hydroxybutyrate-co-3-hydroxypropionate)	Glucose	[137]
Recombinant *Pseudomonas putida*	Random copolymer	Poly(3-hydroxybutyrate-co-3-hydroxyhexanoate-co-3-hydroxyoctanoate)	Sodium heptanoate/Oleic acid	[138]
P. entomophila LAC32	Random copolymer	P(3-hydroxybutyrate-co-3-hydroxydecanaote)	Glucose/decanoic acid/dodecanoic acid	[139]
		P(3-hydroxybutyrate-co-3-hydroxydodecanaote)		
Pseudomonas putida KT2442	Random copolymer	Poly (3-hydroxybutyrate-co-3-hydroxyhexanaote)	Sodium butyrate/Sodium hexanoate	[140]
Pseudomonas putida KT2442	Random copolymer	Poly (3-hydroxybutyrate-co-mcl 3HA)	scl-fatty acid/mcl-fatty acid	[141]

Table 5. Cont.

Microorganisms	Polymer Type	PHAs Composition	Carbon Source	Reference
P. mendocina NK-01	Random copolymer	P(3-hydroxyoctanoate-co-3-hydroxydecanoate-co-3-hydroxydodecanoate), P(3-hydroxyhexanaote-co-3-hydroxyoctanaote-co-3-hydroxydecanaote-co-3-hydroxydodecanoic acid)	sodium octanoate/Sodium decanoate/dodecanoic acid	[142]
Recombinant Pseudomonas entomophila LAC23	Random copolymer	P (3-hydroxydodecanoate-co-3-hydroxy-9-decenoate)	dodecanoic acid/9-decenol	[144]
Recombinant Pseudomonas putida	Block copolymer	Poly(hydroxybutyrate-b-polyhydroxyvalerate-hexanoate-heptanaote)	Butyrate/hexanaote	[144]
	Random copolymer	Poly(3-hydroxybutyrate-co-valerate-hexanoate-heptanaote)	Butyrate/hexanaote	
Ralstonia eutropha NCIMB 11599	Triblock copolymer	Poly(3-hydroxybutyrate-co-3-hydroxyvalerate-b-poly(3-hydroxybutyrate)-b-Poly(3-hydroxybutyrate-co-3-hydroxyvalerate)	Glucose/Pentanoic acid	[145]
R. eutropha NCIMB 11599	Block copolymer	Poly(3-hydroxybutyrate-co-3-hydroxyvalerate)-b-poly(3-hydroxybutyrate)	Glucose/Pentanoic acid	[146]
Recombinant Pseudomonas putida KT2442	Diblock copolymer	Poly(3-hydroxybutyrate-b-poly-3-hydroxyhexanoate)	Sodium butyrate/sodium hexanoate	[147]
Burkholderia sacchari DSM 17165	Block copolymer	Poly(3-hydroxybutyrate-b-3-hydroxyvalerate)	Xylose/levulinic acid	[148]
Mixed culture containing Azohydromonas lata DSM 1122 and Burkholderiasacchari DSM 17165	Random copolymer and block copolymer	Poly(3-hydroxybutyric-co-3-hydroxyvalerate-co-4-hydroxyvalerate) P(3-hydroxybutyrate-b-3-hydroxyvalerate)	Glucose/levulinic acid	[150]
Pseudomonas putida Gpo1	Functional polymer (cationic PHA)	Poly[(β-hydroxy-octanoate)-co-(β-hydroxy-11-(bis(2-hydroxyethyl)-amino)-10-hydroxyundecanoate)]	Sodium octanoate	[151]
Recombinant Escherichia coli and Pseudomonas putida	Functional polymer	3-hydroxydecanoic acid	Fructose	[152]
Recombinant Escherichia coli	Polylactic acid random copolymer	P(lactic acid-co-3-hydroxybutyrate-co-3-hydroxypropionate)	Glucose/glycerol	[161]
Recombinant Escherichia coli	Polylactic acid random copolymer	Poly(glycolate-co-lactate-co-3-hydroxybutyrate)	Glucose	[162]
Recombinant Escherichia coli	Polylactic acid random copolymer	Poly(glycolate-co-lactate-co-3-hydroxybutyrate-co-4-hydroxybutyrate)	Glucose	[163]
Recombinant Escherichia coli	Polylactic acid random copolymer	Poly (lactate-co-3-hydroxybutyrate)	Xylose/acetate	[164]

Table 5. Cont.

Microorganisms	Polymer Type	PHAs Composition	Carbon Source	Reference
Recombinant *Escherichia coli*	Polylactic acid random copolymer	Poly(lactate-co-3-hydroxybutyrate)	Xylose-based hydrolysate	[165]
Recombinant *Escherichia coli*	Polylactic acid random copolymer	Poly(D-lactate-co-glycolate-co-4-hydroxybutyrate)	Glucose/xylose	[166]
Recombinant *Escherichia coli*	Polylactic acid random copolymer	Poly(lactate-co-3-hydroxybutyrate)	Glucose	[167]

$Y_{p/s}$: PHB yield to substrate (g PHB produced per g substrate consumed); %-PHA: Final intracellular PHA content (g/g) of DCW; NA: Data not available.

5. Enhancement of PHA Yield by β-Oxidation Inhibition

The mcl-PHAs usually occur as copolymers because the substrates used for biosynthesis are subjected to β-oxidation, resulting in the production of a mixture of repeat units that differ in chain length. In other words, even when the substrate is one structure, the resultant mcl-PHA is heterogeneous due to β-oxidation. To avoid β-oxidation of fatty acids, two methodologies have been developed; the first is to suppress or remove β-oxidation genes. Alternatively, inhibitors can be used to suppress enzymes catalyzing β-oxidation. Genetically modified organisms were found to be effective in obtaining a dominant repeat unit structure [168]. Jiang et al., also reported that cofeeding acrylic acid is a useful strategy to increase the direct incorporation of a selected substrate. In one example, a fed-batch cultivation of *P. putida* KT2440 on a substrate mixture consisting of glucose: nonanoic acid: acrylic acid (1: 1.25: 0.05, mass ratio) resulted in a total biomass, %-PHA of CDW content and productivity of 71.4 g/L, 75.5% (89 mol% 3HN) and 1.8 g/L/h, respectively [169]. In the absence of acrylic acid, the content of 3HN was reduced to 65 mol%. Also, β-oxidation-deleted mutants of *P. putida* or *P. entomophila* were effective in preparing mcl-PHAs that closely approached being homopolymers of a selected repeat unit [134,170].

Gao et al., deleted the β-oxidation gene of *P. putida* KT2440 towards synthesizing mcl-PHAs that closely approximated a homopolymer composition. In one example, cultivation of the β-oxidation deleted recombinant *P. putida* KT2440, using decanoic acid as the mcl-PHA producing substrate along with co-substrates glucose and acetic acid (2:8 g/g), the total biomass, %-PHA in the CDW and overall productivity reached 18 g/L, 59% and 0.32 g/L/h, respectively [171]. Remarkably, the mcl-PHA formed consisted of only 3HD units. In contrast, the wild strain cultivated under identical conditions, resulted in a total biomass, %-PHA content in the CDW and overall productivity of 39 g/L, 67% and 0.84 g/L/h, respectively [171]. Hence, the PHA yield was higher for the wild-type strain. However, the comonomer composition was 3HD: 3HO: 3HHx units in a molar ratio of 74:14:12.

Zhao et al., reported the construction of a recombinant *P. mendocina* by deletion of multiple genes associated with the β-oxidation pathway. The objective was to prepare mcl-PHA in higher yields that consisted of predominantly 3HD and 3HDD repeat units. Relative to the wild type strain, the recombinant *P. mendocina* had about a 5-fold increase in mcl-PHAs produced from sodium octanoate and sodium decanoate [142]. Using dodecanoic acid as the feedstock the mcl-PHA yield increased by 10-fold. The resulting mcl-PHAs have nearly uniform repeat unit structures that showed higher melting point transitions, mechanical and crystallization properties [142]. This approach demonstrated the potential of developing recombinant *P. mendocina* strains that both provide increased mcl-PHA production but also uniform compositions.

Oliveira et al., investigated mcl-PHA formation by wild-type and recombinant β-oxidation deleted strains of *P. putida* KT2440. Cultivations were conducted by a fed-batch operational mode using sugarcane biorefinery-derived hydrolyzed sucrose and decanoic acid as carbon sources. Using linear phase feeding strategy, *P. putida* KT2440 reached a total biomass, %-PHA content in the CDW and overall productivity of 53.4 g/L, 33%

and 0.4 g/L/h, respectively [172]. The composition of the mcl-PHA formed consisted of C10:C8:C6 repeat units in a molar ratio of 84:14:2. However, using the same cultivation conditions, the β-oxidation deleted *P. putida* strain reached a total biomass, %-PHA of CDW and overall productivity of 24.6 g/L, 42% and 0.25 g/L/h, respectively with C10:C8:C6 mol% composition of 95:5:0 [172]. Hence, while the β-oxidation knockout mutant had lower productivity, it provided a means to produce an mcl-PHA that closely approaches a 3HD homopolymer.

A β-oxidation pathway modified mutant of *R. eutropha* was explored for P(3HB-*co*-3-HHx) production from soybean oil [173]. Deletion of fadB1 (enoyl-CoA hydratase/3HA-CoA dehydrogenase) in recombinant *R. eutropha* strains with other genes encoding (R)-enoyl-CoA hydratases, a 6–21% increase in 3-HHx content in the copolymer was observed [173]. This was attributed to an increased availability of mcl-2-enoyl-CoAs by partial impairment of β-oxidation. This work provides a useful strategy to increase the 4HHx content in copolyesters synthesized from fatty acids.

6. PHA Production Using Mixed Cultures

Mixed microbial consortia (MMC) provide advantageous routes for PHA production as cultivations are conducted in open systems, circumventing the need to maintain sterility, which can reduce operating costs [174–179] (Figure 6). As will be further elaborated below, agro-industrial and municipal waste streams offer low-cost substrates and sources of PHA producing microbes. Volatile fatty acids (VFAs) are a valuable PHA substrate produced by acidogenic fermentation during anaerobic digestion. Afterwards, activated (microbe containing) waste sludge generally undergoes a series of feast/famine feed cycles (visualized within the dashed border of Figure 6) to maximize the population of PHA-producing microbes. Other conditions, such as pH, cultivation time, and temperature are modified to further maximize PHA production. Lastly, the maximum PHA-producing MMC is utilized in subsequent PHA fermentation processes.

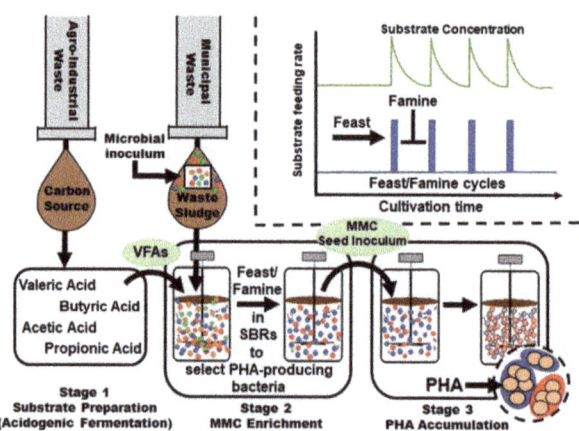

Figure 6. The use of mixed microbial consortia (MMC) for PHA production.

Harnessing MMCs for PHA production requires a first step, often carried out in sequential batch cultivations where selection and enrichment occur resulting in a microbial consortium with a high capacity for PHA formation. This normally required the application of transient conditions. Then, cultivation conditions are applied that further maximizes PHA formation [180,181].

MMC technology often utilizes volatile fatty acids, VFAs (i.e., acetic, propionic, butyric, and valeric acids), as substrates for PHA production [182–184]. These substrates are available via anaerobic digestion processes of wet wastes such as that from biomass fractions, foods, wastewater sludge, fats, animal waste, and more. VFA are advantageous over glucose and other carbohydrate feed sources since they are highly oxidized providing relatively higher equivalent carbon and energy [185]. The relative composition of VFA will determine the PHA composition as, for example, high propionic acid contents will be metabolized to 3HV units whereas even-carbon chain length VFAs will likely form P3HB [186,187].

The diversity of microorganisms in MMCs provides multiple PHA production pathways. A batch reactor that is acetate-fed run with a biomass residence time of one day, and 12 h cycles of feast-famine (F-F) conditions, gave an MMC enriched in PHA producers [180]. Subsequently, fed-batch cultivation experiment using the MMC mixed culture under growth limiting conditions resulted in high PHA storage cell accumulation (89% of the CDW) and high P(3HB) production rates (~1.2 g/g/h) [180]. The dominant microorganism in the cultivation was a *Gammaproteobacterium* that proved to have low similarity to known bacteria.

Jiang et al., used lactate as well as lactate/acetate mixtures to obtain PHB producing MMC's and investigated the use of lactate and a mixture of lactate and acetate for enrichment of PHB producing mixed cultures [188]. Operational conditions that previously resulted in a performing strain for PHB production from acetate was adopted for this work [189]. The enrichments from acetate and lactate used microbes from activated sludge. As above, the dominant microorganism from the enrichment on lactate was a *Gammaproteobacterium* that reached in 6 h PHB contents in its CDW of 90% [189]. When an acetate/lactate mixture was used during the enrichment, the dominant microorganisms were the bacteria *Thauera selenatis* and *Plasticicumulans acidivorans* that reached PHB contents in the CDW of 84% in 8 h [189]. Previous work with strains *Thauera selenatis* and *Plasticicumulans acidivorans* showed that they were capable of high PHB accumulation.

Cui et al., performed enrichments to obtain PHA forming MMCs using starch, glucose and acetate as substrates. Aerobic extended-time dynamic feeding intervals were used. The organisms were exposed to long-term aerobic dynamic feeding periods. Following 350 enrichment cycle intervals under F-F regimes, the MMCs enriched by feeding starch, glucose and acetate accumulated 27%, 61% and 65% PHA in their CDWs, respectively [190]. Sequencing studies revealed that, in addition to PHA forming bacteria, microbes that are non-PHA producing also survived. Using acetate for enrichment, the dominant PHA forming genera were *Stappia* and *Pseudomonas*. In contrast, enrichments conducted using glucose resulted in the PHA forming genera *Vibrio, Piscicoccus* and *Oceanicella*. For enrichments on starch, the sole PHA forming Genus was *Vibrio*.

Palmeiro-Sánchez et al., explored how imposing recurrent sodium chloride concentrations (NaOH, 0.8 g Na$^+$/L) would influence the resulting MMC cultures ability to produce PHA. Enrichments were on a mixture of C2-C5 aliphatic VFAs where the major constituents were acetic and propionic acids (54 and 27 mol%, respectively) [191]. PHA production by the MMC reached 53% of the CDW and the corresponding copolymer composition was P(3HB-co-27 mol%3HV) [191]. A comparative study where the NaOH was not imposed during enrichments resulted in a relatively lower ability of the resulting MMC to produce PHA.

Cavaillé et al., (2016) developed an MMC that consisted of PHA forming microbes from activated sludge. Cultivations were performed under non-sterile conditions using acetic acid as the carbon sources under phosphorous limitation. A stable continuous process for PHA formation was reached where the CDW of effluent cells reached 74% [192].

Thermophilic and thermotolerant strains can be used to develop MMC's for PHA formation to reduce essential heating and cooling operations that are energy intense processes [193]. For such cultivations, adequate oxygen must be accessible to microbes at working temperatures. Benefits of working with thermotolerant MMC enrichments include faster diffusion, higher solubility of substrates, higher PHA production rates and a lower risk of culture contamination [194].

Toward reducing the costs of PHA formation, MCCs were carried out using low cost agro-industrial wastes that include paper mill effluents, cannery effluents, municipal sludge, saponified sunflower oil, fermented sugar cane molasses, industrial and domestic wastewaters and oil mill effluents [195]. Johnson et al., studied MMCs with high storage capacity for PHA production. Acetate was fed for 24 h in the batch reactor to accumulate biomass with F-F cycles of 12 h. This strategy results in accumulation of 89% PHA in the CDW in 7.6 h with an overall PHA productivity of 1.2 g/g/h [196].

Lorini et al., reported on application of sequencing batch reactors for cultivation of mixed culture with uncoupled carbon and nitrogen feeding. The different organic load rates studied (4.25 to 12.72 g COD/L·d) results in PHA productivity of 0.1 g PHA/L/h at optimum conditions [197]. Silva et al., explored how nitrogen feeding would influence PHA formation by MMCs. A mixture of C2 and C3 VFAs were used as carbon sources at an organic load of 8.5 g COD/L·d with concurrent feeding of $(NH_4)_2SO_4$. This strategy resulted in the accumulation of PHA with up to $20 \pm 1\%$ w/w HV) [198]. Liu et al., reported mcl-PHA formation by *Pseudomonas-Saccharomyces*, using xylose as carbon source, results in 152.3 mg/L PHA [199]. The presence of *S. cerevisiae* in the consortium improved the sedimentation of cell mass.

7. Industrial/Agro-Industrial Waste for Production of scl-and mcl-PHA

Despite the promising properties of PHAs, they remain uncompetitive with conventional plastics due to both high cost and performance shortfalls. Significant improvements in performance can be realized by polymer formulation during processing [200–202], but this is beyond the scope of this review. The cost of carbon substrates constitutes ~50% of the total manufacturing cost [203]. Therefore, a major research effort is underway focused on reducing the production cost by utilizing low or no-cost waste materials as carbon sources [20,204,205].

The yield and monomer composition during PHA formation varies depending on the type of microbial strain and carbon sources. The use of commercial sugars in PHA production by bacterial fermentation leads to high-cost processes. Therefore, to achieve cost effective PHA production, in addition to reducing energy utilization and efficient downstream processes, it is also critical to move to low-no cost feedstocks such as cellulose, lignocelluloses, hemicelluloses and sugars derived therefrom. The sustainable sources of feedstock for PHA production are depicted in Figure 7.

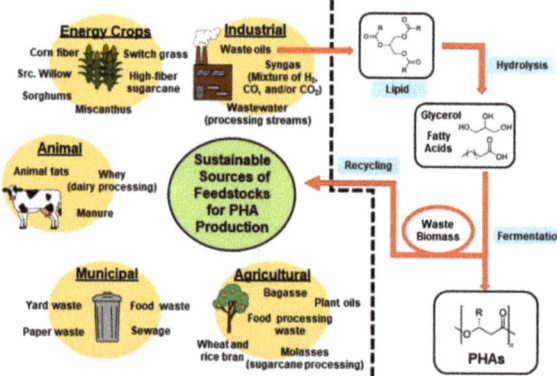

Figure 7. Sustainable sources and conversion of feedstocks for PHA production. Provided within the dashed area is a simplified representation of the conversion of waste oils (lipids) to PHAs. The waste lipid is hydrolyzed to yield substrates for fermentation: fatty acids and glycerol. The substrates are purified prior to use in fermentations. Recovered waste biomass can be recycled as a carbon source for future fermentations.

In addition to the above, large waste quantities are generated from numerous sources such as agriculture (including lignocellulosic materials), food, industrial and municipalities. An opportunity exists to create value from these zero cost feedstocks that would otherwise be landfilled. Wastes may be enriched in valuable carbon sources for PHA production such as carbohydrates of various complexity and fatty acids. Pretreatments that may involve biochemical, chemical or physical processes can significantly enhance the contents of wastes to more readily metabolizable substrates. The representative literature on HCDC using renewable feedstock are presented in Table 6.

7.1. Lignocellulosic Feedstock

A great opportunity exists to exploit the estimated 200 billion tons global production of lignocellulosic feedstocks [206,207]. It is well known that efficient utilization of lignocellulosic feedstocks comes with numerous challenges since these materials were designed by nature to be difficult to access by microbes to enable their long lifetime as plant structural materials.

Furthermore, phenolic compounds generated by lignin can inhibit PHA producing microbial growth and metabolic processes [208]. Fortunately, due to the numerous products envisioned by conversions of highly abundant lignocellulose by integrated biorefinery processes, there have been continuous improvements in pretreatment processes that use, for example, physicochemical, alkaline and enzyme-mediated reactions to achieve high sugar yields for production of biofuels and other valuable products [209]. For example, xylose is highly abundant in hemicellulose that can be liberated by efficient enzymatic or chemical processes for use as primary feedstocks for scl-PHA formation by *P. cepacia* ATCC 17759 [210].

About 25% w/w of wheat straw (WSH) consists of hemicellulose that are rich (80 dry wt) in pentoses [211]. Cesario et al., reported fed batch cultivations of *B. sacchari* DSM 17165 for PHA production from wheat straw hydrolysate that provides primarily arabinose, xylose and glucose [211]. *B. sacchari* was shown to efficiently metabolize these sugars. To promote polymer accumulation, KH_2PO_4 limitation was imposed by reducing its initial concentration to 3 g/L. Fed-batch *B. sacchari* cultivations did not exhibit catabolite repression that can occur when fed wheat straw hydrolysate sugars and reached a total biomass, %-PHB in the CDW and an overall productivity of 146 g/L, 72% and 1.6 g/L/h, respectively [211].

Cesario et al., adopted a fed-batch feeding strategy for the formation of P(3HB-*co*-4HB) by *B. sacchari* [212]. These authors used as carbon source a xylose-rich hydrolysate of lignocellulose and, γ-butyrolactone served as a 4HB unit precursor. This strategy results in a total biomass, %-P(3HB-*co*-6 mol%4HB) in the CDW and an overall productivity of 88 g/L, 27% P(3HB-*co*-6 mol%4HB) and 0.5 g/L/h, respectively [212]. These authors claimed that this is the first demonstration where *B. sacchari*, a strain that produces PHA from xylose-rich lignocellulosic hydrolysates, formed P(3HB-*co*-4HB) by using as co-substrate γ-butyrolactone. Thus, to effectively produce PHA from hemicellulose, the corresponding microbes must have the metabolic machinery to utilized hemicellulose derived pentoses and hexoses. The composition of wheat straw hydrolysate is 35.5% of glucose, 24.7% of xylose, 3.3% of arabinose, 16.4% of lignin, 8.8% of ash and 1.7% of raw proteins [213].

A promising feedstock from lignocellulose or derived sugars is levulinic acid (4-oxovaleric acid) as it can be produced with increasing efficiency from lignocellulose and various waste materials [214]. Levulinic acid has 5-carbons and, as such, is a good precursor for formation of 3HV units in PHAs. While it can serve as a primary substrate for cell growth of some PHA producing organisms, it may also cause inhibitory effects such that its concentrations in cultivations need to be kept below 2 g/L [215,216].

Xu et al., used a media consisting of wheat hydrolysate. By fed-batch cultivation of *W. eutropha* they reached a total biomass, %-PHB in the CDW and an overall productivity of 175 g/L, 93% PHB (i.e., 162.8 g/L) and 0.88 g/L/h, respectively [217].

Sathiyanarayanan et al., explored the sponge-associated endosymbiont *B. subtilis* MSBN17 for PHB accumulation from pulp industry waste (PIW) as the major substrate and tamarind kernel flour as a co-substrate. PIW was provided to cultures at 12, 28 and 40 h time intervals. Under optimized conditions, the total biomass, %-PHB in the CDW and the overall productivity of *B. subtilis* MSBN17 was 24.2 g/L, 51.6% and 0.26 g/L/h, respectively [218].

Dietrich et al., reported a HCDC of *Paraburkholderia sacchari* IPT 101 LMG 19450 for PHB production from hard wood hydrolysate as the sole carbon source. These authors compared a synthetic hydrolysate that consisted of equivalent substrates (acetate, xylose, glucose) with a wood hydrolysate. PHB formation was performed during 52 h HCDCs [219]. Interestingly, the PHB concentration (22 g/L) was relatively lower using the synthetic hydrolysate medium. This can be explained by the presence of additional nutrients in wood hydrolysate and the corresponding specific growth rates that were higher on the wood hydrolysate (0.36 vs. 0.33 h^{-1}). The cultivation on wood hydrolysate, the total biomass, %-PHB in the CDW and an overall productivity were 59.5 g/L, 58% and 0.72 g/L/h, respectively [219]. In comparison, the synthetic hydrolysate resulted in %-PHB in the CDW and an overall productivity of 55% and 0.46 g/L/h, respectively [219]. This report on the conversion of hard wood hydrolysate is promising but, as in the above work by Sathiyanarayanan et al. on PIW, further process and potentially cell engineering will be needed to achieve industrially relevant product titers and productivity.

7.2. Waste Glycerol

Crude glycerol from biodiesel production that, annually, is about 150 million gallons, generates 50 million kg of crude glycerol [220]. Problems encountered with the use of crude glycerol are co-generated impurities (i.e., fatty acids, salts, methanol) that lower its value. With continuous improvements in triglyceride production by oleaginous yeasts from lignocellulosic sugars, it is anticipated that triglyceride production from non-food sources will be available in increasing amounts as will crude glycerol [221]. This has spurred research to utilize crude glycerol as a carbon source for the biotechnological production citric acid, 2,3-butane diol, PHB, 1,3-propane diol and more [222–224].

Cavalheiro et al., investigated the conversion of crude glycerol to P3HB copolymers with 4HB and 3HV repeat units [23]. As discussed above, incorporating γ-butyrolactone as a co-feedstock will lead to the formation of 4HB units, whereas the co-substrate propionic acid can be metabolized to both 3HV and 4HB units. Incorporation of 4HB monomers was promoted by γ-butyrolactone. After optimizing dissolved oxygen in cultivation media, using crude glycerol and γ-butyrolactone as cosubstrates in cultivations of *C. necator* DSM 545 resulted in a total biomass, %-PHA in the CDW and overall productivity of 30.2 g/L, 36.1% P(3HB-17.6 mol%4HB) and 0.17 g/L/h, respectively [23]. In contrast, by using crude glycerol, γ-butyrolactone and propionic acid as co-feedstocks, the total biomass, %-PHA in the CDW and overall productivity reached 45.25 g/L, 36.9% of P(3HB-43.6 mol%4HB-6 mol%3HV) and 0.25 g/L/h, respectively [23]. Production of PHA from crude glycerol helps reduce production costs with concomitant glycerol valorization. Using the same production strain, Cavalheiro et al., studied PHB production from crude glycerol during fed-batch cultivations of *C. necator* DSM 545. They reported a total biomass, %-PHB in the CDW and an overall productivity of 68.8 g/L, 50% and 1.1 g/L/h, respectively [225].

Mozumder et al., reported a three-stage control strategy that was claimed to be organic substrate-independent for automated substrate feeding in a two-phase fed-batch culture [41]. This sensitive and robust feeding strategy was applied to *C. necator* DSM 545 cultivations for conversion of crude glycerol to PHB at experimentally determined optimal substrate concentrations. To reach maximal cell biomass, the feeding strategy combined exponential feeding and alkali-addition monitoring. Subsequently, the substrate feeding rate was kept constant while nitrogen feeding was stopped, resulting in PHB accumulation. This resulted in a total biomass, %-PHB in the CDW and overall productivity of 104.7 g/L and 62.7% and 1.36 g/L/h [41]. This work is certainly a promising example where high PHB titers (66 g/L) were achieved using crude glycerol as the sole carbon source.

Salakkam and Webb investigated whether biodiesel production by-products rapeseed meal and crude glycerol could function synergistically for PHB production by *C. necator* DSM 4058. Rapeseed meal processing provides a media component rich in free amino nitrogen. Adopting a fed-batch cultivation strategy for the co-substrates crude glycerol and processed rapeseed meal, without further nutrient supplements, resulted in a total biomass, %-P3HB in the CDW and overall productivity of 28.9 g/L, 85.8% and 0.21 g/L/h, respectively [226].

A strategy to improve PHA production economics is to use or develop production strains that also generate high value coproducts. Kumar et al., used glycerol as a carbon source for cultivations of *Paracoccus* sp. LL1 that produced carotenoids along with PHA. An enhancement of total biomass was achieved through cell retention, resulting in 24.2 g/L biomass that consisted of 39.3% PHA and 7.14 mg/L carotenoids [227]. Volova et al., reported a fed-batch cultivation of *C. eutrophus* B-10646 for PHB production from glycerol of different purification degrees (99.5%, 99.7% and 82.1% purity) in 30 L and 150 L fermenters. When glycerol of purity 99.3% was used in a 30 L fermenter, the maximum total biomass concentration and PHB content in the CDW was 69.3 g/L and 72.4%, respectively [228]. A further increase in glycerol purity to 99.7% resulted in similar results at 30 L. Interestingly, the use of crude glycerol of purity 82.1% did not significantly decrease total biomass concentration (69.3 g/L) or PHB content in the CDW (78.1%). When PHB biosynthesis by *C. eutrophus* B-10646 on glycerol of purity 99.7% was scaled-up from 30 L runs to 150 L, the total biomass, %-PHB of the CDW and overall productivity reached 110 g/L, 78% and 1.83 g/L/h, respectively [228]. Given the promising results at 30 L with crude glycerol, it would be interesting to see if similar improvements in PHB production efficiency from crude glycerol would be obtained using the same process conditions at 150 L. Zhu et al., reported on cultivation of *B. cepacia* ATCC 17759 for biosynthesis of PHB using crude glycerol. The fermentation process, when scaled to 200 L, gave 23.6 g/L biomass of which 31% was P3HB [229].

Kachrimanidou et al., reported cultivations of *C. necator* DSM 7237 operated in fed batch mode that utilized sunflower meal and crude glycerol as co-substrates for P3HB formation. When levulinic acid and sunflower meal hydrolysate were used as co-substrates, the product form was P(3HB-*co*-3HV) [230]. Sunflower meal served as a source of both inorganic phosphorus and free amino nitrogen that were critical components to achieve substantial PHA formation and cell growth. Cultivations conducted in fed-batch mode resulted in a total biomass, %-PHB of the CDW and overall productivity of 37 g/L, 72.9% and 0.28 g/L/h, respectively [230]. Continuous feeding of levulinic acid results in a total biomass, %-PHB of the CDW and overall productivity of 35.2 g/L, 66.4% P(3HB-*co*-22.5 mol%3HV) content and 0.24 g/L/h [230]. This further reinforces the role of levulinic acid as a precursor substrate for 3HV formation and incorporation in PHAs.

7.3. Sugar-Cane Molasses as Carbon Source

Molasses is a sugar-rich byproduct from sugar beet and sugarcane refining into sugar [231]. It has been extensively studied as a substrate for industrial-scale fermentation processes due to its abundance and low price. Molasses contains predominantly 50% sucrose with lower quantities of glucose and fructose, about 4% protein, trace elements calcium, magnesium, potassium, and iron and vitamins H or B7 [232]. Jiang et al., used sugarcane molasses as substrates for PHA formation by *P. fluorescens* A2a5. In a 5 L bioreactor, the total biomass, %-P3HB in the CDW and overall productivity reached 32 g/L, 68.75% and 0.23 g/L/h, respectively [233].

Kulpreecha et al., reported the use of urea and sugarcane molasses as nitrogen and carbon sources, respectively, for cultivations of *B. megaterium* BA-019. Cultivations were conducted in fed-batch mode under pH-stat feeding control. Under conditions where the molasses feeding solution was 400 g/L and the C/N molar ratio was 10:1, the total biomass, %-PHB in the CDW and overall productivity reached 72.6 g/L, 42% and 1.27 g/L/h, respectively [38]. In a later study, Kanjanachumpol et al., used the same production organism and substrates but changed the following process variables: C/N ratio was increased to 12.5/1 and an intermittent feeding strategy was implemented [234]. As a result, the total biomass, %-PHB in the CDW and overall productivity reached 90.7 g/L, 45.8% and 1.73 g/L/h, respectively [234].

Sugarcane vinasse is the final by-product of biomass distillation from ethanol production using substrates such as sugar crops (beet and sugarcane). It consists of primarily acid-insoluble nitrogen [235], and is rich in melanoidins and phenolics [236]. Dalsasso et al., reported the fed-batch cultivation of *C. necator* for PHB production using sugarcane vinasse and molasses as mixed substrates. Addition of vinasse to the molasses cultivation medium resulted in an increase from 0.19 to 0.36 h^{-1} in the maximum specific growth rate [237]. This was attributed to the presence of organics in vinasse that were rapidly consumed. Molasses was added at two-time intervals and the cultivation gave 20.9 g/L biomass of which 56% of the CDW is PHB [237]. The nitrogen content in both molasses and vinasse (around 1.8 g/L) caused an uncharacteristic rise during the P3HB formation stage in the biomass concentration.

7.4. Green Grass as Carbon Source

Recently, researchers investigated the potential of transforming green grass into substrates for the formation of mcl-PHA production. Davis et al., reported on mcl-PHA formation from perennial ryegrass biomass. The pretreatment of grass biomass results in highly digestible (~75%) substrate rich in C5 and C6 sugars for mcl-PHA formation. While the authors provided information on the fact that *P. putida* W619 and *P. fluorescens* 555 utilized the pretreated perennial ryegrass substrate forming 25-34% mcl-PHA of the CDW, they did not report the total biomass and overall productivity [238].

7.5. Starch as Carbon Source

Many researchers have explored starch as low-cost carbon source in industrial fermentation processes. While starch is not a waste source as it can be directly used by humans in foods; the high productivity of corn by US farmers has created excess starch that is available for fermentation to produce chemicals. As the demands grow for starch in fermentation to produce chemicals or the population increases, it will be necessary to avoid further consumption of starch that is needed to feed the population. That said, some bacterial strains do not produce α-amylase; hence this enzyme needs to be added externally to hydrolyze starch. Haas et al., reported fed-batch HCDC of *R. eutropha* NCIMB 11599 using saccharified (converted to sugars) waste potato starch as the carbon source. They reported a total biomass, %-PHB in the CDW and overall productivity of 179 g/L, 52.5% and 1.47 g/L/h, respectively [239]. While residual maltose accumulated in the bioreactor, it did not cause a noticeable inhibition of cell growth or metabolic processes leading to PHB. Chen et al., explored extruded starch for production of PHA using extremely halophilic Archaeon *H. mediterranei*. Starch was extruded with yeast extract at a weight ratio of 1/1.7 to attain a favorable carbon-to-nitrogen ratio in cultivations. Fermentation were operated in fed-batch mode using pH-stat control. The resulting total biomass, %-PHA in the CDW and overall productivity reached 39.4 g/L, 50.8% P(3HB-*co*-10.4 mol%3HV) and 0.29 g/L/h [240].

7.6. Whey, Wheat, and Rice Bran as Carbon Sources

Under the section of PHA production using genetically modified organisms (Section 3), we presented work on whey utilization by recombinant *E. coli* [92,241]. Park et al., used the recombinant *E. coli* strain CGSC 4401 that harbors biosynthetic PHA biosynthetic genes of *A. latus* [242]. In contrast to the above work by Ahn et al. [92], they conducted studies at low lactose feed rates and did not use concentrated oxygen for oxygen supply to cultures. By maintaining lactose concentrations (<2 g/L), low cell growth and PHA contents resulted (12 g/L and 9%, respectively) [237]. However, at 20 g/L lactose, the total biomass concentration, %-PHB in the CDW, and overall productivity reached 51 g/L, 70%, and 1.35 g/L/h, respectively [242]. This fermentation, when scaled to 300 L, resulted in a total biomass, %-PHB in the CDW and overall productivity of 30 g/L, 67% and 1 g/L/h, respectively [242].

Wheat bran, that contains substantial quantities of hemicellulose and cellulose, is a rich source of proteins, carbohydrates, and other minerals. Annamalai and Sivakumar used *R. eutropha* NCIMB 11599 as the P3HB production strain and wheat bran hydrolysate as the source of carbon and other nutrients. This work resulted in a total biomass, %-PHB in the CDW and overall productivity of 24.5 g/L, 62.5%, and 0.255 g/L/h, respectively [243]. An important observation by these workers is that wheat bran hydrolysate did not cause any apparent toxicity to the PHB production strain.

Huang et al., reported the production of PHA by *H. mediterranei* using as carbon sources corn starch and rice bran that was first extruded to enhance the ease by which cells can utilize these substrates [244]. This strain functions in hyper-saline conditions that virtually circumvents contamination problems. Furthermore, PHA downstream processing is simplified as the haloarchaea is readily lysed in distilled water giving PHA pellets that can be recovered by centrifugation at low speeds. Rice bran and starch, extruded in a 1:8 (g/g) ratio, was used as the fermentations source of carbon nutrient. The authors employed a pH-stat control strategy to maintain pH at 6.9 to 7.1 in a 5 L fermenter. Cultivations were operated in repeated fed-batch mode and the pH was maintained between 6.9–7.1 using a pH-stat. They reported a total biomass and %-PHB in the CDW of 140 g/L and 55.6% [244]. The development of processes for PHA formation by halophiles such as *H. mediterranei* and increasing rice bran and starch access by cells using extrusion are notable achievements by the authors.

7.7. Waste Vegetable Oils and Plant Oils as Carbon Sources

Triglycerides are important substrates for PHA production as they have a high density of carbon by weight, are directly metabolized to acetate (an excellent PHA substrate), are used to form 3HHx units in copolymers and can form mcl-PHAs. However, triglyceride derived fatty acids produced from edible oils is not economically feasible. Also, the widespread use of edible oils as fermentation feedstocks, oleochemicals and biodiesel production can lead to a food crisis [245]. Waste frying oil has been reported as a useful substrate by many bacterial strains for PHA formation. In cases where triglycerides are used directly, their rate of hydrolysis to fatty acids and glycerol may limit cell growth and product formation. Approaches to mediate these drawbacks are: (i) conversion of triglycerides to fatty acids or their methyl esters prior to cultivations, (ii) amending media with lipases and (iii) emulsification of triglycerides with non-toxic surfactant to increase their availability [246].

Tufail et al., investigated the utility of waste frying oil as a substrate for PHA formation. Cultivations were conducted in shake flasks containing 2% waste frying oil. Oil droplets were formed by sonication and subsequent shaking. The maximum PHA formed from waste frying oil (53.2% PHA and total biomass 23.7 g/L) was by *P. aeruginosa* (KF270353), cultured for 72 h at 100 rpm [247].

Obruca et al., investigated PHA formation by *C. necator* H16 using waste frying rapeseed oil as the carbon source. Cultivations operated in fed-batch mode resulted in a total biomass, %-PHA in the CDW and overall productivity of 138 g/L, 76% (i.e., 105 g/L PHB) and 1.46 g/L/h, respectively [248]. Furthermore, the substrate conversion efficiency reached 0.83 g PHB per g oil. This is an extraordinary result showing the value that waste rapeseed oil can provide in the cultivation of prepared PHAs. By addition to cultivations of 1% v/v propanol, biomass and PHA formation was increased. Also, the PHA formed contained 8 mol% 3HV units in the copolymer structure [248].

Cruz et al., focused on processes that will lead to efficient conversions by *C. necator* DSM 428 of waste cooking oil (WCO) to PHA. Operational strategies to increase cultivation efficiencies included exponential substrate feeding and use of DO-stat for process control. Utilizing exponential feeding, the total biomass, %-PHB content in the CDW and overall productivity reached 21.3 g/L, 84% and 0.1875 g/L/h, respectively [249]. A substantial improvement in the productivity of PHB formation (to 0.525 g/L/h) resulted from controlling pH with ammonium hydroxide as well as using a DO-stat for process regulation [249].

Ruiz et al., used *P. putida* KT2440 as the PHA production strain and fatty acids derived from WCO as the carbon source. The implemented feed strategy delayed the stationary phase to enhance biomass and PHA formation. Use of intermittent feeding and conditions that led to HCDC, *P. putida* KT2440 reached a total biomass, %-mcl-PHA content in the CDW and overall productivity of 159.4 g/L, 36.4% (i.e., 57 g/L mcl-PHA) and 1.93 g/L/h, respectively [250]. This bioprocess provides a strong foundation that, with further process optimization, could lead to a commercially viable process for conversion of WCO to PHA.

Ruiz et al., using *P. chlororaphis* 555 as the production strain, developed a HCDC bioreactor-based process to convert WCOs from restaurants to mcl-PHA. The composition of the WCO is palmitic acid (7.9%), stearic acid (42.3%), oleic acid (42.3%), linoleic acid (32.2%), and others (0.7%) [251]. In batch bioreactor experiments, when 60 g/L of WCO was supplied during a 30 h fermentation, the total biomass, %-mcl-PHA content in the CDW and overall productivity reached 45.5 g/L, 19.8%, and 0.30 g/L/h [251]. The authors then implemented a pulse feeding strategy controlled by the DO-stat. That is, increase in the DO above 20% prompted substrate feeding. They also implemented nutrient limitation (phosphorous) during PHA formation. The resulting cultivation in which 140 g WCO/L was supplied over 48 h resulted in a total biomass, %-PHA content, and overall productivity of 73 g/L, 19%-P(20 mol%3HDD-37 mol%3HD-36 mol%3HO-7 mol%3HHx) and 0.29 g/L/h, respectively [251].

Thuoc et al., reported the use of glycerol and waste fish oil as substrates for PHA formation. The PHA-forming bacterium *Salinivibrio* sp. M318 was isolated from fermenting shrimp paste. By operating cultivations in fed batch mode, the total biomass, %-PHB and overall volumetric productivity reached 69.1 g/L, 51.5% and 0.46 g/L/h, respectively [252]. The isolation and development of PHA-forming microbes that can use waste materials from aquaculture is important given the large volume of these waste materials.

Da Cruz Pradella et al., explored pulsed feeding of soybean oil to attain high yield and productivity of PHB in *C. necator* DSM 545. By using pulsed feeding strategy, the total biomass, %-PHB and overall volumetric productivity reached 83 g/L, 80% and 2.5 g/L/h, respectively [253]. Shang et al., reported mass production of mcl-PHA from corn oil hydrolysate in *Pseudomonas putida* KT2442. By using fed-batch cultivation, the total biomass, %-PHB and overall volumetric productivity reached 103 g/L, 27.1% and 0.61 g/L/h, respectively [254].

7.8. Wastewater for PHA Production

Wastewater is a potential source of carbon and other nutrients for PHA production. Ryu et al., focused on swine wastes water augmented with yeast extract, inorganic salts and glucose as substrates for *A. vinelandii* UWD cultivations. Supplementing batch cultures of swine wastewater with 30 g/L glucose, resulted in about a six-fold increase in both cell growth and PHA formation [255]. These increases varied based on the extent of swine wastewater dilution. At 50% dilution, the total biomass reached 9.4 g/L that contained 58%-PHA [255]. This work is an example of where, to achieve suitable production of PHA from a waste source, supplementation with another carbon source is required.

Yan et al., assessed the potential use of activated sludge as a source of carbon and nutrients for PHA formation. Substrates were obtained from full-scale wastewater plants that include those from cheese, starch, municipal and pulp/paper manufacturing facilities. These wastewaters were used as sources of PHA producing microbes for shake flask experiments. VFAs from wastewaters were believed to function as carbon substrates. Activated sludge from pulp and paper mills gave maximal PHA formation (43% of suspended solid dry weight) [256].

Table 6. Production of scl- and mcl-PHA using renewable resources.

Microorganisms	PHA Type	Carbon Source	Biomass (g/L)	%-PHA	Overall Productivity (g/L/h)	Carbon Conversion Efficiency $(Y_{p/s})$ g/g	Reference
Cupriavidus necator DSM 545	P(3HB-co-4HB)	Waste glycerol/γ-butyrolactone	30.19	36.1	0.17	0.06	[23]
	P(3HB-4HB-3HV)	Waste glycerol/γ-butyrolactone/propionic acid	45.25	36.9	0.25	0.08	
Bacillus megaterium BA-019	PHB	Sugarcane molasses	72.6	42.1	1.27	NA	[38]
C. necator DSM 545	PHB	Waste glycerol	104.7	62.7	1.36	NA	[41]
Zobellella denitrificans MW1	PHB	Crude glycerol	81	66.9	1.09	0.25	[48]
Burkholderia sacchari DSM 17165	PHB	Wheat straw hydrolysate	146	72	1.6	0.22	[211]
Burkholderia sacchari DSM 17165	P(3HB-co-4HB)	Wheat straw hydrolysate/-butyrolactone	88	27	0.5	NA	[212]
Wautersia eutropha NCIMB 11599	PHB	Wheat hydrolysate	175.05	93	0.89	0.47	[217]
Paraburkholderia sacchari IPT 101 LMG 19450	PHB	Hardwood hydrolysate	59.5	58	0.72	0.15	[219]
Cupriavidus necator DSM 4058	PHB	Crude glycerol/rapeseed meal	28.86	85.75	0.21	0.32	[226]
Paracoccus sp. LL1	P(3HB-co-3HV)	Crude glycerol	24.2	39.3	NA	0.136	[227]
Cupriavidus eutrophus B-10646	PHB	Glycerol (99.3% purity) (30 L fermenter)	69.3	72.4	NA	NA	[228]
		Glycerol (99.7% purity) (30 L fermenter)	69.4	73.3	NA	NA	
		(Glycerol 82.1% purity) (30 L fermenter)	69.3	78.1	NA	NA	
		Glycerol (99.7% purity) (150 L fermenter)	110	78	1.83	NA	
Burkholderia cepacia ATCC 17759	PHB	Crude glycerol	23.6	31	NA	NA	[229]
C. necator DSM 7237	PHB	Crude glycerol and sunflower meal hydrolysate	37	72.9	0.28	0.32	[230]
	P(3HB-co-3HV)	Sunflower meal hydrolysate and levulinic acid	35.2	66.4	0.24	0.28	
P. fluorescens A2a5	PHB	Cane liquor medium	32	68.75	0.23	NA	[233]
Bacillus megaterium BA-019	PHB	Sugarcane molasses	90.71	45.85	1.73	NA	[234]

Table 6. Cont.

Microorganisms	PHA Type	Carbon Source	Biomass (g/L)	%-PHA	Overall Productivity (g/L/h)	Carbon Conversion Efficiency $(Y_{p/s})$ g/g	Reference
Cupriavidus necator	PHB	Sugarcane vinnase and molasses	20.89	56	NA	NA	[237]
R. eutropha NCIMB 11599	PHB	Waste potato starch	179	52.51	1.47	0.22	[239]
R. eutropha NCIMB 11599	PHB	Wheat bran hydrolysate	24.5	62.5	0.25	0.32	[243]
Haloferax mediterranei ATCC 33500	P(3HB-co-3HV)	Extruded rice bran/extruded cornstarch/yeast extract	140	55.6	3.2	NA	[244]
H. mediterranei ATCC 33500	P(3HB-co-3HV)	Enzymatic extruded starch	39.4	50.8	0.29	NA	[244]
P. aeruginosa STN-10	P(3HB-co-3HV)	Waste frying oil	44.71	53	0.33	NA	[247]
Cupriavidus necator H16	P(3HB-co-3HV)	Waste rapeseed oil/propanol	138	76	1.46	0.83	[248]
Cupriavidus necator DSM 428	PHB	Used cooking oil (Exponential profile)	21.3	84	0.1875	0.65	[249]
	PHB	Used cooking oil (DO stat strategy)	27.2	77	0.525	0.52	
Pseudomonas putida KT2440ATCC 47054	PHD/PHO	Hydrolyzed waste cooking oil	159.3	36.4	1.93	0.76	[250]
P. chlororaphis 555	PHD/PHDD/PHO/PHHx	Waste cooking oil	73	19	0.29	0.11	[251]
Salinivibrio sp. M318	PHB	Waste fish oil and glycerol	61.1	51.5	0.46	0.32	[252]
C. necator DSM 545	PHB	Soybean oil	83	80	2.5	0.83	[253]
P. putida KT2442	PHO/PHD	Corn oil hydrolysate	103	27.1	0.61	NA	[254]

$Y_{p/s}$: PHB yield to substrate (g PHB produced per g substrate consumed); %-PHA: Final intracellular PHA content (g/g) of DCW; NA: Data not available, PHB: Polyhydroxybutyrate, P(3HB-co-4HB): Poly(3-hydroxybutyrate-co-4-hydroxybutyrate), P(3HB-co-3HV): Poly (3-hydroxybutyrate-co-3-hydroxyvalerate), P(3HB-4HB-3HV): Poly(3-hydroxybutyrate-4-hydroxybutyrate-3-hydroxyvalerate), PHO: Polyhydroxyoctanaote, PHD: Polyhydroxydecanoate.

8. Global PHA Producer Companies at Pilot and Industrial Scale

Pilot- and industrial-scale PHA manufacturers along with product trade names, PHA type, raw material, production capacity and estimated prices are listed in Table 7. At present, numerous companies are pursuing the manufacture of PHB and its copolymers at pilot and industrial scales. Some PHA-producing companies are focused on PHA commercialization for high-value biomedical applications. These include Terra Verdae Bioworks (Edmonton, AB, Canada), Tepha Inc. (Lexington, MA, USA), and PolyFerm Canada (Kingston, ON, Canada). Their products range from heart valves, scaffolds, biodegradable sutures, and materials for controlled delivery. These product sectors are smaller in volume but have high profit margins. Further discussion of these companies and their activities is given below.

Danimer Scientific (previously MHG), USA, produces PHA under tradename of Nodax™ (Bainbridge, GA, USA). They entered the PHA space in 2007 by purchasing intellectual property from Procter and Gamble Co (Cincinnati, OH, USA). Danimer produced a copolymer consisting of 3HB units and (R)-3-hydroxyalkanoate comonomer units with mcl-PHA repeat units [257]. Nodax™ PHA is 100% renewable and is produced from readily available feedstock. The company claims that custom formulation of Nodax™ PHA can provide application-specific tailored plastic resins [258].

PHB industrial (Brazil), produced and marketed biodegradable plastic, PHB and P(3HB-co-3HV) under the trade name of Biocycle®. The production of these polymers was inaugurated in September 2000 by PHB Industrial (Sao Paulo, Brazil) [259]. The industry operated on a pilot scale until 2015 as well as exporting products to Japan. However, the industrial plant is now defunct.

The mission of Bio-on, founded in 2007, was to be integrated with the agri-food industries that they believed would provide the technologies needed to produce PHAs. Unfortunately, Bio-on announced bankruptcy in 20 December 2019 [260]. Bio-on then founded LUX-ON whose focus was to develop PHA manufacturing using carbon dioxide as the feedstock. Their technology also harvests renewable solar energy to power bioproduction process. Bio-on bioplastic currently uses feedstocks that include, sugar cane molasses, sugar cane, food wastes, WCO, glycerol and carbohydrates [261].

TianAn Biopolymers (Ningbo, China), Ecomann Biotechnology (Shenzhen, China), and Tianjin GreenBio Materials (Tianjin, China), and were said to possess a cumulative capacity of 15,000 tonnes/year to produce the PHA materials discussed below [262,263].

TianAn Biopolymers (Ningbo, China) is the world's largest producer (10,000 tons/year) of P(3HB-co-3HV), marketed as ENMAT™. They produce the copolymer by fermentation of *C. necator* on D-glucose and propionic acid. They claim that P(3HB-co-3HV) serves as the primary component in materials for injection molding, thermoforming, blown films and extrusions. They also supply PHB (ENMAT Y3000 and Y3000P) along with P(3HB-co-3HV)/PLA blends (ENMAT F9000P) [262].

Tianjin GreenBio Material Co. is the first company in China to produce 10,000 tonnes of PHA per year. They developed the pellets (SoGreen 2013) for blown film processing that are fully biodegradable. They also developed PHA foam pellets that can be made into fully-biodegradable foams for the food service industry as well as appliance packaging. The PHA they produce consists of P(3HB-co-4HB) copolymer [263].

In 2007, Biomer Biotechnology Co. (Germany) manufactured at both the pilot and research scales 10 tons/year in under the name Biomer™ Biopolyester [264]. To the best of our knowledge, the current status of this company is unclear.

Kaneka Corporation (Japan) manufactures P(3HB-co-3HHx), under the tradename Kaneka PHBH® and AONILEX®, the reported production capacity in 2007 was 100 tons/year. It is produced by a microorganism fermentation process, in which plant oils and its fatty acids are used as the primary raw material. As of 2019, Kaneka Corporation was producing 5000 tons per year [265].

Yield10 was launched by Metabolix, Inc. in 2015 and is traded on the Nasdaq. Yield10 Bioscience produces PHB and its copolymers under the trade name of Mirel™ [266]. Commercialization of Mirel™ resulted when Archer Daniels Midland Company and Metabolix entered into the joint venture called Telles.

Tepha's TephaFLEX® PHA consists of 4HB units (i.e., P4HB). After purification, P4HB is processed into medical products that include films, sutures and textile products [267]. The hydrolysis of P4HB leads to 4HB that is a normal constituent of the mammalian body. Tepha's TephELAST is more elastic than TephaFLEX and is also being applied to develop medical devices. These polymers are produced at both pilot and research scales.

Mitsubishi Gas Chemical (Japan) produces P(3HB) under the trade name Biogreen® at both research and pilot scales [268].

Polyferm Canada produces mcl-PHA under the tradename of VersaMer™ from naturally selected organisms and feedstocks such as vegetable oils and sugars [269]. They are currently developing applications for medical devices, sealants, adhesives, plastic additives and more.

Metabolix, a Massachusetts-based company, completed the sale (US$10 million) of its intellectual property concerning polyhydroxyalkanoate (PHA) biopolymers, to an affiliate firm CJ CheilJedang Corp (Seoul, South Korea). The sale includes production and application patents as well as microbes used in Metabolix's production processes [270].

Newlight Technologies, using a microorganism isolated from the Pacific Ocean, developed a proprietary biocatalyst for PHA production from carbon dioxide and methane. Newlight takes methane from a dairy farm in California that it mixes with air to form a polymer they trademarked as AirCarbon [271,272]. Information on the production capacity and composition of PHA produced by Newlight technologies is currently not available. However, as a validation of the potential of Newlight technologies, Newlight signed a deal in 2015 with Vinmar International, a petrochemical distributor [273]. Vinmar agreed to purchase over the next 20 years 450 million tons of AirCarbon PHA. Furthermore, once Newlight constructs its production facility (23 million tons/year), Vinmar has committed to purchase 100% of the AirCarbon PHA produced at this site.

Table 7. Pilot and industrial-scale PHA manufacturers currently active worldwide.

Company	Trade Name	PHA Type	Raw Material	Production Capacity (tons per year)	Price per kg	References
Danimer Scientific (Bainbridge, GA, USA)	Nodax™	PHBH	NA	20	NA	[257,258]
PHB Industrial (Sao Paulo, Brazil)	Biocycle®	PHB, P(3HB-co-3HV)	Sugarcane	600	NA	[259]
Bio-On, (Bologna, Italy)	Minerv-PHA™	PHA	Sugar beet and sugar cane molasses, other waste-derived feedstocks.	10,000	NA	[261]
TianAn GreenBio Materials Co. Biopolymer, (Ningbo, China)	ENMAT™, SoGreen™	PHB, P(3HB-co-4HB)	Dextrose	10,000	NA	[262]
GreenBio-DSM (TEDA Tianjin, China)	Ecoflex blend Enmat®	P(3HB-co-4HB), P(3HB-co-3HV) + Ecoflex blend Enmat®	NA	10,000	€3.26	[263]
Biomer Biotechnology Co. (Schwalbach, Germany)	Biomer® biopolyesters	PHB	Glucose from corn starch	50	€3.00–5.00	[264]
Kaneka Corporation (Minato-ku, Tokyo, Japan)	PHBH™	P(3HB-co-HHx)	Vegetable oil	5000	NA	[265]
Yield10 Bioscience, (formerly Metabolix, Inc.), (Woburn, MA, USA)	Mirel™	PHB copolymers (e.g., P(3HB-co-HHx-co-HO)	Corn sugar	50,000	NA	[266]
Tepha Inc (Lexington, MA, USA)	TephaFlex® TephElast®	P4HB, P3HB-co-4HB	NA	NA	NA	[267]

NA: Data not available.

9. Downstream Processing of PHA

Downstream processing constitutes a critical step in PHA manufacturing with regards to product purity, cost and environmental impact. There are two basic approaches to PHA recovery after fermentation: (i) dissolving the biomass with acids, alkali, surfactants, and enzymes leaving the granules for isolation or (ii) direct solvent extraction of PHAs from the cells that produce it [195].

Methods for the release of PHA granules from cells that largely circumvent solvent utilization, such as enzymatic digestion of non-PHA biomass, are of great interest. With enzyme digestion and other methods that seek to remove biomass from granules, one must consider the extent that outer granule components are removed. Indeed, it is well known that specific proteins are localized on the PHB granule surface. Such proteins include enzymes that function for PHA synthases, PHA depolymerization, polymer surface-displayed proteins (e.g., phasins) and other proteins that regulate granule subcellular localization [274]. In selecting or developing PHA isolation methods, one must balance economic, safety, environmental impact, energy input, recovery yield, PHA purity and ease of scale-up [275]. Other factors include the PHA producing strain, PHA composition and effects on PHA molecular weight [276]. For example, the PHA composition will influence its solubility in solvents (e.g., mcl-PHAs are highly soluble in numerous solvents including acetone where PHB is not). Another important factor in PHA granule isolation is that the methods used must cause minimal change in the polymers' molecular weight. This is due to that, if the molecular weight of the final melt-processed product falls below the critical chain entanglement molecular weight, the product's mechanical properties will be compromised. Different downstream processing strategies for PHA isolation/purification are depicted in Figure 8.

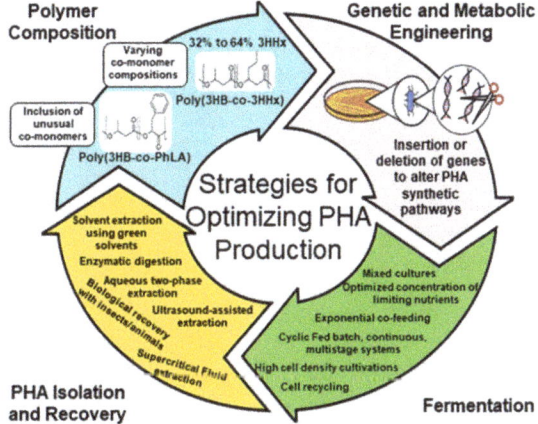

Figure 8. Key areas of PHA production to optimize to obtain high-yields of PHA. To compete with petroleum-based plastics whilst remaining as environmentally friendly as possible, PHA production requires constant innovation and optimization in four major areas. The cyclic arrows illustrate that any modification to one of the four categories will affect the following stage of production. The strategies presented herein have been proposed due to the published success regarding improved PHA production yields or PHA properties.

9.1. Solvent Extraction

This is a common method used due to its ease of operation and simplicity. Solvents rupture the cell increasing cell membrane permeability. Subsequently, solvent access to PHA granules results in solubilization of the polymer. To isolate the polymer, it is precipitated using a non-solvent. Acetone is a preferred solvent for mcl-PHA extraction while chilled methanol or ethanol are preferred non-solvents. Solvent extraction is advantaged in that it results in high-purity PHA with little or no change in PHA molecular weight [277].

9.1.1. Halogenated Solvents

Examples of halogenated solvents used for PHA extraction include 1,2-dichloroethane, methylene chloride and chloroform [278]. Studies show that PHA recovery yields as well as PHA purity are high using these solvents [279]. The most used solvent for extraction is chloroform that, unfortunately, is toxic and increases process cost. PHAs from chloroform extraction have low contents of endotoxin, which is a high priority for PHAs used in medical applications. Several countries have banned the use of chlorinated solvents in consumer products.

9.1.2. Non-Halogenated Solvents

Considering the problems encountered using chloroform or other halogenated solvents, several commercial producers have published patents on chlorinated solvent alternatives [280–286]. While non-halogenated solvents for polymer extraction may be less harmful to personnel carrying out PHA purification, their sustainability requires careful life-cycle analysis. Also, the inherent process requirements including solvent recycling requires technoeconomic analysis.

Extraction of scl-PHA that, due to their structures (i.e., copolyesters) have low crystallinity, may result in acetone, ethyl acetate, methyl-isobutyl ketone, and cyclohexanone. Extraction of mcl-PHAs can be accomplished at room temperature using diethyl ether, tetrahydrofuran and acetone [276].

As mentioned above, an important consideration during PHA isolation, especially where the final application is as a medical product, is the extent that endotoxins (e.g., lipopolysaccharides) are effectively removed. One approach that has shown promise is extraction methods that are temperature-controlled as demonstrated by Ferrur et al., for the recovery of P(3HO-*co*-3HHx) from *P. putida* GPo1 [287]. Solvents such as 2-propanol and *n*-hexane are particularly useful, giving PHA purities >97% (w/w) and low endotoxin levels (i.e., 10–15 endotoxin units, EUs) per g PHO. Furthermore, redissolution of PHO at 45 °C in 2-propanol and subsequent precipitation of PHO at 10 °C gave a high purity product (nearly 100%) with correspondingly low endotoxicity (2 EU/g PHO) [287].

Koller et al., explored the potential of applying elevated temperatures and pressures to enhance the ability of acetone to extract scl-PHAs. The results of this approach were compared with solvent extraction at ambient temperature and pressure with chloroform. Extraction of scl-PHA using acetone performs similarly to chloroform in terms of extraction yield (96.8% by both methods) and scl-PHA purity (98.4 vs. 97.7%) [276]. However, extraction using acetone is much more rapid than using chloroform (20 min vs.12 h).

Non-halogenated solvents including acetic acid, methanol, hexane, dimethyl formamide, dimethyl sulfoxide and ethylene carbonate were assessed for PHB extraction from *C. necator* cells [278]. Maximum PHB recovery yield and product purities (98.6 and 98%, respectively) was achieved by using ethylene carbonate at 150 °C for 60 min [278]. Some solvents function to extract lipids from cell debris [288].

9.1.3. Green Solvents

Green processes for PHA extraction will require low energy input, mild conditions, avoid toxic chemicals, and deliver PHA products in high purity and extraction yield [289]. Advantaged solvents include dimethyl carbonate that is fully biodegradable and those that can be produced from biomass and/or bioprocesses from readily renewable resources. Examples of the latter include ethanol and esters of acetate and lactate.

Yabueng et al., explored the use of green solvents for PHB recovery from *C. necator* strain A-04 [290]. Solvents studied included ethyl lactate, 1,3-dioxolane, 2-methyltetrahydrofuran and 1,3-propanediol. The results obtained were compared with that by chloroform extraction. Water was mixed with 1,3-dioxolane to give a water-miscible system. Dried *C. necator* cells containing PHB were extracted at 80 °C for 6 h in 1,3-dioxolane. Subsequently, added water resulted in the phase separation of P3HB and 1,3-dioxolane. This method resulted in isolation of 97.9% ± 1.8% pure PHB with a 92.7% ± 1.4% recovery yield [290]. Analysis of recovered PHB showed that its molecular weight was unaffected.

Characteristic features of ionic liquids are their low vapor pressures and their ability to be tuned to dissolve a wide range of water-insoluble materials [291]. However, one must take care that the produced PHA does not contain residual ionic liquids that may be toxic. Furthermore, recyclability of the ionic liquid is required for economically viable processes.

Dubey et al., reported ionic liquid-based extraction of PHB from *Halomonas hydrothermalis* [292]. The ionic liquid, 1-ethyl-3-methylimidazolium diethylphosphate (EMIM DEP), due to its strong H-bond basicity ($\beta = 1.07$), easily dissolved *Halomonas hydrothermalis* without water removal (i.e., wet biomass). EMIM DEP disrupts the cell membrane leading to PHB solubilization. Subsequently, PHB was separated by precipitation using methanol as the non-solvent. However, the recovery of PHB was low (60%) as was the percent purity (86%) [292]. Other process concerns are the ability to sufficiently clean and recover EMIM DEP for re-use as it is expensive.

9.2. Ultrasound-Assisted Extraction

To reduce PHA recovery cost, it is essential to reach high PHA recovery yields and PHA purities in short time periods while minimizing solvent use [293]. Ultrasound irradiation increases the efficiency of mass transfer during extraction processes. Ishak et al., applied ultrasound irradiation to mcl-PHA containing cells that were suspended in a good and marginal non-solvent mixture [293]. The term marginal non-solvent is meant to describe a substance that, when alone, functions as a non-solvent. However, when that solvent is mixed with a good solvent (i.e., when alone it dissolves the PHA) in a suitable ratio, the PHA remains in solution [294]. By increasing the concentration of the marginal solvent, the PHA will precipitate. These workers interrogated the effects of sonication volumetric energy dissipation, solvent/marginal non-solvent ratio, and time on mcl-PHA extraction efficiency. By choosing heptane as the marginal non-solvent, the application of ultrasound resulted in improved PHA extraction yield. By optimizing process conditions, ultrasound irradiation did not negatively affect mcl-PHA molecular weight and efficiently removes endotoxins [277].

Ultrasound-assisted extraction processes were shown to provide higher extraction yields relative to conventional processes. Furthermore, the application of ultrasound during PHA extraction allows the reduction of solvent quantities, the use of safe solvents and decreased process times [295].

9.3. Supercritical Fluid Extraction

Supercritical fluids (SCFs) are substances at a pressure and temperature that is above its critical point. Under these conditions is any substance at a temperature and pressure that is above its critical point, where separate gas and liquid phases are non-existent. Advantageously, diffusion and solvation of SCFs have properties akin to a gas and liquid, respectively. Similar to a gas, SCFs can diffuse through materials and as well as function as liquids by dissolving polymers [279].

The most frequently used SCF is that of carbon dioxide (sCO$_2$). Work has demonstrated the ability of sCO$_2$ to function in biomolecule purification and separation processes. Further advantages of sCO$_2$ is that it is non-toxic, non-flammable and inert chemically [296]. Also, sCO$_2$ functions at moderate temperatures (31 °C) and pressures (74 bar). Removal of sCO$_2$ subsequent to its use is accomplished by vaporization at reduced pressure. This technique causes cell disruption that allows isolation of PHA in purities that vary between 86 and 99% purity. Raza et al., claimed that, under optimized conditions, supercritical fluid extraction is highly effective providing impurity-free PHAs for medical applications [297]. Supercritical CO$_2$ provides a green solvent alternative that deserves further investigation to optimize process efficiency [298]. Primary disadvantages of this method include high capital and maintenance costs.

Hampson and Ashby used sCO$_2$ to isolate/purify mcl-PHAs produced by *Pseudomonas resinovorans*. The process used consists of treating the dry cell biomass with sCO$_2$ to extract lipid impurities which accounted for 2–11% of the non-PHA biomass [299]. However, these authors then turned to chloroform for isolation/purification of the mcl-PHA. The resulting extraction yield was relatively low (42.4%) [299]. However, the introduction of an initial sCO$_2$ step did decrease the quantity of chlorinated solvent used in the second step.

Hejazi et al., also explored the potential use of sCO$_2$ for separation of non-PHA biomass from the product. They varied multiple process parameters that include temperature, pressure, time and even use a co-solvent modifier. The optimized process was conducted at 40 °C, 200 atm and 100 min using low quantities of methanol. PHB recovery from *R. eutropha* cells reached 89% [300]. This percent recovery approaches that of other methods described above while it has an improved environmental footprint.

9.4. Aqueous Two-Phase Extraction

Purification by an aqueous two-phase extraction (ATPE) involves the formation of two coexisting immiscible phases when one polymer and an inorganic salt are mixed beyond the critical concentration in water. Special features of ATPE is that it consists of higher water content (up to 90% *w/w*) affording environmentally friendly conditions for biopolymer separation. Moreover, the ATPE phase-forming components can be non-toxic and compare well to the solvents used in conventional extraction processes. In addition, ATPE reduces the number of purifications step and has high purification capabilities. ATPE has been claimed to provide a readily scalable high-efficiency economic process for PHA purification [13,301,302].

Leong et al., reported optimization of PHB extraction from *R. eutropha*. Using PEG 8000/sodium sulphate at pH 6 and 0.5 M NaCl resulted in a recovery yield of 65% [301]. To enhance the biopolymer recovery, cloud-point extraction (CPE) was used [302]. This allowed reuse of the phase-forming component by using a thermo-responsive polymer (i.e., ethylene oxide-propylene oxide, EOPO). EOPO consists of polyethylene oxide (PEO) and polypropylene oxide (PPO) blocks that are most often synthesized to give PPO-PEO-PPO triblocks or PEO-PEO deblocks. CPE of PHB from *R. eutropha* cells was performed with 20% *w/w* EOPO (Mol. wt. 3900 g/mol), NaCl (10 mM) and a phase-separation temperature of 60 °C. The recovery yield was 94.8% with a 1.42 purification factor [302]. The authors demonstrated the EOPO 3900 could be reused at least two times without a significant effect on the recovery yield or the purification factor. Subsequently, Leong et al., reinvestigated the conditions for isolation/purification of PHB from *C. necator* [303]. The process was modified such that the wt-% of EOPO 3900 and ammonium sulfate were 14 and 14, respectively, at pH 6. Under these conditions, extra centrifugation steps were circumvented and the recovery yield and purification factor reached 72.2% and 1.61-fold, respectively [303].

Subsequently, Leong et al., adopted an ATPE process [304]. As above, the removal of non-PHA biomass for to isolate/purify a PHA focused on PHB from *C. necator* H16. Optimized process conditions used were EOPO 3900 (5%), pH 6 and a fermentation temperature of 30 °C. The authors reported a purification factor and recovery yield of 1.36-fold and 97.6%, respectively [304].

9.5. Enzymatic and Chemical Digestion Method

The objective of enzyme and chemical digestion processes is to conserve intact PHA granules by dissolution of non-PHA cell mass (NPCM). Approaches have focused on enzymatic and chemical (acid/alkali) processes. The principle of these approaches is to disrupt microbial cell walls and thereby release PHAs from cells [305,306].

Early work explored removal of NPCM by treating cells with strong oxidizing agents such as sodium hydroxide and sodium hypochlorite. Critical to this approach is careful control of oxidizing agent concentration, temperature and reaction time as extreme conditions result in oxidation of both NPCM and PHAs (i.e., reduced molecular weight). Given the lack of selectivity by strong oxidants, research focused on increasingly selective agents such as acids, anionic surfactants and proteolytic enzymes.

Dong et al., studied combinations of surfactant and sodium hypochlorite for recovery of PHA from *Azotobacter chroococcum* G-3 [307]. The biomass was first pretreated by freezing which liberated the PHA granules that were isolated by centrifugation. Another approach combined sodium hypochlorite and surfactant treatments. In one example, PHA-containing cells were incubated for 15 min with sodium dodecyl sulfate 10 g/L (SDS). This treatment solubilized the lipids and proteins. Subsequently, incubation of the NPCM with sodium hypochlorite (30%, 3 min) eliminated peptidoglycans and other impurities. The combined treatment with surfactant and sodium hypochlorite resulted in a PHA recovery and a final purity of 86.6% and 98%, respectively [307].

Proteolytic enzymes catalyze hydrolytic reactions that target proteins. In 2006, Kapritchkoff et al., used enzymes to lyse *R. eutropha* DSM545 and thereby purify PHB [308]. Preliminary studies revealed that the most promising enzymes for this process are lysozyme, bromelain, and trypsin. The best result used bromelain (14.1 U/mL) at pH 9 and 50 °C which gave PHB in 88.8% purity [308]. Subsequently, pancreatin was assessed for cell lysis and removal of NPCM. This resulted in isolation of 90% pure PHB with a 3-fold reduction in cost relative to bromelain and no loss in PHB molecular weight [308]. The selectivity and mild conditions by which enzymes function provide important advantages relative to other methods that makes it likely that enzymes will continue to be evaluated and process improvements will be made [309]. It also may be that combinations of enzymes as well as enzyme-surfactant systems may further boost the efficiency of PHA purification processes.

Alcalase and lysozyme are both effective in digesting cellular biomass. Yasotha et al., attempted to optimize mcl-PHAs recovery/purification by digesting denatured proteins with alcalase (a protease), SDS to enhance solubilization of NPCM, EDTA (Ethylenediaminetetraacetic acid) for divalent metal complexation and lysozyme to catalyze the decomposition of peptidoglycans in cell walls [310]. Analysis of experiments revealed that alcalase contributed most (71% relative to other process factors) to NPCM breakdown and mcl-PHA isolation/purification. Crossflow ultrafiltration effectively removed NPCM and facilitated mcl-PHA granule recovery (90% efficiency) and in high purity (92.6%) [310].

Kachrimanidou et al., reported the isolation/purification of P(3HB-*co*-3HV) by an enzyme process. They used solid-state fermentation of the *Aspergillus oryzae* to develop a mixture of crude enzymes to catalyze *C. necator* lysis to liberate the PHA copolyesters. The enzymatic process was conducted at 48 °C without pH control. The results of this work are impressive as poly(3HB-*co*-3HV) recovery yield and purity reached 98% and 96.7%, respectively [311].

To isolate/purify from *C. necator* cells amorphous PHB granules, Martino et al., used alcalase (0.3 AU/g), EDTA (0.01 g/g), and SDS (0.3 g/g) P3HB granules were recovered with purities >90% without crystallization [312]. In other words, amorphous PHB granules formed during cultivation remained amorphous under the imposed recovery conditions.

Israni et al., harnessed the high lytic activity of *Streptomyces albus* Tia1 for isolation/purification of PHA from PHA-producing cells [313]. In one approach, *S. albus* and PHA-producing cells were co-inoculated. In the second approach, the lytic enzymes from *S. albus* were introduced at the end of PHA cultivations for PHA recovery. Conditions resulting in the maximum activity for cell lysis of PHA were the addition of a concentrated *S. albus* culture filtrate (33.3 mL) from a 100 mL cultivation to *B. megaterium* (220 mg) with incubation at 40 °C and pH 6. By co-inoculation of *S. albus* and PHA producing *B. megaterium*, the PHA yield reached 0.55 g/g which was similar to treatment with sodium hypochlorite for PHA recovery [313]. In contrast, comparison of the enzyme based with co-inoculation showed that the former resulted in a 1.74-fold increase in PHA yield [313]. The authors attributed this difference to use by *S. albus* of released PHA as a carbon source.

Although enzymatic digestion methods are selective, energy efficient and result in good recovery, it is generally agreed that improvement in process economics is required [288].

9.6. New Biological Recovery Methods

Less traditional methods for PHA biological recovery include the use of insects and animals. These methods eliminate the use of solvents and hazardous chemicals resulting in ecofriendly processes. The bases for these approaches are that insects and animals will utilize lyophilized bacterial biomass but not PHA. Murugan et al., reported that the larval form of the mealworm beetle, *Tenebrio molitor*, can selectively consume NPCM of *C. necator* without breaking down PHA granules that it excretes in its feces [314]. The resulting feces containing PHA were purified using detergent, water and heat giving PHA granules that are almost 100% pure. Furthermore, the molecular weight of the larva-processed isolated PHA was identical to the same product isolated by chloroform extraction [314]. Nevertheless, there are questions about the scalability of such processes.

Kunasundari et al., reported P3HB isolation using laboratory rats [315]. *C. necator* H16 cells where P3HB was 39% of its CDW was fed to rats. The test rats excreted pellets of PHB with 82–97% purity in the form of feces [315]. The polymer so obtained was further purified using detergent, water and heat. The result was highly purified P3HB granules. Molecular weights of rat-processed P3HB was nearly identical to that isolated by chloroform extraction [315].

10. Conclusions: Challenges and Future Perspectives

PHAs have a long history of study both by science/engineering academicians, industrial, and national laboratory researchers who have been working towards successful commercialization of PHA-based materials. These bioplastics offer an appealing platform of polymer compositions that can be integrated into biorefinery processes. It is this vision that has resulted in a history of over 80 years of fundamental and applied studies as well as multiple investments for scale-up and commercialization of PHA products.

The present report describes PHA biosynthesis strategies to obtain desired copolymer compositions where process efficiency was emphasized. The design of microbial fermentation processes depends on the production organism, feedstock selection as well as the cultivation conditions. Implementation of metabolic engineering and evolutionary approaches will lead to continual improvements in PHA production efficiency, copolymer structural diversity and cost-performance metrics. Process development must carefully consider the PHA production organism, consistent availability of feedstocks and metrics such as energy input, carbon conversion efficiency, substrate cost, cell growth rate and PHA production rate. While PHA titer and productivity are important measures of success, many published works do not provide carbon utilization rates which is a critical determinant of a successful outcome. Furthermore, techno-economic analysis and life-cycle analysis are rarely included in published works.

The development of organisms that are metabolically engineered to better produce PHAs from waste materials such as molasses, whey, spent cooking oils, and various hydrolyzed lignocellulosic raw materials will likely play a critical role in the development of robust and cost-effective processes. While much progress has been made, metabolic engineering can avoid the use of feedstocks such as γ-butyrolactone, levulinic acid and fatty acids to produce copolymers with 4HB, 3HV and 3HHx units. Metabolic engineering can introduce metabolic routes to desired monomers from unrelated carbon sources. Halophilic PHA producers have great potential as they allow PHA production to be conducted under non-sterile conditions in media resembling seawater. Furthermore, placing these organisms in DI results in cell lysis easing downstream processing. More attention should be paid to increasing the metabolic potential of these strains to produce the desired copolymer products at high titers and productivity.

There is no question that a longstanding Achilles heel to PHA commercialization has been the need for cost-efficient downstream processing methods. Effective methods will likely not be generic to an organism or product but will be customized and optimized for selected systems. Residual impurities in PHAs can result in melt-processed products of reduced molecular weight and coloration. While enzymatic processes have shown promise, their efficiency must be improved to increase PHA purity and reduce enzyme costs. Routes to recycle enzymes used in removing non-PHA biomass are important as a route to reduce costs. The use of impure enzyme cocktails generated by microbes is interesting but challenging since the cocktail composition must be optimized and result in consistent outcomes.

What is beyond the scope of this review is PHA physical–mechanical properties. While its academically interesting to pursue PHAs with new compositions of matter, it is crucial that application scientists focus on the many tools of material science which includes developing optimal blend compositions and composites that improve mechanical and rheological properties while retaining chemical and biological recycling.

Although there are likely many cynics who, based on the commercial history of PHAs believe it foolish to consider further investments in PHA technologies, in the view of the authors of this review biotechnological developments in concert with work by fermentation engineers, materials scientists and breakthroughs in lignocellulose processing will lead to a future where PHAs will provide sustainable commodity materials for a range of applications.

Author Contributions: Conceptualization, D.K.B. and R.A.G.; writing—original draft preparation, D.K.B. and C.B.E.; writing—review and editing, R.A.G.; supervision, R.A.G.; project administration, R.A.G.; funding acquisition, R.A.G. All authors have read and agreed to the published version of the manuscript.

Funding: The authors are grateful for funding received from the National Science Foundation Partnerships for International Research and Education (PIRE) Program (Award #1243313).

Data Availability Statement: Data supporting reported results can be found in publicly archived datasets such as SciFinder and Scopus.

Acknowledgments: We thank the Center for Biotechnology and Interdisciplinary Science (CBIS) at RPI for use of their Core facilities.

Conflicts of Interest: The authors declare no conflict of interest.

References

1. Lemoigne, M. Dehydration and polymerization product of β-oxybutyric acid. *Bull. Soc. Chim. Biol.* **1926**, *8*, 770–782.
2. Muhammadi, S.; Muhammad, A.; Shafqat, H. Bacterial polyhydroxyalkanoates-eco-friendly next generation plastic: Production, biocompatibility, biodegradation, physical properties and applications. *Green Chem. Lett. Rev.* **2015**, *8*, 56–77. [CrossRef]
3. Blunt, W.; Levin, D.B.; Cicek, N. Bioreactor operating strategies for improved polyhydroxyalkanoate (PHA) productivity. *Polymers* **2018**, *10*, 1197. [CrossRef]
4. Kessler, B.; Witholt, B. Production of microbial polyesters. Fermentation and downstream processes. *Adv. Biochem. Eng. Biotechnol.* **2001**, *71*, 159–182. [PubMed]
5. Sudesh, K.; Abe, H.; Doi, Y. Synthesis, structure and properties of polyhydroxyalkanoates: Biological polyesters. *Prog. Polym. Sci.* **2000**, *25*, 1503–1555. [CrossRef]
6. Lee, S.Y.; Chang, H.N. Production of poly (hydroxyalkanoic acid). *Adv. Biochem. Eng. Biotechnol.* **1995**, *52*, 27–58. [PubMed]
7. Muhr, A.; Rechberger, E.M.; Salerno, A.; Reiterer, A.; Schiller, M.; Kwicien, M.; Adamus, G.; Kowalczuk, M.; Strohmeier, K.; Schober, S.; et al. Biodegradable latexes from animal-derived waste: Biosynthesis and characterization of mcl-PHA accumulated by *P. citronellolis*. *React. Funct. Polym.* **2013**, *73*, 1391–1398. [CrossRef]
8. Muhr, A.; Rechberger, E.M.; Salerno, A.; Reiterer, A.; Malli, K.; Strohmeier, K.; Schober, S.; Mittelbach, M.; Koller, M. Novel description of mcl-PHA biosynthesis by *Pseudomonas chlororaphis* from animal-derived waste. *J. Biotechnol.* **2013**, *165*, 45–51. [CrossRef]
9. Koller, M.; Rodriguez-Contreras, A. Techniques for tracing PHA-producing organisms and for qualitative and quantitative analysis of intra- and extracellular PHA. *Eng. Life Sci.* **2015**, *15*, 558–581. [CrossRef]
10. Singh, M.; Kumar, P.; Ray, S.; Kalia, V.C. Challenges and opportunities for customizing polyhydroxyalkanoates. *Indian J. Microbiol.* **2015**, *55*, 235–249. [CrossRef] [PubMed]

11. Koller, M. Advances in polyhydroxyalkanoate (PHA) production. *Bioengineering* **2017**, *4*, 88. [CrossRef] [PubMed]
12. Chen, G.-Q.; Jiang, X.-R. Engineering bacteria for enhanced polyhydroxyalkanoates (PHA) biosynthesis. *Synth. Syst. Biotechnol.* **2017**, *2*, 192–197. [CrossRef]
13. Kourmentza, C.; Plácido, J.; Venetsaneas, N.; Burniol-Figols, A.; Varrone, C.; Gavala, H.N.; Reis, M.A.M. Recent advances and challenges towards sustainable polyhydroxyalkanoate (PHA) production. *Bioengineering* **2017**, *4*, 55. [CrossRef]
14. Fernández-Dacosta, C.; Posada, J.A.; Kleerebezem, R.; Cuellar, M.C.; Ramirez, A. Microbial community-based polyhydroxyalkanoates (PHAs) production from wastewater: Techno-economic analysis and ex-ante environmental assessment. *Bioresour. Technol.* **2015**, *185*, 368–377. [CrossRef]
15. Lee, S.Y.; Wong, H.H.; Choi, J.; Lee, S.H.; Lee, S.C.; Han, C.S. Production of medium-chain-length polyhydroxyalkanoates by high-cell-density cultivation of *Pseudomonas putida* under phosphorus limitation. *Biotechnol. Bioeng.* **2000**, *68*, 466–470. [CrossRef]
16. Park, S.J.; Ahn, W.S.; Green, P.R.; Lee, S.Y. Production of poly (3-hydroxybutyrate-co-3-hydroxyhexanoate) by metabolically engineered *Escherichia coli* strains. *Biomacromolecules* **2001**, *2*, 248–254. [CrossRef]
17. Kim, D.Y.; Hyung, W.K.; Moon, G.C.; Young, H.R. Biosynthesis, modification, and biodegradation of bacterial medium-chain-length polyhydroxyalkanoates. *J. Microbiol.* **2007**, *45*, 87–97. [PubMed]
18. Zinn, M.; Witholt, B.; Egli, T. Occurrence, synthesis and medical application of bacterial polyhydroxyalkanoate. *Adv. Drug. Del. Rev.* **2001**, *53*, 5–21. [CrossRef]
19. Nitschke, M.; Costa, S.G.V.A.O.; Contiero, J. Rhamnolipids and PHAs: Recent reports on *Pseudomonas*-derived molecules of increasing industrial interest. *Process Biochem.* **2011**, *46*, 621–630. [CrossRef]
20. Reddy, C.S.K.; Ghai, R.; Rashmi, T.; Kahia, V.C. Polyhydroxyalkanoates: An overview. *Bioresour. Technol.* **2003**, *87*, 137–146. [CrossRef]
21. Riesenberg, D.; Guthke, R. High-cell-density cultivation of microorganisms. *Appl. Microbiol. Biotechnol.* **1999**, *51*, 422–430. [CrossRef]
22. Turner, C.; Gregory, M.E.; Thornhill, N. Closed-loop control of fed-batch cultures of recombinant *Escherichia coli* using on-line HPLC. *Biotechnol. Bioeng.* **1994**, *44*, 819–829. [CrossRef]
23. Cavalheiro, J.M.B.T.; Raposo, S.R.; Almeida, M.C.M.D.; Sevrin, M.T.C.C.; Grandfils, C.; Fonseca, M.M.R. Effect of cultivation parameters on the production of poly (3-hydroxybutyrate-co-4-hydroxybutyrate) and poly (3-hydroxybutyrate-4-hydroxybutyrate-3-hydroxyvalerate) by *Cupriavidus necator* using waste glycerol. *Bioresour. Technol.* **2012**, *111*, 391–397. [CrossRef] [PubMed]
24. Ienczak, J.L.; Quines, L.K.; Melo, A.A.; Brandellero, M.; Mendes, C.R.; Schmidell, W.; Aragao, G.M.F. High cell density strategy for poly (3-hydroxybutyrate) production by *Cupriavidus necator*. *Braz. J. Chem. Eng.* **2011**, *28*, 585–596. [CrossRef]
25. Yee, L.; Blanch, H.W. Recombinant protein expression in high cell density fed-batch cultures of *Escherichia coli*. *Biotechnol. Adv.* **1992**, *10*, 1550–1556. [CrossRef] [PubMed]
26. Davis, R.; Duane, G.; Kenny, S.T.; Cerrone, F.; Guzik, M.W.; Babu, R.P.; Casey, E.; O'Connor, K.E. 2015. High cell density cultivation of *Pseudomonas putida* KT2440 using glucose without the need for oxygen enriched air supply. *Biotechnol. Bioeng.* **2015**, *112*, 725–733. [CrossRef]
27. Maclean, H.; Sun, Z.; Ramsay, J.; Ramsay, B. Decaying exponential feeding of nonanoic acid for the production of medium-chain-length poly(3-hydroxyalkanoates) by *Pseudomonas putida* KT2440. *Can. J. Chem.* **2008**, *86*, 564–569. [CrossRef]
28. Huschner, F.; Grousseau, E.; Brigham, C.J.; Plassmeier, J.; Popovic, M.; Rha, C.; Sinskey, A.J. Development of a feeding strategy for high cell and PHA density fed-batch fermentation of *Ralstonia eutropha* H16 from organic acids and their salts. *Process Biochem.* **2015**, *50*, 165–172. [CrossRef]
29. Ryu, H.W.; Hahn, S.K.; Chang, Y.K.; Chang, H.N. Production of poly(3-hydroxybutyrate) by high cell density fed-batch culture of *Alcaligenes eutrophus* with phosphate limitation. *Biotechnol. Bioeng.* **1997**, *55*, 25–32. [CrossRef]
30. Aragao, G.M.F.; Lindley, N.D.; Uribelarrea, J.L.; Pareilleux, A. Maintaining a controlled residual growth capacity increases the production of polyhydroxyalkanoate copolymers by *Alcaligenes eutrophus*. *Biotechnol. Lett.* **1996**, *18*, 937–942. [CrossRef]
31. Lee, W.H.; Azizan, M.N.M.; Sudesh, K. Effect of culture conditions of poly(3-hydroxybutyrate-co-4-hydroxybutyrate) synthesized by *Comamonas acidovorans*. *Polym. Degrad. Stab.* **2004**, *84*, 129–134. [CrossRef]
32. Kaur, G.; Roy, I. Strategies for large-scale production of polyhydroxyalkanoates. *Chem. Biochem. Eng. Q.* **2015**, *29*, 157–172. [CrossRef]
33. Singh, G.; Kumari, A.; Mittal, A.; Yadav, A.; Aggarwal, N.K. Poly β-hydroxybutyrate production by *Bacillus subtilis* ng 220 using sugar industry waste water. *BioMed Res. Int.* **2013**, 952641. [CrossRef]
34. Rai, R.; Yunos, D.M.; Boccaccini, A.R.; Knowles, J.C.; Barker, I.A.; Howdle, S.M.; Tredwell, G.D.; Keshavarz, T.; Roy, I. Poly-3-hydroxyoctanoate P(3HO), a medium chain length polyhydroxyalkanoate homopolymer from *Pseudomonas mendocina*. *Biomacromolecules* **2011**, *12*, 2126–2136. [CrossRef]
35. Pirt, S.J. The dynamics of microbial processes: A personal view. In *Microbial Growth Dynamics*; Poole, R.K., Bazin, M.J., Keevil, C.M., Eds.; IRL: Tokyo, Japan, 1990; Volume 28, pp. 1–16.
36. Ienczak, J.L.; Schmidell, W.; de Aragao, G.M.F. High-cell-density culture strategies for polyhydroxyalkanoate production: A review. *J. Ind. Microbiol. Biotechnol.* **2013**, *40*, 275–286. [CrossRef]
37. McNeil, B.; Harvey, L.M. *Fermentation, a Practical Approach*; IRL: Tokyo, Japan, 1990.
38. Kulpreecha, S.; Boonruangthavorn, A.; Meksiriporn, B.; Thongchul, N. Inexpensive fed-batch cultivation for high poly(3-hydroxybutyrate) production by a new isolate of *Bacillus megaterium*. *J. Biosci. Bioeng.* **2009**, *107*, 240–245. [CrossRef]

39. Shang, L.; Jiang, M.; Chang, H.N. Poly(3-hydroxybutyrate) synthesis in fed-batch culture of *Ralstonia eutropha* with phosphate limitation under different glucose concentrations. *Biotechnol. Lett.* **2003**, *25*, 1415–1419. [CrossRef] [PubMed]
40. Yamane, T.; Fukunaga, M.; Lee, Y.W. Increased PHB production by high-cell-density fed-batch culture of *Alcaligenes latus*, a growth associated PHB producer. *Biotechnol. Bioeng.* **1996**, *50*, 197–202. [CrossRef]
41. Mozumder, M.S.I.; De Wever, H.; Volcke, E.I.P.; Garcia-Gonzalez, E. A robust fed-batch feeding strategy independent of the carbon source for optimal polyhydroxybutyrate production. *Process. Biochem.* **2014**, *49*, 365–373. [CrossRef]
42. Miranda, D.S.D.M.; Koller, M.; Puppi, D.; Morelli, A.; Chiellini, F.; Braunegg, G. Fed-batch synthesis of poly(3-hydroxybutyrate) and poly(3-hydroxybutyrate-co-4-hydroxybutyrate) from sucrose and 4-hydroxybutyrate precursors by *Burkholderia sacchari* strain DSM 17165. *Bioengineering* **2017**, *4*, 36. [CrossRef]
43. Haas, C.; El-Najjar, T.; Virgolini, N.; Smerilli, M.; Neureiter, M. High cell-density production of poly (3-hydroxybutyrate) in a membrane bioreactor. *New Biotechnol.* **2017**, *37*, 117–122. [CrossRef] [PubMed]
44. Racine, F.M.; Saha, B.C. Production of mannitol by *Lactobacillus intermedius* NRRL B-3693 in fed-batch and continuous cell-recycle fermentations. *Process. Biochem.* **2007**, *42*, 1609–1613. [CrossRef]
45. Ienczak, J.L.; Schmidt, M.; Quines, L.K.; Zanfonato, K.; da Cruz Pradella, J.G.; Schmidell, W.; de Aragao, G.M.F. Poly(3-hydroxybutyrate) production in repeated fed-batch with cell recycle using a medium with low carbon source concentration. *Appl. Biochem. Biotechnol.* **2016**, *178*, 408–417. [CrossRef]
46. Schmidt, M.; Ienczak, J.L.; Quines, L.K.; Zanfonato, K.; Schmidell, W.; de Aragão, G.M.F. Poly (3-hydroxybutyrate-co-3-hydroxyvalerate) production in a system with external cell recycle and limited nitrogen feeding during the production phase. *Biochem. Eng. J.* **2016**, *112*, 130–135. [CrossRef]
47. Rodríguez-Contreras, A.; Koller, M.; Miranda-de Sousa Dias, M.; Calafell-Monfort, M.; Braunegg, G.; Marqués-Calvo, M.S. Influence of glycerol on poly (3-hydroxybutyrate) production by *Cupriavidus necator* and *Burkholderia sacchari*. *Biochem. Eng. J.* **2015**, *94*, 50–57. [CrossRef]
48. Ibrahim, M.H.; Steinbüchel, A. Poly(3-hydroxybutyrate) production from glycerol by *Zobellella denitrificans* MW1 via high-cell-density fed-batch fermentation and simplified solvent extraction. *Appl. Environ. Microbiol.* **2009**, *75*, 6222–6231. [CrossRef]
49. Chanprateep, S.; Buasri, K.; Muangwong, A.; Utiswannakul, P. Biosynthesis and biocompatibility of biodegradable poly (3-hydroxybutyrate-co-4-hydroxybutyrate). *Polym. Degrad. Stab.* **2010**, *95*, 2003–2012. [CrossRef]
50. Le Meur, S.; Zinn, M.; Egli, T.; Thöny-Meyer, L.; Ren, Q. Improved productivity of poly (4-hydroxybutyrate) (P4HB) in recombinant *Escherichia coli* using glycerol as the growth substrate with fed-batch culture. *Microb. Cell Fact* **2014**, *13*, 131. [CrossRef]
51. Stanley, A.; Punil Kumar, H.N.; Mutturi, S.; Vijayendra, S.V.N. Fed-batch strategies for production of PHA using a native isolate of *Halomonas venusta* KT832796 strain. *Appl. Biochem. Biotechnol.* **2017**, *184*, 935–952. [CrossRef]
52. Rocha, R.C.S.; Silva, L.F.; Taciro, M.K.; Pradella, J.G.C. Production of poly(3-hydroxybutyrate-co-3-hydroxyvalerate) P(3HB-co-3HV) with a broad range of 3HV content at high yields by *Burkholderia sacchari* IPT 189. *World J. Microbiol. Biotechnol.* **2008**, *24*, 427–431. [CrossRef]
53. Chen, G.Q.; Zhang, G.; Park, S.J.; Lee, S.Y. Industrial scale production of poly(3-hydroxybutyrate-co-3-hydroxyhexanoate). *Appl. Microbiol. Biotechnol.* **2001**, *57*, 50–55.
54. Blunt, W.; Dartiailh, C.; Sparling, R.; Gapes, D.J.; Levin, D.B.; Cicek, N. Development of High cell density cultivation strategies for improved medium chain length polyhydroxyalkanoate productivity using *Pseudomonas putida* LS46. *Bioengineering* **2019**, *6*, 89. [CrossRef]
55. Gao, J.; Ramsay, J.A.; Ramsay, B.A. Fed-batch production of poly-3-hydroxydecanoate from decanoic acid. *J. Biotechnol.* **2016**, *218*, 102–107. [CrossRef]
56. Sun, Z.; Ramsay, J.; Guay, M.; Ramsay, B.A. Enhanced yield of medium-chain-length polyhydroxyalkanoates from nonanoic acid by co-feeding glucose in carbon-limited fed-batch culture. *J. Biotechnol.* **2009**, *143*, 262–267. [CrossRef]
57. Cerrone, F.; Duane, G.; Casey, E.; Davis, R.; Belton, I.; Kenny, S.T.; Guzik, M.W.; Woods, T.; Babu, R.P.; O'Connor, K. Fed-batch strategies using butyrate for high cell density cultivation of *Pseudomonas putida* and its use as a biocatalyst. *Appl. Microbiol. Biotechnol.* **2014**, *98*, 9217–9228. [CrossRef]
58. Sun, Z.; Ramsay, J.A.; Guay, M.; Ramsay, B.A. Carbon-limited fed-batch production of medium-chain-length polyhydroxyalkanoates from nonanoic acid by *Pseudomonas putida* KT2440. *Appl. Microbiol. Biotechnol.* **2007**, *74*, 69–77. [CrossRef] [PubMed]
59. Diniz, S.; Taciro, M.K.; Gomez, J.G.C.; Pradella, J.G.C. High-cell-density cultivation of *Pseudomonas putida* IPT 046 and medium-chain-length polyhydroxyalkanoate production from sugarcane carbohydrates. *Appl. Biochem. Biotechnol. Part A Enzym. Eng. Biotechnol.* **2004**, *119*, 51–69. [CrossRef]
60. Kim, B.S. Production of medium chain length polyhydroxyalkanoates by fed-batch culture of *Pseudomonas oleovorans*. *Biotechnol. Lett.* **2002**, *24*, 125–130. [CrossRef]
61. Kim, G.J.; Lee, I.Y.; Yoon, S.C.; Shin, Y.C.; Park, Y.H. Enhanced yield and a high production of medium-chain-length poly(3-hydroxyalkanoates) in a two-step fed-batch cultivation of *Pseudomonas putida* by combined use of glucose and octanoate. *Enzyme Microb. Technol.* **1997**, *20*, 500–505. [CrossRef]
62. Andin, N.; Longieras, A.; Veronese, T.; Marcato, F.; Molina-Jouve, C.; Uribelarrea, J.-L. Improving carbon and energy distribution by coupling growth and medium chain length polyhydroxyalkanoate production from fatty acids by *Pseudomonas putida* KT2440. *Biotechnol. Bioprocess Eng.* **2017**, *22*, 308–318. [CrossRef]

63. Dey, P.; Rangarajan, V. Improved fed-batch production of high-purity PHB (poly-3 hydroxy butyrate) by *Cupriavidus necator* (MTCC 1472) from sucrose-based cheap substrates under response surface-optimized conditions. *3 Biotech* **2017**, *7*, 310. [CrossRef] [PubMed]
64. Penloglou, G.; Vasileiadou, A.; Chatzidoukas, C.; Kiparissides, C. Model-based intensification of a fed-batch microbial process for the maximization of polyhydroxybutyrate (PHB) production rate. *Bioproc. Biosyst. Eng.* **2017**, *40*, 1247–1260. [CrossRef]
65. Jung, K.; Hazenberg, W.; Prieto, M.; Witholt, B. Two-stage continuous process development for the production of medium-chain-length poly(3-hydroxyalkanoates). *Biotech Bioengineering* **2001**, *71*, 19–24. [CrossRef]
66. Atlic, A.; Koller, M.; Scherzer, D.; Kutschera, C.; Grillo-Fernandes, E.; Horvat, P.; Chiellini, E.; Braunegg, G. Continuous production of poly([R]-3-hydroxybutyrate) by *Cupriavidus necator* in a multistage bioreactor cascade. *Appl. Microbiol. Biotechnol.* **2011**, *91*, 295–304. [CrossRef]
67. Horvat, P.; Vrana, S.I.; Lopar, M.; Atlić, A.; Koller, M.; Braunegg, G. Mathematical modelling and process optimization of a continuous 5-stage bioreactor cascade for production of poly [-(R)-3-hydroxybutyrate] by *Cupriavidus necator*. *Bioproc. Biosyst. Eng.* **2013**, *36*, 1235–1250. [CrossRef]
68. Du, G.; Chen, J.; Yu, J.; Lun, S. Continuous production of poly- 3-hydroxybutyrate by *Ralstonia eutropha* in a two-stage culture system. *J. Biotechnol.* **2001**, *88*, 59–65. [CrossRef]
69. Du, G.; Chen, J.; Yu, J.; Lun, S. Kinetic studies on poly-3-hydroxybutyrate formation by *Ralstonia eutropha* in a two-stage continuous culture system. *Process Biochem.* **2001**, *37*, 219–227. [CrossRef]
70. Khanna, S.; Srivastava, A.K. Continuous production of poly-β-hydroxybutyrate by high-cell-density cultivation of *Wautersia eutropha*. *J. Chem. Tech. Biotechnol.* **2008**, *83*, 799–805. [CrossRef]
71. Egli, T. On multiple-nutrient-limited growth of microorganisms, with special reference to dual limitation by carbon and nitrogen substrates. *Antonie Van Leeuwenhoek* **1991**, *60*, 225–234. [CrossRef]
72. Zinn, M.; Weilenmann, H.-U.; Hany, R.; Schmid1, M.; Egli, T. Tailored synthesis of poly([R]-3-hydroxybutyrate-co-3-hydroxyvalerate) (PHB/HV) in *Ralstonia eutropha* DSM 428. *Acta Biotechnol.* **2003**, *23*, 309–316. [CrossRef]
73. Yu, S.T.; Lin, C.C.; Too, J.R. PHBV production by *Ralstonia eutropha* in a continuous stirred tank reactor. *Process Biochem.* **2005**, *40*, 2729–2734. [CrossRef]
74. Huijberts, G.N.M.; Eggink, G. Production of poly(3-hydroxyalkanoates) by *Pseudomonas putida* KT2442 in continuous cultures. *Appl. Microbiol. Biotechnol.* **1996**, *46*, 233–239. [CrossRef]
75. Tan, D.; Xue, Y.S.; Aibaidula, G.; Chen, G.Q. Unsterile and continuous production of polyhydroxybutyrate by *Halomonas* TD01. *Bioresour. Technol.* **2011**, *102*, 8130–8136. [CrossRef] [PubMed]
76. Macauley-Patrick, S.; Finn, B. Modes of fermenter operation. In *Practical Fermentation Technology*; McNeil, B., Harvey, L.M., Eds.; John Wiley and Sons Ltd.: Chichester, UK, 2008; pp. 69–95.
77. Kosaric, N.; Vardar-Sukan, F. *Fermentation modes of industrial interest, In The Biotechnology of Ethanol: Classical and Future Applications*; Roehr, M., Ed.; Wiley-VCH GmbH: Weinheim, Germany, 2001; pp. 139–149.
78. Ibrahim, M.H.A.; Steinbüchel, A. High-cell-density cyclic fed-batch fermentation of a poly(3-hydroxybutyrate)-accumulating thermophile, *Chelatococcus* sp. strain MW10. *Appl. Environ. Microbiol.* **2010**, *76*, 7890–7895. [CrossRef]
79. Karasavvas, E.; Chatzidoukas, C. Model-based dynamic optimization of the fermentative production of polyhydroxyalkanoates (PHAs) in fed-batch and sequence of continuously operating bioreactors. *Biochem. Eng. J.* **2020**, *162*, 107702.
80. Dufresne, A.; Samain, E. Preparation and characterization of a poly (β- hydroxyalkanoate) latex produced by *Pseudomonas oleovorans*. *Macromolecules* **1998**, *31*, 6426–6433. [CrossRef]
81. Kellerhals, M.B.; Hazenberg, W.M.; Witholt, B. High cell density fermentations of *Pseudomonas oleovorans* for the production of mcl-PHAs in two-liquid-phase media. *Enzyme Microb. Technol.* **1999**, *24*, 111–116. [CrossRef]
82. Ramsay, B.A.; Saracovan, I.; Ramsay, J.A.; Marchessault, R.H. Effect of nitrogen limitation on long-side-chain poly-β-hydroxyalkanoate synthesis by *Pseudomonas resinovorans*. *Appl. Environ. Microbiol.* **1992**, *58*, 744–746. [CrossRef]
83. Oeding, V.; Schlegel, H.G. β-Ketothiolase from *Hydrogenomonas eutropha* H16 and its significance in the regulation of poly- β-hydroxybutyrate metabolism. *Biochem. J.* **1973**, *134*, 239–248. [CrossRef]
84. Tu, W.; Zhang, D.; Wang, H. Polyhydroxyalkanoates (PHA) production from fermented thermal hydrolyzed sludge by mixed microbial cultures: The link between phosphorus and PHA yields. *Waste Manag.* **2019**, *96*, 149–157. [CrossRef]
85. Wen, Q.; Chen, Z.; Tian, T.; Chen, W. Effects of phosphorus and nitrogen limitation on PHA production in activated sludge. *J. Environ. Sci.* **2010**, *22*, 1602–1607. [CrossRef]
86. Portugal-Nunes, D.J.; Pawar, S.S.; Lidén, G.; Gorwa-Grauslund, M.F. Effect of nitrogen availability on the poly-3-d-hydroxybutyrate accumulation by engineered *Saccharomyces cerevisiae*. *AMB Expr.* **2017**, *7*, 35. [CrossRef]
87. Grousseau, E.; Blanchet, E.; Déléris, S.; Albuquerque, M.G.E.; Paul, E.; Uribelarrea, J.-L. Phosphorus limitation strategy to increase propionic acid flux towards 3-hydroxyvaleric acid monomers in *Cupriavidus necator*. *Biores Technol.* **2014**, *153*, 206–215. [CrossRef]
88. Da Cruz Pradella, J.G.; Taciro, M.K.; Mateus, A.Y.P. High-cell-density poly (3-hydroxybutyrate) production from sucrose using *Burkholderia sacchari* culture in airlift bioreactor. *Bioresour. Technol.* **2010**, *101*, 8355–8360. [CrossRef]
89. Grousseau, E.; Blanchet, E.; Déléris, S.; Albuquerque, M.G.E.; Paul, E.; Uribelarrea, J.-L. Impact of sustaining a controlled residual growth on polyhydroxybutyrate yield and production kinetics in *Cupriavidus necator*. *Bioresour. Technol.* **2013**, *148*, 30–38. [CrossRef]

90. Kim, S.W.; Kim, P.; Lee, H.S.; Kim, J.H. High production of polyhydroxybutyrate (PHB) from *Methylobacterium organophilum* under potassium limitation. *Biotechnol. Lett.* **1996**, *18*, 25–30. [CrossRef]
91. Kim, J.S.; Lee, B.H.; Kim, B.S. Production of poly (3-hydroxybutyrate-co-4-hydroxybutyrate) by *Ralstonia eutropha*. *Biochem. Eng. J.* **2005**, *23*, 169–174. [CrossRef]
92. Ahn, W.S.; Park, S.J.; Lee, S.Y. Production of poly(3-hydroxybutyrate) from whey by cell recycle fed-batch culture of recombinant *Escherichia coli*. *Biotechnol. Lett.* **2001**, *23*, 235–240. [CrossRef]
93. Riedel, S.L.; Bader, J.; Brigham, C.J.; Budde, C.F.; Yusof, Z.A.; Rha, C.; Sinskey, A.J. Production of poly(3-hydroxybutyrate-co-3-hydroxyhexanoate) by *Ralstonia eutropha* in high cell density palm oil fermentations. *Biotechnol. Bioeng.* **2012**, *109*, 74–83. [CrossRef]
94. Chen, G.-Q.; Jiang, X.-R. Engineering microorganisms for improving polyhydroxyalkanoate biosynthesis. *Curr. Opin. Biotechnol.* **2018**, *53*, 20–25. [CrossRef]
95. Madison, L.L.; Huisman, G.W. Metabolic engineering of poly(3-hydroxyalkanoates): From DNA to plastic. *Microbiol. Mol. Biol. Rev.* **1999**, *63*, 21–53. [CrossRef]
96. Liu, F.; Jian, J.; Shen, X.; Chung, A.; Chen, J.; Chen, G.Q. Metabolic engineering of *Aeromonas hydrophila* 4AK4 for production of copolymers of 3-hydroxybutyrate and medium-chain length 3-hydroxyalkanoate. *Bioresour. Technol.* **2012**, *102*, 8123–8129. [CrossRef] [PubMed]
97. Hoffer, P.; Vermette, P.; Groleau, D. Production and characterization of polyhydroxyalkanoates by recombinant *Methylobacterium extorquens*: Combining desirable thermal properties with functionality. *Biochem. Eng. J.* **2011**, *54*, 26–33. [CrossRef]
98. Huisman, G.W.; Wonin, E.; Koning, G.; Preusting, H.; Witholt, B. Synthesis of poly(3-hydroxyalkanoates) by mutant and recombinant *Pseudomonas* strains. *Appl. Microbiol. Biotechnol.* **1992**, *38*, 1–5. [CrossRef]
99. Matsusaki, H.; Abe, H.; Doi, Y. Biosynthesis and properties of poly(3-hydroxybutyrate-*co*-3-hydroxyalkanoates) by recombinant strains of *Pseudomonas* sp. 61-3. *Biomacromolecules* **2000**, *1*, 17–22. [CrossRef] [PubMed]
100. Hu, D.; Chung, A.L.; Wu, L.P.; Zhang, X.; Wu, Q.; Chen, J.C.; Chen, G.Q. Biosynthesis and characterization of polyhydroxyalkanoate block copolymer P3HB-*b*-P4HB. *Biomacromolecules* **2011**, *12*, 3166–3173. [CrossRef]
101. Fidler, S.; Dennis, D. Polyhydroxyalkanoate production in recombinant *Escherichia coli*. *FEMS Microbiol. Rev.* **1992**, *103*, 231–236. [CrossRef]
102. Hahn, S.K.; Chang, Y.K.; Lee, S.Y. Recovery and characterization of poly (3-hydroxybutyric acid) synthesized in *Alcaligenes eutrophus* and recombinant *Escherichia coli*. *Appl. Environ. Microbiol.* **1995**, *61*, 34–39. [CrossRef]
103. Nduko, J.M.; Suzuki, W.; Matsumoto, K.; Kobayashi, H.; Ooi, T.; Fukuoka, A.; Taguchi, S. Polyhydroxyalkanoates production from cellulose hydrolysate in *Escherichia coli* LS5218 with superior resistance to 5-hydroxymethylfurfural. *J. Biosci. Bioeng.* **2012**, *113*, 70–72. [CrossRef]
104. Chen, G.Q. A microbial polyhydroxyalkanoates (PHA) based bio- and material industry. *Chem. Soc. Rev.* **2009**, *38*, 2434–2446. [CrossRef] [PubMed]
105. Ren, Q.; Roo, G.D.; Beilen, J.B.V.; Zinn, M.; Kessler, B.; Witholt, B. Poly (3- hydroxyalkanoate) polymerase synthesis and in vitro activity in recombinant *Escherichia coli* and *Pseudomonas putida*. *Appl. Gen. Mol. Biotechnol.* **2005**, *69*, 286–292. [CrossRef] [PubMed]
106. Suriyamongkol, P.; Weselake, R.; Narine, S.; Moloney, M.; Shah, S. Biotechnological approaches for the production of polyhydroxyalkanoates in microorganisms and plants-A review. *Biotechnol. Adv.* **2007**, *25*, 148–175. [CrossRef]
107. Wang, Q.; Yang, P.; Xian, M.; Liu, H.; Cao, Y.; Yang, Y.; Zhao, G. Production of block copolymer poly (3-hydroxybutyrate)-block-poly (3-hydroxypropionate) with adjustable structure from an inexpensive carbon source. *ACS Macro Lett.* **2013**, *2*, 996–1000. [CrossRef]
108. Kahar, P.; Agus, J.; Kikkawa, Y.; Taguchi, K.; Doi, Y.; Tsuge, T. Effective production and kinetic characterization of ultra-high-molecular-weight poly[(R)-3-hydroxybutyrate] in recombinant *Escherichia coli*. *Polym. Degrad. Stab.* **2005**, *87*, 161–169. [CrossRef]
109. Nikel, P.I.; Giordano, A.M.; Almeida, A.; Godoy, M.S.; Pettinari, M.J. Elimination of D-lactate synthesis increase poly(3-hydroxybutyrate) and ethanol synthesis from glycerol and affects cofactor distribution in recombinant *Escherichia coli*. *Appl. Environ. Biol.* **2010**, *76*, 7400–7406. [CrossRef]
110. Nikel, P.I.; Almeida, A.; Melillo, E.C.; Galvagno, M.A.; Pettinari, M.J. New recombinant *Escherichia coli* strain tailored for the production of poly(3-hydroxybutyrate) from agroindustrial byproducts. *Appl. Environ. Biol.* **2006**, *72*, 3949–3954. [CrossRef]
111. Agus, J.; Kahar, P.; Abe, H.; Doi, Y.; Tsuge, T. Altered expression of polyhydroxyalkanoate synthase gene and its effect on poly[(R)-3-hydroxybutyrate] synthesis in recombinant *Escherichia coli*. *Polym. Degrad. Stab.* **2006**, *91*, 1645–1650. [CrossRef]
112. Wang, F.; Lee, S.Y. High cell density culture of metabolically engineered *Escherichia coli* for the production of poly (3-hydroxybutyrate) in a defined medium. *Biotechnol. Bioeng.* **1998**, *58*, 325–328. [CrossRef]
113. Wong, H.H.; Lee, S.Y. Poly(3-hydroxybutyrate) production from whey by high cell density cultivation of recombinant *Escherichia coli*. *Appl. Microbiol. Biotechnol.* **1998**, *50*, 30–33. [CrossRef] [PubMed]
114. Morr, C.V. Whey proteins: Manufacture. *Dev. Dairy Chem.* **1989**, *4*, 245–284.
115. Riedel, S.L.; Jahns, S.; Koenig, S.; Bock, M.C.; Brigham, C.J.; Bader, J.; Stahl, U. Polyhydroxyalkanoates production with *Ralstonia eutropha* from low quality waste animal fats. *J. Biotechnol.* **2015**, *214*, 119–127. [CrossRef] [PubMed]

116. Sato, S.; Maruyama, H.; Fujiki, T.; Matsumoto, K. Regulation of 3-hydroxyhexanoate composition in PHBH synthesized by recombinant *Cupriavidus necator* H16 from plant oil by using butyrate as a co-substrate. *J. Biosci. Bioeng.* **2015**, *120*, 246–251. [CrossRef] [PubMed]
117. Povolo, S.; Toffano, P.; Basaglia, M.; Casella, S. Polyhydroxyalkanoates production by engineered *Cupriavidus necator* from waste material containing lactose. *Bioresour. Technol.* **2010**, *101*, 7902–7907. [CrossRef]
118. Kim, D.Y.; Kim, Y.B.; Rhee, Y.H. Evaluation of various carbon substrates for the biosynthesis of polyhydroxyalkanoates bearing functional groups by *Pseudomonas putida*. *Int. J. Biol. Macromol.* **2000**, *28*, 23–29. [CrossRef]
119. Cai, L.; Yuan, M.Q.; Liu, F.; Jian, J.; Chen, G.Q. Enhanced production of medium-chain-length polyhydroxyalkanoates (PHA) by PHA depolymerase knockout mutant of *Pseudomonas putida* KT2442. *Bioresour. Technol.* **2009**, *100*, 2265–2270. [CrossRef] [PubMed]
120. Le Meur, S.; Zinn, M.; Egli, T.; Thöny-Meyer, L.; Ren, Q. Production of medium-chain-length polyhydroxyalkanoates by sequential feeding of xylose and octanoic acid in engineered *Pseudomonas putida* KT2440. *BMC Biotechnol.* **2012**, *12*, 53.
121. Kahar, P.; Tsuge, T.; Taguchi, K.; Doi, Y. High yield production of polyhydroxyalkanoates from soybean oil by *Ralstonia eutropha* and its recombinant strain. *Polym. Degrad. Stab.* **2004**, *83*, 79–86. [CrossRef]
122. Jing, H.; Yuan-Zheng, Q.; Dai-Cheng, L.; Guo-Qiang, C. Engineered *Aeromonas hydrophila* for enhanced production of poly(3-hydroxybutyrate-co-3-hydroxyhexanoate) with alterable monomers composition. *FEMS Microbiol. Lett.* **2004**, *239*, 195–201.
123. Qiu, Y.Z.; Han, J.; Guo, J.J.; Chen, G.Q. Production of poly (3-hydroxybutyrate-co-3-hydroxyhexanoate) from gluconate and glucose by recombinant *Aeromonas hydrophila* and *Pseudomonas putida*. *Biotechnol. Lett.* **2005**, *27*, 1381–1386. [CrossRef]
124. Ouyang, S.; Han, J.; Qiu, Y.; Qin, L.; Chen, S.; Wu, Q.; Leski, M.L.; Chen, G. Poly(3-hydroxybutyrate-co-3-hydroxyhexanoate) production in recombinant *Aeromonas hydrophila* 4ak4 harboring phba, phbb and vgb genes. *Macromol. Symp.* **2005**, *224*, 21–34. [CrossRef]
125. Budde, C.F.; Riedel, S.L.; Willis, L.B.; Rha, C.; Sinskey, A.J. Production of poly (3-hydroxybutyrate-co-3-hydroxyhexanoate) from plant oil by engineered *Ralstonia eutropha* strains. *Appl. Environmen. Microbiol.* **2011**, *77*, 2847–2854. [CrossRef] [PubMed]
126. Thinagaran, L.; Sudesh, K. Evaluation of sludge palm oil as feedstock and development of efficient method for its utilization to produce polyhydroxyalkanoate. *Waste Biomass Valoriz.* **2017**, *10*, 709–720. [CrossRef]
127. Choi, J.; Lee, S.Y.; Kyuboem, H. Cloning of the *Alcaligenes latus* polyhydroxyalkanoate biosynthesis genes and use of these genes for enhanced production of poly(3-hydroxybutyrate) in *Escherichia coli*. *Appl. Environ. Microbiol.* **1998**, *64*, 4897–4903. [CrossRef] [PubMed]
128. Choi, J.; Lee, S.Y.; Shin, K.; Lee, W.G.; Park, S.J.; Chang, H.N.; Chang, Y.K. Pilot scale production of poly(3-hydroxybutyrate-co-3-hydroxy-valerate) by fed-batch culture of recombinant *Escherichia coli*. *Biotechnol. Bioprocess Eng.* **2002**, *7*, 371–374. [CrossRef]
129. Chen, X.; Yin, J.; Ye, J.; Zhang, H.; Che, X.; Ma, Y.; Li, M.; Wu, L.P.; Chen, G.Q. Engineering *Halomonas bluephagenesis* TD01 for non-sterile production of poly(3-hydroxybutyrate-co-4-hydroxybutyrate). *Bioresour. Technol.* **2017**, *244*, 534–541. [CrossRef]
130. Poblete-Castro, I.; Rodriguez, A.L.; Lam, C.M.; Kessler, W. Improved production of medium-chain-length polyhydroxyalkanoates in glucose-based fed-batch cultivations of metabolically engineered *Pseudomonas putida* strains. *J. Microbiol. Biotechnol.* **2014**, *24*, 59–69. [CrossRef]
131. Yang, J.E.; Park, S.J.; Kim, W.J.; Kim, H.J.; Kim, B.J.; Lee, H.; Shin, J.; Lee, S.Y. One-step fermentative production of aromatic polyesters from glucose by metabolically engineered *Escherichia coli* strains. *Nat. Commun.* **2018**, *9*, 79. [CrossRef] [PubMed]
132. Chen, G.-Q.; Hajnal, I. The 'PHAome'. *Trends Biotechnol.* **2015**, *33*, 559–564. [CrossRef]
133. Wang, Y.; Chung, A.; Chen, G.Q. Synthesis of medium-chain-length polyhydroxyalkanoate homopolymers, random copolymers, and block copolymers by an engineered strain of *Pseudomonas entomophila*. *Adv. Healthc. Mater.* **2017**, *2017*, 6.
134. Li, S.; Cai, L.; Wu, L.; Zeng, G.; Chen, J.; Wu, Q.; Chen, G.Q. Microbial synthesis of functional homo-, random and block polyhydroxyalkanoates by β-oxidation deleted *Pseudomonas entomophila*. *Biomacromolecules* **2014**, *15*, 2310–2319. [CrossRef] [PubMed]
135. Wang, H.H.; Zhou, X.R.; Liu, Q.; Chen, G.Q. Biosynthesis of polyhydroxyalkanoate homopolymers by *Pseudomonas putida*. *Appl. Microbiol. Biotechnol.* **2011**, *89*, 1497–1507. [CrossRef]
136. Liu, Q.; Luo, G.; Zhou, X.R.; Chen, G.Q. Biosynthesis of poly(3-hydroxydecanoate) and 3-hydroxydodecanoate dominating polyhydroxyalkanoates by β-oxidation pathway inhibited *Pseudomonas putida*. *Metab. Eng.* **2011**, *13*, 11–17. [CrossRef] [PubMed]
137. Meng, D.C.; Wang, Y.; Wu, L.P.; Shen, R.; Chen, J.C.; Wu, Q.; Chen, G.Q. Production of poly(3-hydroxypropionate) and poly(3-hydroxybutyrate-co-3-hydroxypropionate) from glucose by engineering *Escherichia coli*. *Metab. Eng.* **2015**, *29*, 189–195. [CrossRef]
138. Cheema, S.; Bassas-Galia, M.; Sarma, P.M.; Lal, B.; Arias, S. Exploiting metagenomic diversity for novel polyhydroxyalkanoate synthases: Production of a terpolymer poly(3-hydroxybutyrate-*co*-3-hydroxyhexanoate-*co*-3-hydroxyoctanoate) with a recombinant *Pseudomonas putida* strain. *Bioresour. Technol.* **2012**, *103*, 322–328. [CrossRef]
139. Li, M.; Chen, X.; Che, X.; Zhang, H.; Wu, L.P.; Du, H.; Chen, G.Q. Engineering *Pseudomonas entomophila* for synthesis of copolymers with defined fractions of 3-hydroxybutyrate and medium-chain-length 3-hydroxyalkanoates. *Metab. Eng.* **2019**, *52*, 253–262. [CrossRef] [PubMed]
140. Tripathi, L.; Wu, L.P.; Dechuan, M.; Chen, J.; Wu, Q.; Chen, G.Q. *Pseudomonas putida* KT2442 as a platform for the biosynthesis of polyhydroxyalkanoates with adjustable monomer contents and compositions. *Bioresour Technol.* **2013**, *142*, 225–231. [CrossRef]

141. Ouyang, S.P.; Liu, Q.; Fang, L.; Chen, G.Q. Construction of pha operon defined knockout mutants of *Pseudomonas putida* KT2442 and its applications in polyhydroxyalkanoates production. *Macromol. Biosci.* **2007**, *7*, 227–233. [CrossRef]
142. Zhao, F.; He, F.; Liu, X.; Shi, J.; Liang, J.; Wang, S.; Yang, C.; Liu, R. Metabolic engineering of *Pseudomonas mendocina* NK-01 for enhanced production of medium-chain-length polyhydroxyalkanoates with enriched content of the dominant monomer. *Int. J. Biol. Macromol.* **2020**, *154*, 1596–1605. [CrossRef] [PubMed]
143. Guzik, M.W.; Narancic, T.; Ilic-Tomic, T.; Vojnovic, S.; Kenny, S.T.; Casey, W.T.; Duane, G.F.; Casey, E.; Woods, T.; Babu, R.P.; et al. Identification and characterization of an acyl-CoA dehydrogenase from *Pseudomonas putida* KT2440 that shows preference towards medium to long chain length fatty acids. *Soc. Gen. Microbiol.* **2014**, *160*, 1760–1771. [CrossRef]
144. Li, S.Y.; Dong, C.L.; Wang, S.Y.; Ye, H.M.; Chen, G.Q. Microbial production of polyhydroxyalkanoate block copolymer by recombinant *Pseudomonas putida*. *Appl. Microbiol. Biotechnol.* **2011**, *90*, 659–669. [CrossRef]
145. Nakaoki, T.; Yasui, J.; Komaeda, T. Biosynthesis of P3HBV-b-P3HB-b-P3HBV triblock copolymer by *Ralstonia eutropha*. *J. Polym. Environ.* **2019**, *27*, 2720–2727. [CrossRef]
146. Nakaoki, T.; Yamagishi, R.; Ishii, D. Biosynthetic Process and characterization of poly (3-hydroxybutyrate-co-3-hydroxyvalerate)-block-poly (3-hydroxybutyrate) by *R. Eutropha*. *J. Polym. Environ.* **2015**, *23*, 487–492. [CrossRef]
147. Tripathi, L.; Wu, L.P.; Chen, J.; Chen, G.Q. Synthesis of diblock copolymer poly-3-hydroxybutyrate-block-poly-3-hydroxyhexanoate [PHB-b-PHHx] by a beta-oxidation weakened *Pseudomonas putida* KT2442. *Microb. Cell Fact* **2012**, *11*, 44. [CrossRef]
148. Ashby, R.D.; Solaiman, D.; Nuñez, A.; Strahan, G.D.; Johnston, D.B. *Burkholderia sacchari* DSM 17165: A source of compositionally-tunable block-copolymeric short-chain poly(hydroxyalkanoates) from xylose and levulinic acid. *Bioresour. Technol.* **2018**, *253*, 333–342. [CrossRef]
149. Ashby, R.D.; Solaiman, D.K.Y. Levulinic Acid: A valuable platform chemical for the fermentative synthesis of poly(hydroxyalkanoate) biopolymers. *Green Polym. Chem. New Prod. Process Applic.* Chapter **2018**, *21*, 339–354.
150. Ashby, R.D.; Solaiman, D.K.; Strahan, G.D. The Use of *Azohydromonas lata* DSM 1122 to produce 4-hydroxyvalerate-containing polyhydroxyalkanoate terpolymers, and unique polymer blends from mixed-cultures with *Burkholderia sacchari* DSM 17165. *J. Polym. Environ.* **2019**, *27*, 198–209. [CrossRef]
151. Sparks, J.; Scholz, C. Synthesis and characterization of a cationic poly(beta-hydroxyalkanoate). *Biomacromolecules* **2008**, *9*, 2091–2096. [CrossRef]
152. Zheng, Z.; Gong, Q.; Chen, G.Q. A novel method for production of 3-hydroxydecanoic acid by recombinant *Escherichia coli* and *Pseudomonas putida*. *Chin. J. Chem. Eng.* **2004**, *12*, 550–555.
153. Hazer, B. Simple synthesis of amphiphilic poly (3-hydroxy alkanoate)s with pendant hydroxyl and carboxylic groups via thiol-ene photo click reactions. *Polym. Degrad. Stab.* **2015**, *119*, 159–166. [CrossRef]
154. Yu, L.P.; Yan, X.; Zhang, X.; Chen, X.B.; Wu, Q.; Jiang, X.R.; Chen, G.Q. Biosynthesis of functional polyhydroxyalkanoates by engineered *Halomonas bluephagenesis*. *Metab. Eng.* **2020**, *59*, 119–130. [CrossRef]
155. Yuan, M.Q.; Shi, Z.Y.; Wei, X.X.; Wu, Q.; Chen, S.F.; Chen, G.Q. Microbial production of medium-chain length 3-hydroxyalkanoic acids by recombinant *Pseudomonas putida* KT2442 harboring genes fadL, fadD and phaZ. *FEMS Microbiol. Lett.* **2008**, *283*, 167–175. [CrossRef] [PubMed]
156. Ma, L.; Zhang, H.; Liu, Q.; Chen, J.; Zhang, J.; Chen, G.Q. Production of two monomers containing medium-chain-length polyhydroxyalkanoates by β-oxidation impaired mutant of *Pseudomonas putida* KT2442. *Bioresour. Technol.* **2009**, *100*, 4891–4894. [CrossRef] [PubMed]
157. Cereghino, J.L.; Cregg, J.M. Heterologous protein expression in the methylotrophic yeast *Pichia pastoris*. *FEMS Microbiol Rev.* **2000**, *24*, 45–66. [CrossRef] [PubMed]
158. Poirier, Y.; Erard, N.; MacDonald-Comber Petétot, J. Synthesis of polyhydroxyalkanoate in the peroxisome of *Pichia pastoris*. *J. FEMS Microbiol. Lett.* **2002**, *207*, 97–102. [CrossRef]
159. Vijayasankaran, N.; Carlson, R.; Srienc, F. Synthesis of poly [(r)-3-hydroxybutyric acid)] in the cytoplasm of *Pichia pastoris* under oxygen limitation. *Biomacromolecules* **2005**, *6*, 604–611. [CrossRef] [PubMed]
160. Yamada, M.; Matsumoto, K.; Uramoto, S.; Motohashi, R.; Abe, H.; Taguchi, S. Lactate fraction dependent mechanical properties of semitransparent poly(lactate-co-3-hydroxybutyrate) s produced by control of lactyl-CoA monomer fluxes in recombinant *Escherichia coli*. *J. Biotechnol.* **2011**, *154*, 255–260. [CrossRef] [PubMed]
161. Ren, Y.; Meng, D.; Wu, L.; Chen, J.; Wu, Q.; Chen, G.Q. Microbial synthesis of a novel terpolyester P(LA-co-3HB-co-3HP) from low-cost substrates. *Microb. Biotechnol.* **2017**, *10*, 371–380. [CrossRef]
162. Li, Z.J.; Qiao, K.; Shi, W.; Pereira, B.; Zhang, H.; Olsen, B.D.; Stephanopoulos, G. Biosynthesis of poly(glycolate-co-lactate-co-3-hydroxybutyrate) from glucose by metabolically engineered *Escherichia coli*. *Metab. Eng.* **2016**, *35*, 1–8. [CrossRef]
163. Li, Z.J.; Qiao, K.; Che, X.M.; Stephanopoulos, G. Metabolic engineering of *Escherichia coli* for the synthesis of the quadripolymer poly(glycolate-co-lactate-co-3-hydroxybutyrate-co-4-hydroxybutyrate) from glucose. *Metab. Eng.* **2017**, *44*, 38–44. [CrossRef]
164. Salamanca-Cardona, L.; Scheel, R.A.; Bergey, N.S.; Stipanovic, A.J.; Matsumoto, K.; Taguchi, S.; Nomura, C.T. Consolidated bioprocessing of poly(lactate-co-3-hydroxybutyrate) from xylan as a sole feedstock by genetically-engineered *Escherichia coli*. *J. Bioscie. Bioeng.* **2016**, *122*, 406–414. [CrossRef]
165. Takisawa, K.; Ooi, T.; Matsumoto, K.; Kadoya, R.; Taguchi, S. Xylose-based hydrolysate from eucalyptus extract as feedstock for poly(lactate-co-3-hydroxybutyrate) production in engineered *Escherichia coli*. *Process Biochem.* **2017**, *54*, 102–105. [CrossRef]

166. Choi, S.Y.; Chae, T.U.; Shin, J.; Im, J.A.; Lee, S.Y. Biosynthesis and characterization of poly (D-lactate-co-glycolate-co-4-hydroxybutyrate). *Biotechnol. Bioeng.* **2020**, *117*, 2187–2197. [CrossRef] [PubMed]
167. Goto, S.; Suzuki, N.; Matsumoto, K.; Taguchi, S.; Tanaka, K.; Matsusaki, H. Enhancement of lactate fraction in poly(lactate-co-3-hydroxybutyrate) synthesized by *Escherichia coli* harboring the D-lactate dehydrogenase gene from *Lactobacillus acetotolerans* HT. *J. Gen. Appl. Microbiol.* **2019**, *65*, 204–208. [CrossRef] [PubMed]
168. Park, S.J.; Park, J.P.; Lee, S.Y.; Doi, Y. Enrichment of specific monomer in medium chain length polyhydroxyalkanoate by amplification of fadD and fadE genes in recombinant *Eschericia coli*. *Enzyme Microb. Technol.* **2003**, *33*, 62–70. [CrossRef]
169. Jiang, X.J.; Sun, Z.; Ramsay, J.A.; Ramsay, B.A. Fed-batch production of MCL-PHA with elevated 3-hydroxynonanoate content. *AMB Express* **2013**, *3*, 50. [CrossRef]
170. Chung, A.L.; Jin, H.L.; Huang, L.J.; Ye, H.M.; Chen, J.C.; Wu, Q.; Chen, G.Q. Biosynthesis and characterization of poly(3-hydroxydodecanoate) by β-oxidation inhibited mutant of *Pseudomonas entomophila* L48. *Biomacromolecules* **2011**, *12*, 3559–3566. [CrossRef]
171. Gao, J.; Vo, M.T.; Ramsay, J.A.; Ramsay, B.A. Overproduction of mcl-PHA with high 3-hydroxydecanoate content. *Biotechnol. Bioeng.* **2018**, *115*, 390–400. [CrossRef]
172. Oliveira, G.H.D.; Zaiat, M.; Rodrigues, J.A.D.; Ramsay, J.A.; Ramsay, B.A. Towards the production of mcl-pha with enriched dominant monomer content: Process development for the sugarcane biorefinery context. *J. Polym. Environ.* **2020**, *28*, 844–853. [CrossRef]
173. Insomphun, C.; Mifune, J.; Orita, I.; Numata, K.; Nakamura, S.; Fukui, T. Modification of β-oxidation pathway in *Ralstonia eutropha* for production of poly(3-hydroxybutyrate-co-3-hydroxyhexanoate) from soybean oil. *J. Biosci. Bioeng.* **2014**, *117*, 184–190. [CrossRef]
174. Oliveira, C.S.S.; Silva, C.E.; Carvalho, G.; Reis, M.A. Strategies for efficiently selecting PHA producing mixed microbial cultures using complex feedstocks: Feast and famine regime and uncoupled carbon and nitrogen availabilities. *New Biotechnol.* **2016**, *37*, 69–79. [CrossRef] [PubMed]
175. Shen, X.W.; Jian, Y.Y.J.; Wu, Q.; Chen, J.Q. Production and characterization of homopolymer poly(3-hydroxyvalerate) (PHV) accumulated by wild type and recombinant *Aeromonas hydrophila* strain 4AK4. *Bioresour. Technol.* **2009**, *100*, 4296–4299. [CrossRef] [PubMed]
176. Albuquerque, M.G.E.; Torres, C.A.V.; Reis, M.A.M. Polyhydroxyalkanoate (PHA) production by a mixed microbial culture using sugar molasses: Effect of the influent substrate concentration on culture selection. *Water Res.* **2010**, *44*, 3419–3433. [CrossRef]
177. Koller, M. Characterization of polyhydroxyalkanoates. In *Recent Advances in Biotechnology (Volume 2) Microbial Biopolyester*; Bentham Science Publishers: Sharjah, United Arab Emirates, 2016; p. 283.
178. Magdouli, S.; Brar, S.K.; Blais, J.F.; Tyagi, R.D. How to direct the fatty acid biosynthesis towards polyhydroxyalkanoates production. *Biomass Bioenergy* **2015**, *74*, 268–279. [CrossRef]
179. Anjum, A.; Zuber, M.; Zia, K.M.; Noreen, A.; Anjum, M.N.; Tabasum, S. Microbial production of polyhydroxyalkanoates (PHAs) and its copolymers: A review of recent advancements. *Int. J. Biol. Macromol.* **2016**, *89*, 161–174. [CrossRef]
180. Johnson, K.; Jiang, Y.; Kleerebezem, R.; Muyzer, G.; Van Loosdrecht, M.C.M. Enrichment of a mixed bacterial culture with a high polyhydroxyalkanoate storage capacity. *Biomacromolecules* **2009**, *10*, 670–676. [CrossRef] [PubMed]
181. Serafim, L.S.; Lemos, P.C.; Albuquerque, M.G.E.; Reis, M.A.M. Strategies for PHA production by mixed cultures and renewable waste materials. *Appl. Microbiol. Biotechnol.* **2008**, *81*, 615–628. [CrossRef]
182. Dionisi, D.; Majone, M.; Papa, V.; Beccari, M. Biodegradable polymers from organic acids by using activated sludge enriched by aerobic periodic feeding. *Biotechnol. Bioeng.* **2004**, *85*, 569–579. [CrossRef]
183. Beccari, M.; Dionisi, D.; Giuliani, A.; Majone, M.; Ramadori, R. Effect of different carbon sources on aerobic storage by activated sludge. *Water Sci. Technol.* **2002**, *45*, 157–168. [CrossRef]
184. Lemos, P.C.; Serafim, L.S.; Reis, M.A.M. Synthesis of polyhydroxyalkanoates from different short-chain fatty acids by mixed cultures submitted to aerobic dynamic feeding. *J. Biotechnol.* **2006**, *122*, 226–238. [CrossRef] [PubMed]
185. Solaiman, D.K.Y.; Ashby, R.D.; Foglia, T.A.; Marmer, W.N. Conversion of agricultural feedstock and coproducts into poly(hydroxyalkanoates). *Appl. Microbiol. Biotechnol.* **2006**, *71*, 783–789. [CrossRef]
186. Shen, L.; Hu, H.; Ji, H.; Cai, J.; He, N.; Li, Q.; Wang, Y. Production of poly(hydroxybutyrate-hydroxyvalerate) from waste organics by the two-stage process: Focus on the intermediate volatile fatty acids. *Bioresour. Technol.* **2014**, *166*, 194–200. [CrossRef]
187. Marang, L.; Jiang, Y.; van Loosdrecht, M.C.M.; Kleerebezem, R. Butyrate as preferred substrate for polyhydroxybutyrate production. *Bioresour. Technol.* **2013**, *142*, 232–239. [CrossRef]
188. Jiang, Y.; Marang, L.; Kleerebezem, R.; Muyzer, G.; van Loosdrecht, M.C.M. Polyhydroxybutyrate production from lactate using a mixed microbial culture. *Biotechnol. Bioeng.* **2011**, *108*, 2022–2035. [CrossRef]
189. Jiang, Y.; Hebly, M.; Kleerebezem, R.; Muyzer, G.; van Loosdrecht, M.C.M. Metabolic modeling of mixed substrate uptake for polyhydroxyalkanoate (PHA) production. *Water Res.* **2011**, *45*, 1309–1321. [CrossRef] [PubMed]
190. Cui, Y.-W.; Zhang, H.-Y.; Lu, P.-F.; Peng, Y.-Z. Effects of carbon sources on the enrichment of halophilic polyhydroxyalkanoate-storing mixed microbial culture in an aerobic dynamic feeding process. *Sci. Rep.* **2016**, *6*, 30766. [CrossRef]
191. Palmeiro-Sánchez, T.; Fra-Vázquez, A.; Rey-Martínez, N.; Campos, J.L.; Mosquera-Corral, A. Transient concentrations of NaCl affect the PHA accumulation in mixed microbial culture. *J. Hazard. Mater.* **2016**, *306*, 332–339. [CrossRef] [PubMed]

192. Cavaillé, L.; Albuquerque, M.; Grousseau, E.; Lepeuple, A.S.; Uribelarrea, J.L.; Hernandez-Raquet, G.; Paul, E. Understanding of polyhydroxybutyrate production under carbon and phosphorus-limited growth conditions in non-axenic continuous culture. *Bioresour. Technol.* **2016**, *201*, 65–73. [CrossRef]
193. Ceyhan, N.; Ozdemir, G. Poly-β-hydroxybutyrate (PHB) production from domestic wastewater using *Enterobacter aerogenes* 12Bi strain. *Afr. J. Microbiol. Res.* **2011**, *5*, 690–702.
194. Valentino, F.; Morgan-Sagastume, F.; Campanari, S.; Villano, M.; Werker, A.; Majone, M. Carbon recovery from wastewater through bioconversion into biodegradable polymers. *New Biotechnol.* **2017**, *37*, 9–23. [CrossRef] [PubMed]
195. Kosseva, M.R.; Rusbandi, E. Trends in the biomanufacture of polyhydroxyalkanoates with focus on downstream processing. *Int. J. Biol. Macromol.* **2018**, *107*, 762–778. [CrossRef]
196. Johnson, K.; van Loosdrecht, M.C.M.; Kleerebezem, R. Influence of ammonium on the accumulation of polyhydroxybutyrate (PHB) in aerobic open mixed cultures. *J. Biotechnol.* **2010**, *147*, 73–79. [CrossRef]
197. Lorini, L.; di Re, F.; Majone, M.; Valentino, F. High rate selection of PHA accumulating mixed cultures in sequencing batch reactors with uncoupled carbon and nitrogen feeding. *New Biotechnol.* **2020**, *56*, 140–148. [CrossRef] [PubMed]
198. Silva, F.; Campanari, S.; Matteo, S.; Valentino, F.; Majone, M.; Villano, M. Impact of nitrogen feeding regulation on polyhydroxyalkanoates production by mixed microbial cultures. *New Biotechnol.* **2017**, *37*, 90–98. [CrossRef] [PubMed]
199. Liu, C.; Qi, L.; Yang, S.; He, Y.; Jia, X. Increased sedimentation of a *Pseudomonas–Saccharomyces* microbial consortium producing medium chain length polyhydroxyalkanoates. *Chin. J. Chem. Eng.* **2019**, *27*, 1659–1665. [CrossRef]
200. Keskin, G.; Kızıl, G.; Bechelany, M.; Pochat-Bohatier, C.; Öner„ M. IPotential of polyhydroxyalkanoate (PHA) polymers family as substitutes of petroleum-based polymers for packaging applications and solutions brought by their composites to form barrier materials. *Pure Appl. Chem.* **2017**, *89*, 1841–1848.
201. Sun, J.; Shen, J.; Chen, S.; Cooper, M.A.; Fu, H.; Wu, D.; Yang, Z. Nanofiller reinforced biodegradable PLA/PHA composites: Current status and future trends. *Polymers* **2018**, *10*, 505. [CrossRef]
202. Wang, S.; Chen, W.; Xiang, H.; Yang, J.; Zhou, Z.; Zhu, M. Modification and potential application of short-chain-length polyhydroxyalkanoate (SCL-PHA). *Polymers* **2016**, *8*, 273. [CrossRef]
203. Kim, B.S. Production of poly(3-hydroxybutyrate) from inexpensive substrates. *Enzyme Microb. Technol.* **2000**, *27*, 774–777. [CrossRef]
204. Alvi, S.; Thomas, S.; Sandeep, K.P.; Kalaalvirikkal, N.J.V.; Yaragalla, S. *Polymers for Packaging Applications*; Apple Academic Press: Point Pleasant, NJ, USA, 2014.
205. Lee, W.H.; Loo, C.Y.; Nomura, C.T.; Sudesh, K. Biosynthesis of polyhydroxyalkanoate copolymers from mixtures of plant oils and 3-hydroxyvalerate precursors. *Bioresour. Technol.* **2008**, *99*, 6844–6851. [CrossRef]
206. Al-Battashi, H.S.; Annamalai, N.; Sivakumar, N.; Al-Bahry, S.; Tripathi, B.N.; Nguyen, Q.D.; Gupta, V.K. Lignocellulosic biomass (LCB): A potential alternative biorefinery feedstock for polyhydroxyalkanoates production. *Rev. Environ. Sci. Biotechnol.* **2019**, *18*, 183–205. [CrossRef]
207. Obruca, S.; Benesova, P.; Marsalek, L.; Marova, I. Use of lignocellulosic materials for PHA production. *Chem. Biochem. Eng. Q.* **2015**, *29*, 135–144. [CrossRef]
208. Li, M.; Wilkins, M.R. Recent advances in polyhydroxyalkanoate production: Feedstocks, strains and process developments. *Int. J. Biologic. Macromol.* **2020**, *156*, 691–703. [CrossRef]
209. Kim, J.S.; Lee, Y.Y.; Kim, T.H. A review on alkaline pretreatment technology for bioconversion of lignocellulosic biomass. *Bioresour. Technol.* **2016**, *199*, 42–48. [CrossRef]
210. Ramsay, J.A.; Hassan, M.C.A.; Ramsay, B.A. Hemicellulose as a potential substrate for production of poly (β-hydroxyalkanoates). *Can. J. Microbiol.* **1995**, *41*, 262–266. [CrossRef]
211. Cesário, M.T.; Raposo, R.S.; de Almeida, M.C.M.D.; van Keulen, F.; Ferreira, B.S.; da Fonseca, M.M.R. Enhanced bioproduction of poly-3-hydroxybutyrate from wheat straw lignocellulosic hydrolysates. *New Biotechnol.* **2014**, *3*, 104–113. [CrossRef]
212. Cesário, M.T.; Raposo, R.S.; de Almeida, M.C.M.D.; van Keulen, F.; Ferreira, B.S.; Telo, J.P.; da Fonseca, M.M.R. Production of poly(3-hydroxybutyrate-*co*-4-hydroxybutyrate) by *Burkholderia sacchari* using wheat straw hydrolysates and gamma-butyrolactone. *Int. J. Biol. Macromol.* **2014**, *71*, 59–67. [CrossRef] [PubMed]
213. Baroi, G.N.; Skiadas, I.V.; Westermann, P.; Gavala, H.N. Continuous fermentation of wheat straw hydrolysate by *Clostridium tyrobutyricum* with in-situ acids removal. *Waste Biomass Valoriz.* **2015**, *6*, 317–326. [CrossRef]
214. Antonetti, C.; Licursi, D.; Fulignati, S.; Valentini, G.; Raspolli Galletti, A.M. New frontiers in the catalytic synthesis of levulinic acid: From sugars to raw and waste biomass as starting feedstock. *Catalysts* **2016**, *6*, 196. [CrossRef]
215. Habe, H.; Sato, S.; Morita, T.; Fukuoka, T.; Kirimura, K.; Kitamoto, D. Bacterial production of short-chain organic acids and trehalose from levulinic acid: A potential cellulose-derived building block as a feedstock for microbial production. *Bioresour. Technol.* **2015**, *177*, 381–386. [CrossRef]
216. Jaremko, M.; Yu, J. The initial metabolic conversion of levulinic acid in *Cupriavidus necator*. *J. Biotechnol.* **2011**, *155*, 293–298. [CrossRef] [PubMed]
217. Xu, Y.; Wang, R.-H.; Kautinass, A.A.; Webb, C. Microbial biodegradable plastic production from a wheat-based biorefining strategy. *Process Biochem.* **2010**, *45*, 153–163. [CrossRef]
218. Sathiyanarayanan, G.; Saibaba, G.; Seghal, K.G.; Selvin, J. A statistical approach for optimization of polyhydroxybutyrate production by marine *Bacillus subtilis* MSBN17. *Int. J. Biol. Macromol.* **2013**, *59*, 170–177. [CrossRef] [PubMed]

219. Dietrich, K.; Oliveira-Filho, E.R.; Dumont, M.J.; Gomez, J.G.C.; Taciro, M.K.; DaSilva, L.F.; Orsat, V.; Del Rio, L.F. Increasing PHB production with an industrially scalable hardwood hydrolysate as a carbon source. *Ind. Crops Prod.* **2020**, *154*, 112703. [CrossRef]
220. Yang, F.; Hanna, M.A.; Sun, R. Value-added uses for crude glycerol-a byproduct of biodiesel production. *Biotechnol. Biofuels.* **2012**, *5*, 13. [CrossRef]
221. Koutinas, A.A.; Papanikolaou, S. Biodiesel production from microbial oil. In *Handbook of Biofuels Production—Processes and Technologies*; Luque, R., Campelo, J., Clark, J.H., Eds.; Woodhead Publishing Limited: Sawston, UK, 2011; pp. 177–198.
222. Chatzifragkou, A.; Papanikolaou, S. Effect of impurities in biodiesel-derived waste glycerol on the performance and feasibility of biotechnological processes. *Appl. Microbiol. Biotechnol.* **2012**, *95*, 13–27. [CrossRef]
223. Chatzifragkou, A.; Papanikolaou, S.; Kopsahelis, N.; Kachrimanidou, V.; Dorado, M.P.; Koutinas, A. Biorefinery development through utilization of biodiesel industry by-products as sole fermentation feedstock for 1,3-propanediol production. *Bioresour. Technol.* **2014**, *159*, 167–175. [CrossRef] [PubMed]
224. Kachrimanidou, V.; Kopsahelis, N.; Chatzifragkou, A.; Papanikolaou, S.; Yanniotis, S.; Kookos, I.; Koutinas, A.A. Utilization of by-products from sunflower-based biodiesel production processes for the production of fermentation feedstock. *Waste Biomass Valoriz.* **2013**, *4*, 529–537. [CrossRef]
225. Cavalheiro, J.M.B.T.; de Almeida, M.C.M.D.; Grandfils, C.; da Fonseca, M.M.R. Poly(3-hydroxybutyrate) production by *Cupriavidus necator* using waste glycerol. *Process Biochem.* **2009**, *44*, 509–515. [CrossRef]
226. Salakkam, A.; Webb, C. Production of poly(3-hydroxybutyrate) from a complete feedstock derived from biodiesel by-products (crude glycerol and rapeseed meal). *Biochem. Eng. J.* **2018**, *137*, 358–364. [CrossRef]
227. Kumar, P.; Jun, H.B.; Kim, B.S. Co-production of polyhydroxyalkanoates and carotenoids through bioconversion of glycerol by *Paracoccus* sp. strain LL1. *Int. J. Biol. Macromol.* **2018**, *107*, 2552–2558. [CrossRef]
228. Volova, T.; Demidenko, A.; Kiselev, E.; Baranovskiy, S.; Shishatskaya, E.; Zhila, N. Polyhydroxyalkanoate synthesis based on glycerol and implementation of the process under conditions of pilot production. *Appl. Microbiol. Biotechnol.* **2019**, *103*, 225–237. [CrossRef]
229. Zhu, C.; Nomura, C.T.; Perrotta, J.A.; Stipanovic, A.J.; Nakas, J.P. Production and characterization of poly-3-hydroxybutyrate from biodiesel-glycerol by *Burkholderia cepacia* ATCC 17759. *Biotechnol. Progress* **2010**, *26*, 424–430.
230. Kachrimanidou, V.; Kopsahelis, N.; Papanikolaou, S.; Kookos, I.K.; De Bruyn, M.; Clark, J.H.; Koutinas, A.A. Sunflower-based biorefinery: Poly(3-hydroxybutyrate) and poly(3-hydroxybutyrate-co-3-hydroxyvalerate) production from crude glycerol, sunflower meal and levulinic acid. *Bioresour. Technol.* **2014**, *172*, 121–130. [CrossRef] [PubMed]
231. Albuquerque, M.G.E.; Eiroa, M.; Torres, C.; Nunes, B.R.; Reis, M.A.M. Strategies for the development of a side stream process for polyhydroxyalkanoate (PHA) production from sugar cane molasses. *J. Biotechnol.* **2007**, *130*, 411–421. [CrossRef]
232. Shasaltaneh, M.D.; Moosavi-Nejad, Z.; Gharavi, S.; Fooladi, J. Cane molasses as a source of precursors in the bioproduction of tryptophan by *Bacillus subtilis. Iran. J. Microbiol.* **2013**, *5*, 285–292. [PubMed]
233. Jiang, Y.; Song, X.; Gong, L.; Li, P.; Dai, C.; Shao, W. High poly(hydroxybutyrate) production by *Pseudomonas fluorescens* A2a5 from inexpensive substrates. *Enzyme Microb. Technol.* **2008**, *42*, 167–172. [CrossRef]
234. Kanjanachumpol, P.; Kulpreecha, S.; Tolieng, V.; Thongchul, N. Enhancing polyhydroxybutyrate production from high cell density fed-batch fermentation of *Bacillus megaterium* BA-019. *Bioprocess Biosyst. Eng.* **2013**, *36*, 1463–1474. [CrossRef]
235. Parnaudeau, V.; Condom, N.; Oliver, R.; Cazevieille, P.; Recous, S. Vinasse organic matter quality and mineralization potential, as influenced by raw material, fermentation and concentration processes. *Bioresour. Technol.* **2008**, *99*, 1553–1562. [CrossRef]
236. FitzGibbon, F.; Singh, D.; McMullan, G.; Marchant, R. The effect of phenolic acids and molasses spent wash concentration on distillery wastewater remediation by fungi. *Process Biochem.* **1998**, *33*, 799–803. [CrossRef]
237. Dalsasso, R.R.; Pavan, F.A.; Bordignon, S.E.; de Aragão, G.M.F.; Poletto, P. Polyhydroxybutyrate (PHB) production by *Cupriavidus necator* from sugarcane vinasse and molasses as mixed substrate. *Process Biochem.* **2019**, *85*, 12–18. [CrossRef]
238. Davis, R.; Kataria, R.; Cerrone, F.; Woods, T.; Kenny, S.; O'Donovan, A.; Guzik, M.; Shaikh, H.; Duane, G.; Gupta, V.K.; et al. Conversion of grass biomass into fermentable sugars and its utilization for medium chain length polyhydroxyalkanoate (mcl-PHA) production by *Pseudomonas* strains. *Bioresour. Technol.* **2013**, *150*, 202–209. [CrossRef]
239. Haas, R.; Jin, B.; Zepf, F.T. Production of poly(3-hydroxybutyrate) from waste potato starch. *Biosci. Biotechnol. Biochem.* **2008**, *72*, 253–256. [CrossRef]
240. Chen, C.W.; Don, T.M.; Yen, H.F. Enzymatic extruded starch as a carbon source for the production of poly(3-hydroxybutyrate-co-3-hydroxyvalerate) by *Haloferax mediterranei. Process Biochem.* **2006**, *41*, 2289–2296. [CrossRef]
241. Ahn, W.S.; Park, S.J.; Lee, S.Y. Production of poly(3-Hydroxybutyrate) by fed-batch culture of recombinant *Escherichia coli* with a highly concentrated whey solution. *Appl. Environ. Microb.* **2000**, *66*, 3624–3627. [CrossRef]
242. Park, S.J.; Park, J.P.; Lee, S.Y. Production of poly(3-hydroxybutyrate) from whey by fed-batch culture of recombinant *Escherichia coli* in a pilot-scale fermenter. *Biotechnol. Lett.* **2002**, *24*, 185–189. [CrossRef]
243. Annamalai, N.; Sivakumar, N. Production of polyhydroxybutyrate from wheat bran hydrolysate using *Ralstonia eutropha* through microbial fermentation. *J. Biotechnol.* **2016**, *237*, 13–17. [CrossRef]
244. Huang, T.Y.; Duan, K.J.; Huang, S.Y.; Chen, C.W. Production of polyhydroxyalkanoates from inexpensive extruded rice bran and starch by *Haloferax mediterranei. J. Ind. Microbiol. Biotechnol.* **2006**, *33*, 701–706. [CrossRef]
245. Demirbas, A.; Bafail, A.; Ahmad, W.; Sheikh, M. Biodiesel production from non-edible plant oils. *Energ. Explor. Exploit* **2016**, *34*, 290–318. [CrossRef]

246. Abid, S.; Raza, Z.A.; Hussain, T. Production kinetics of polyhydroxyalkanoates by using *Pseudomonas aeruginosa* gamma ray mutant strain EBN-8 cultured on soybean oil. *3 Biotech* **2016**, *6*, 142. [CrossRef] [PubMed]
247. Tufail, S.; Munir, S.; Jamil, N. Variation analysis of bacterial polyhydroxyalkanoates production using saturated and unsaturated hydrocarbons. *Braz. J. Microbiol.* **2017**, *48*, 629–636. [CrossRef]
248. Obruca, S.; Marova, I.; Snajdar, O.; Mravcova, L.; Svoboda, Z. Production of poly (3-hydroxybutyrate-co-3-hydroxyvalerate) by *Cupriavidus necator* from waste rapeseed oil using propanol as a precursor of 3-hydroxyvalerate. *Biotechnol. Lett.* **2010**, *32*, 1925–1932. [CrossRef] [PubMed]
249. Cruz, M.V.; Gouveia, A.R.; Dionísio, M.; Freitas, F.; Reis, M.A.M. A process engineering approach to improve production of P(3HB) by *Cupriavidus necator* from used cooking oil. *Int. J. Polym. Sci.* **2019**, *2019*, 1–7. [CrossRef]
250. Ruiz, C.; Kenny, S.T.; Babu, P.R.; Walsh, M.; Narancic, T.; O'Connor, K.E. High cell density conversion of hydrolyzed waste cooking oil fatty acids into medium chain length polyhydroxyalkanoate using *Pseudomonas putida* KT2440. *Catalysts* **2019**, *9*, 468. [CrossRef]
251. Ruiz, C.; Kenny, S.T.; Narancic, T.; Babu, R.; O'Connor, K. Conversion of waste cooking oil into medium chain polyhydroxyalkanoates in a high cell density fermentation. *J. Biotechnol.* **2019**, *306*, 9–15. [CrossRef] [PubMed]
252. Thuoc, D.V.; My, D.N.; Loan, T.T.; Sudesh, K. Utilization of waste fish oil and glycerol as carbon sources for polyhydroxyalkanoate (PHA) production by *Salinivibrio* sp. M318. *Int. J. Biologic. Macromol.* **2019**, *141*, 885–892. [CrossRef] [PubMed]
253. Da Cruz Pradella, J.G.; Ienczak, J.L.; Delgado, C.R.; Taciro, M.K. Carbon source pulsed feeding to attain high yield and high productivity in poly(3-hydroxybutyrate) (PHB) production from soybean oil using *Cupriavidus necator*. *Biotechnol. Lett.* **2012**, *34*, 1003–1007. [CrossRef] [PubMed]
254. Shang, L.; Jiang, M.; Yun, Z.; Yan, H.-Q.; Chang, H.-N. Mass production of medium-chain-length poly(3-hydroxyalkanoates) from hydrolyzed corn oil by fed-batch culture of *Pseudomonas putida*. *World J. Microbiol. Biotechnol.* **2008**, *24*, 2783–2787. [CrossRef]
255. Ryu, H.W.; Cho, K.S.; Goodrich, P.R.; Park, C.H. Production of polyhydroxyalkanoates by *Azotobacter vinelandii* UWD using swine wastewater: Effect of supplementing glucose, yeast extract, and inorganic salts. *Biotechnol. Bioprocess Eng.* **2008**, *13*, 651–658. [CrossRef]
256. Yan, S.; Tyagi, R.D.; Surampalli, R.Y. Polyhydroxyalkanoates (PHA) production using wastewater as carbon source and activated sludge as microorganisms. *Water Sci. Technol.* **2006**, *53*, 175–180. [CrossRef] [PubMed]
257. Noda, I.; Lindsey, S.B.; Caraway, D. Nodax™ class PHA copolymers: Their properties and applications. In *Plastics from Bacteria*; Chen, G.Q., Ed.; Microbiology 2010 Monographs 14; Springer: Berlin/Heidelberg, Germany, 2010.
258. Danimer Scientific Nodax™. Available online: http://danimerscientific.com/pha-begining-of-life/ (accessed on 16 August 2020).
259. Available online: http://ww.biocycle.com.br (accessed on 16 August 2020).
260. Plastics News. Available online: https://www.plasticsnews.com/news/bio-declares-bankruptcy (accessed on 13 October 2020).
261. Bio-On. S.p.A. *Flos to be World's First Company to Use Revolutionary Bioplastic Designed by Bio-On*. Available online: http://www.bio-on.it/project.php?lin=portoghese (accessed on 7 August 2020).
262. TianAn Biopolymer Homepage. Available online: http://www.tianan-enmat.com (accessed on 16 August 2020).
263. Tianjin Gouyun Biomaterials Co. Homepage. Available online: http://www.tjgreenbio.com/en/about.aspx?title=Enterprise%20Introduction&cid=25 (accessed on 16 August 2020).
264. Bio-mer Homepage. Available online: http://www.biomer.de/IndexE.html (accessed on 16 August 2020).
265. Press Release by Kaneka Corporation. Available online: https://www.kaneka.co.jp/en/service/news/nr20191219/ (accessed on 16 August 2020).
266. Yield10 Bioscience Home Page. Available online: https://www.yield10bio.com (accessed on 16 August 2020).
267. Available online:. Available online: https://www.tepha.com/technology/polymer-processing-material-attributes/ (accessed on 16 August 2020).
268. Available online:. Available online: www.mgc.co.jp (accessed on 16 August 2020).
269. Available online:. Available online: http://www.polymerofcanada.com/versamer_phas.html (accessed on 16 August 2020).
270. Online News Article. Available online: https://www.ajudaily.com/view/20160823101739592 (accessed on 16 August 2020).
271. Jennifer, B.; Emily, G. Plastic from thin air. *Popul. Sci.* **2014**, *285*, 24.
272. Daniel, L. Can plastic be made environmentally friendly? *Scientific American*. 2014. Available online: Scientificamerican.com (accessed on 7 April 2021).
273. Grady, B. August 31 2015, Newlight and Vinmar Bet on Carbon-Negative Plastic, GreenBiz. Available online: https://www.greenbiz.com/article/newlight-and-vinmar-bet-carbon-negative-plastic (accessed on 15 August 2020).
274. Bresan, B.; Sznajder, A.; Hauf, W.; Forchhammer, K.; Pfeiffer, D.; Jendrossek, D. Polyhydroxyalkanoate (PHA) granules have no phospholipids. *Sci. Rep.* **2016**, *6*, 26612. [CrossRef]
275. Dietrich, K.; Dumont, M.-J.; Del Rio, L.F.; Orsat, V. Producing PHAs in the bioeconomy-towards a sustainable bioplastic. *Sustain. Prod. Consumpt.* **2017**, *9*, 58–70. [CrossRef]
276. Koller, M.; Niebelschütz, H.; Braunegg, G. Strategies for recovery and purification of poly[(R)-3-hydroxyalkanoates] (PHA) biopolyesters from surrounding biomass. *Eng. Life Sci.* **2013**, *13*, 549–562. [CrossRef]
277. Jacquel, N.; Lo, C.W.; Wei, Y.H.; Wu, H.S.; Wang, S.S. Isolation and purification of bacterial poly(3-hydroxyalkanoates). *Biochem. Eng. J.* **2008**, *39*, 15–27. [CrossRef]

278. Aramvash, A.; Moazzeni Zavareh, F.; Gholami Banadkuki, N. Comparison of different solvents for extraction of polyhydroxybutyrate from *Cupriavidus necator*. *Eng Life Sci.* **2018**, *18*, 20–28. [CrossRef]
279. Pérez-Rivero, C.; López-Gómez, J.P.; Roy, I. A sustainable approach for the downstream processing of bacterial polyhydroxyalkanoates: State-of-the-art and latest developments. *Biochem. Eng. J.* **2019**, *150*, 107283. [CrossRef]
280. Lafferty, R.M.; Heinzle, E. Cyclic Carbonic Acid Esters as Solvents for Poly-(β-Hydroxybutyric Acid), 18 July 1978, US4101533A. Available online: https://patents.google.com/patent/US4101533 (accessed on 20 March 2021).
281. Noda, I.; Schechtman, L.A. Solvent Extraction of Polyhydroxyalkanoates from Biomass. 24 August 1996, US5942597A. Available online: https://patents.google.com/patent/US5942597A (accessed on 15 April 2021).
282. Kurdikar, D.L.; Strauser, F.E.; Solodar, A.J.; Paster, M.D.; Asrar, J. Methods of PHA Extraction and Recovery Using Non-Halogenated Solvents. 28 March 1998, US6043063A. Available online: https://patents.google.com/patent/US6043063A (accessed on 15 March 2021).
283. Horowitz, D.M. Methods for Purifying Polyhydroxyalkanoates. 18 January 2002, US20020058316A1. Available online: https://patents.google.com/patent/US20020058316A1 (accessed on 1 April 2021).
284. Van Walsem, J.; Zhong, L.; Shih, S.S. Polymer Extraction Methods. 23 July 2003, US7252980B2. Available online: https://patents.google.com/patent/US7252980B2 (accessed on 15 April 2021).
285. Kinoshita, K.; Osakada, F.; Ueda, Y.; Narasimhan, K.; Cearley, A.C.; Yee, K.; Noda, I. Method for Producing Polyhydroxyalkanoate Crystal. 26 December 2006, US7153928B2. Available online: https://patents.google.com/patent/US7153928 (accessed on 1 April 2021).
286. Ibrahim, M.H.A.; Takwa, M.; Hatti-Kaul, R. Process for Extraction of Bioplastic and Production of Monomers from the Bioplastic. March 24, 2020, US10597506B2. (accessed on 10 March 2021).
287. Furrer, P.; Panke, S.; Zinn, M. Efficient recovery of low endotoxin medium-chain length poly([R]-3-hydroxyalkanoate) from bacterial biomass. *J. Microbiol. Methods.* **2007**, *69*, 206–213. [CrossRef]
288. Anis, S.N.; Iqbal, N.M.; Kumar, S.; Al-Ashraf, A. Increased recovery and improved purity of PHA from recombinant *Cupriavidus necator*. *Bioengineering* **2013**, *4*, 115–118.
289. Chemat, F.; Vian, M.A.; Cravotto, G. Green extraction of natural products: Concept and principles. *Int. J. Mol. Sci.* **2012**, *13*, 8615–8627. [CrossRef]
290. Yabueng, N.; Napathorn, S.C. Toward non-toxic and simple recovery process of poly (3-hydroxybutyrate) using the green solvent 1, 3-dioxolane. *Process Biochem.* **2018**, *69*, 197–207. [CrossRef]
291. Hecht, S.E.; Niehoff, R.L.; Narasimhan, K.; Neal, C.W.; Forshey, P.A.; Van Phan, D.; Brooker, A.D.M.; Combs, K.H. Extracting Biopolymers from a Biomass Using Ionic Liquids. 26 October 2006, US7763715B2. Available online: https://patents.google.com/patent/US7763715B2 (accessed on 3 April 2021).
292. Dubey, S.; Bharmoria, P.; Singh Gehlot, P.; Agrawal, V.; Kumar, A.; Mishra, S. 1-Ethyl-3-Methylimidazolium Diethylphosphate based extraction of bioplastic "polyhydroxyalkanoates" from bacteria: Green and sustainable approach. *ACS Sustain. Chem. Eng.* **2018**, *6*, 766–773. [CrossRef]
293. Ishak, K.A.; Annuar, M.S.M.; Heidelberg, T.; Gumel, A.M. Ultrasound-assisted rapid extraction of bacterial intracellular medium-chain-length poly(3-hydroxyalkanoates) (mcl-phas) in medium mixture of solvent/marginal non-solvent. *Arab J. Sci. Eng.* **2016**, *41*, 33–44. [CrossRef]
294. Noda, I. Process for Recovering Polyhydroxyalkanotes Using Air Classification. 15 December 1998; US5849854A.
295. Vilkhu, K.; Mawson, R.; Simons, L.; Bates, D. Applications and opportunities for ultrasound assisted extraction in the food industry-A review. *Innov. Food Sci. Emerg. Technol.* **2008**, *9*, 161–169. [CrossRef]
296. Darani, K.K.; Mozafari, M.R. Supercritical fluids technology in bioprocess industries: A review. *J. Biochem. Tech.* **2009**, *2*, 144–152.
297. Raza, Z.A.; Abid, S.; Banat, I.M. Polyhydroxyalkanoates: Characteristics, production, recent developments and applications. *Int. Biodeterior. Biodegrad.* **2018**, *126*, 45–56. [CrossRef]
298. Poliakoff, M.; Licence, P. Supercritical fluids: Green solvents for green chemistry? *Philos. Trans. A Math. Phys. Eng. Sci.* **2015**, *373*. [CrossRef] [PubMed]
299. Hampson, J.W.; Ashby, R.D. Extraction of lipid-grown bacterial cells by supercritical fluid and organic solvent to obtain pure medium chain-length polyhydroxyalkanoates. *J. Am. Oil Chem. Soc.* **1999**, *76*, 1371–1374. [CrossRef]
300. Hejazi, P.; Vasheghani-Farahani, E.; Yamini, Y. Supercritical fluid disruption of *Ralstonia eutropha* for poly(β-hydroxybutyrate) recovery. *Biotechnol. Prog.* **2003**, *19*, 1519–1523. [CrossRef]
301. Leong, Y.K.; Koroh, F.E.; Show, P.L.; Lan, J.C.W.; Loh, H.S. Optimization of extractive bioconversion for green polymer *via* aqueous two-phase system. *Chem. Eng. Trans.* **2015**, *45*, 1495–1500.
302. Leong, Y.K.; Lan, J.C.W.; Loh, H.S.; Ling, T.C.; Ooi, C.W.; Show, P.L. Cloud-point extraction of green-polymers from *Cupriavidus necator* lysate using thermoseparating-based aqueous two-phase extraction. *J. Biosci. Bioeng.* **2016**, *123*, 3270–3375. [CrossRef]
303. Leong, Y.K.; Show, P.L.; Lan, J.C.-W.; Loh, H.-S.; Yap, Y.-J.; Ling, T.C. Extraction and purification of polyhydroxyalkanoates (PHAs): Application of thermoseparating aqueous two-phase extraction. *J. Polym. Res.* **2017**, *24*, 158. [CrossRef]
304. Leong, K.; Show, P.L.; Lan, J.C.W.; Krishnamoorthy, R.; Chu, D.T.; Nagarajan, D.; Yen, H.W.; Chang, J.S. Application of thermoseparating aqueous two-phase system in extractive bioconversion of polyhydroxyalkanoates by *Cupriavidus necator* H16. *Bioresour. Technol.* **2019**, *287*, 121474. [CrossRef]

305. López-Abelairas, M.; García-Torreiro, M.; Lú-Chau, T.; Lema, J.M.; Steinbüchel, A. Comparison of several methods for the separation of poly(3-hydroxybutyrate) from *Cupriavidus necator* H16 cultures. *Biochem. Eng. J.* **2015**, *93*, 250–259. [CrossRef]
306. Gumel, A.M.; Annuar, M.S.M.; Chisti, Y. Recent advances in the production, recovery and applications of polyhydroxyalkanoates. *J. Polym. Environ.* **2013**, *21*, 580–605. [CrossRef]
307. Dong, Z.; Sun, X.; Zhaolin, D.; Xuenan, S.U.N. A new method of recovering polyhydroxyalkanoate from *Azotobacter chroococcum*. *Chinese Sci. Bull.* **2000**, *45*, 252–256. [CrossRef]
308. Kapritchkoff, F.M.; Viotti, A.P.; Alli, R.C.P.; Zuccolo, M.; Pradella, J.G.C.; Maiorano, A.E.; Miranda, E.A.; Bonomi, A. Enzymatic recovery and purification of polyhydroxybutyrate produced by *Ralstonia eutropha*. *J. Biotechnol.* **2006**, *122*, 453–462. [CrossRef]
309. Harrison, S.T.L. Bacterial cell disruption: A key unit operation in the recovery of intracellular products. *Biotechnol. Adv.* **1991**, *9*, 217–240. [CrossRef]
310. Yasotha, K.; Aroua, M.K.; Ramachandran, K.B.; Tan, I.K.P. Recovery of medium-chain-length polyhydroxyalkanoates (PHAs) through enzymatic digestion treatments and ultrafiltration. *Biochem. Eng. J.* **2006**, *30*, 260–268. [CrossRef]
311. Kachrimanidou, V.; Kopsahelis, N.; Vlysidis, A.; Papanikolaou, S.; Kookos, I.K.; Martínez, B.M.; Rondan, M.C.E.; Kautinas, A.A. Downstream separation of poly(hydroxyalkanoates) using crude enzyme consortia produced via solid state fermentation integrated in a biorefinery concept. *Food Bioprod. Process.* **2016**, *100*, 323–334. [CrossRef]
312. Martino, L.; Cruz, M.V.; Scoma, A.; Freitas, F.; Bertin, L.; Scandola, M.; Reis, M.A.M. Recovery of amorphous polyhydroxybutyrate granules from *Cupriavidus necator* cells grown on used cooking oil. *Int. J. Biol. Macromol.* **2014**, *71*, 117–123. [CrossRef] [PubMed]
313. Israni, N.; Thapa, S.; Shivakumar, S. Biolytic extraction of poly(3-hydroxybutyrate) from *Bacillus megaterium* Ti3 using the lytic enzyme of *Streptomyces albus* Tia1. *J. Genet. Eng. Biotechnol.* **2018**, *16*, 265–271. [CrossRef] [PubMed]
314. Murugan, P.; Han, L.; Gan, C.Y.; Maurer, H.J.M.; Sudesh, K. A new biological recovery approach for PHA using mealworm, *Tenebrio molitor*. *J. Biotechnol.* **2017**, *239*, 98–105. [CrossRef] [PubMed]
315. Kunasundari, B.; Arza, C.R.; Maurer, F.H.J.; Murugaiyah, V.; Kaur, G.; Sudesh, K. Biological recovery and properties of poly(3-hydroxybutyrate) from *Cupriavidus necator* H16. *Sep. Purif. Technol.* **2017**, *172*, 1–6. [CrossRef]

Review

Preparation, Marriage Chemistry and Applications of Graphene Quantum Dots–Nanocellulose Composite: A Brief Review

Wan Hazman Danial [1,*], Nur Fathanah Md Bahri [1] and Zaiton Abdul Majid [2]

[1] Department of Chemistry, Kulliyyah of Science, International Islamic University Malaysia, Kuantan 25200, Pahang, Malaysia; fathanahbahri93@gmail.com
[2] Department of Chemistry, Faculty of Science, Universiti Teknologi Malaysia, Johor Bahru 81310, Johor, Malaysia; zaitonmajid@utm.my
* Correspondence: whazman@iium.edu.my

Citation: Danial, W.H.; Md Bahri, N.F.; Abdul Majid, Z. Preparation, Marriage Chemistry and Applications of Graphene Quantum Dots–Nanocellulose Composite: A Brief Review. *Molecules* **2021**, *26*, 6158. https://doi.org/10.3390/molecules26206158

Academic Editor: Sylvain Caillol

Received: 29 August 2021
Accepted: 11 October 2021
Published: 12 October 2021

Publisher's Note: MDPI stays neutral with regard to jurisdictional claims in published maps and institutional affiliations.

Copyright: © 2021 by the authors. Licensee MDPI, Basel, Switzerland. This article is an open access article distributed under the terms and conditions of the Creative Commons Attribution (CC BY) license (https://creativecommons.org/licenses/by/4.0/).

Abstract: Graphene quantum dots (GQDs) are zero-dimensional carbon-based materials, while nanocellulose is a nanomaterial that can be derived from naturally occurring cellulose polymers or renewable biomass resources. The unique geometrical, biocompatible and biodegradable properties of both these remarkable nanomaterials have caught the attention of the scientific community in terms of fundamental research aimed at advancing technology. This study reviews the preparation, marriage chemistry and applications of GQDs–nanocellulose composites. The preparation of these composites can be achieved via rapid and simple solution mixing containing known concentration of nanomaterial with a pre-defined composition ratio in a neutral pH medium. They can also be incorporated into other matrices or drop-casted onto substrates, depending on the intended application. Additionally, combining GQDs and nanocellulose has proven to impart new hybrid nanomaterials with excellent performance as well as surface functionality and, therefore, a plethora of applications. Potential applications for GQDs–nanocellulose composites include sensing or, for analytical purposes, injectable 3D printing materials, supercapacitors and light-emitting diodes. This review unlocks windows of research opportunities for GQDs–nanocellulose composites and pave the way for the synthesis and application of more innovative hybrid nanomaterials.

Keywords: graphene quantum dots; nanocellulose; composite

1. Introduction

Over the past few years, many researchers have developed an interest in investigating carbon-based nanomaterials, such as graphene and graphene quantum dots (GQDs) [1–4]. Graphene is one layer of sp^2-hybridised carbon atoms arranged in a honeycomb lattice [5], whereas GQDs are chopped fragments of a few graphene sheets (<10 nm lateral dimension) and <10 graphene layers which form final particles [6]. GQDs are used in various applications owing to their higher photostability, low cytotoxicity, strong photoluminescence, dispersibility in water, fluorescence and excellent biocompatibility [7,8]. Furthermore, the characteristics of GQDs vary according to morphology, size, doping concentration and type [9]. Owing to these properties and the ability to fine-tune them, GQDs are investigated for different applications in biomedicine [10], catalyst development [11], energy [12] and sensing and photo electronics [6]. Different bottom-up and top-down strategies have been used to produce GQDs, such as organic synthesis [13], hydrothermal [14], microwave irradiation [15], chemical exfoliation [16] and electrochemical exfoliation [3,17].

Furthermore, many researchers have also conducted extensive carbohydrate polymer research using cellulose-based nanomaterials. Cellulose is an abundantly available, naturally occurring organic polymer that is composed of repeating units of β-glucopyranose rings that are covalently linked to one another with a β 1-4 glycosidic bond [18,19]. Cellulose nanowhiskers, or nanocellulose, are cellulose particles present in the form of crystals or fibres [20]. These particles are few micrometres in length and have a diameter <100 nm.

These fibres are lightweight, biodegradable and also have a higher water-binding capacity [21]. Cellulosic nanomaterials display a larger specific surface area; therefore, they can form many hydrogen bonds. This hydrogen bond-forming capability helps the material develop a dense and strong network [22]. Many different kinds of nano-cellulosic materials are described in the literature, such as cellulose nanocrystals (CNCs) [23], nanofibrillated cellulose (NFCs) [24] and bacterial nanocellulose (BNCs) [25]. As nanocellulose can be derived from different cellulosic sources, each displays varying properties and some are derived from natural sources. For instance, the properties of the bacterial nanocelluloses are based on different bacterial sources, whereas the properties of the cellulose nanocrystals and nanofibrillated celluloses are based on sources such as tunicin or plants [26]. Nanocellulose has garnered significant research interest as a promising nanomaterial that can revolutionise multiple fields, such as the pharmaceutical field [27], engineering [28], electronics [29] and health and environmental protection [30].

GQD-nanocellulose is described as a hybrid material that contains nanocellulose and GQDs, which synergistically improves the properties of every individual component, such as their stability and mechanical strength [31]. The addition of GQDs to a nanocomposite material can improve its final tensile strength, stiffness and the toughness of these GQDs–nanocellulose structures, irrespective of their GQD oxidation type and nanocellulose orientation [32]. These structures are designed to exhibit higher conductivity, cycle stability and higher specific capacitance [32]. Therefore, GQDs–nanocellulose composites are designed for use in many applications, such as fluorescence films, bendable and portable paper electronics, and hydrogels. The numerous -COOH and -OH functional groups are present on the surfaces of GQDs result in the formation of hydrogen bonds at the GQDs–nanocellulose interface. This can significantly affect the GQDs–nanocellulose supercells, nanocellulose lattice parameters and the morphological properties of nanocellulose, such as the dihedral angle differences in the hydroxymethyl groups, axial tilt of molecular chains and the flipping motion of terminal groups [33]. In addition, the composites made from these renewable nanomaterials offer a greener approach than petroleum-derived composites and exhibit great potential for various technological applications [34]. Moreover, this nanomaterial combination has recently emerged as a new class of hybrid material due to its exceptional features and notable synergistic effects [35].

In this review, the preparation, marriage chemistry and applications of GQDs–nanocellulose composites are discussed. Although these composites have been investigated in recent years, reviews that focus solely on GQDs–nanocellulose composites have not yet been reported. This review also provides some insight into the development of the fluorescent hydrogel functions of the composites.

1.1. Graphene Quantum Dots (GQDs)

GQDs hold sp^2 hybridised carbon single-layer nanocrystals and they are highly fluorescent, regardless of whether they are in an aqueous or solid state [36,37]. GQDs are a distinct from the type of carbon in carbon nanodots (CNDs) and polymer dots (PDs). This is because all carbon dots possess modified chemical groups, such as oxygen groups, on the surface [38]. However, each of them is of a different size and possesses different properties with which to perform their action. GQDs have an average lattice parameter of 0.24 nm, which corresponds to 100 in plane graphene lattice parameters [39]. On the contrary, CNDs are divided into carbon nanoparticles without a crystal lattice and have a spherical shape [40], while the grafted and cross-linked polymer chains of linear non-conjugated polymers form PDs [36].

GQDs are the simplest carbon dots with connected chemical groups on the surfaces or edges [41]. The surfaces or edges of GQDs contain triple carbene at the zigzag edges and oxygen groups at the graphene core. Additionally, the type of GQD edge plays a significant role in determining the material's optical, electronic and magnetic properties [36]. GQDs are produced using either the top-down or bottom-up method. Both approaches use different parameters to produce GQDs [36]. The top-down method involves direct cutting of graphite or graphene-based materials via acid exfoliation [16], sonochemistry [42], solvothermal synthesis [43], electrochemistry [44–47], or chemical oxidation [48–50]. The advantages of these methods are an abundance of inexpensive precursor (graphite), the high graphitic nature and the formation of GQDs with high oxygen-containing functional groups, which renders good solubility and functionality. However, the drawbacks of these methods include harsh a reaction procedure, as well as the non-uniform size and thickness of the final product [51]. Examples of the bottom-up method of GQD production from molecular precursors include cyclodehydrogenation [52], pyrolysis [53] and solution chemistry [13]. Despite the difficult synthesis processes, the bottom-up approaches provide better control over the size and shape of the GQDs.

Graphene is widely used in many applications, such as electronics, solar cells and Li-ion batteries [54], whereas GQDs have attracted tremendous interest in photoluminescence [55], cell-imaging [56] and drug delivery [57,58], among many things. GQDs have a great advantage, as their properties can be adjusted by changing their band via doping to produce amine-functionalised GQDs [59], nitrogen-doped GQDs (N-GQDs) [60] and sulphur–nitrogen co-doped GQDs (S, N-GQDs) [61]. Furthermore, the photoluminescence colour of GQDs can also be changed from violet to yellow by setting the reactant concentration and temperature during the hydrothermal method [54].

1.2. Nanocellulose

Nanocellulose, a sustainable and renewable nano-structured cellulose, has gained tremendous attention for its potential use in many applications due to its excellent surface chemistry, physical properties and remarkable biological properties [62,63]. Nanocellulose was first prepared in 1947 using sulfuric acid and hydrochloric acid hydrolysis from wood fibres and cotton fibres [64]. As illustrated in Figure 1, there are three types of nanocellulose from different precursors, bacteria nanocellulose, cellulose nanocrystals (CNCs) and cellulose nanofibrils (CNF) [65,66], which can be obtained from cellulose-containing precursors such as plants or bacteria. Plant cellulose is situated within a plant's fibre walls, whereas bacteria produce exopolysaccharides to form microbial cellulose [67]. These bacterial isolates can normally be obtained from rotten vegetables and fruits. Similar to plant-derived cellulose, some parameters or conditions, such as carbon source, nitrogen source, temperature, pH and agitation, are measured to produce a high yield of bacterial cellulose [68,69]. Bacterial nanocellulose is pure in nature and has a low cytotoxicity, high pore distribution and high hydrophilicity due to the presence of OH groups on its surface [70,71]. However, unlike bacterial cellulose, plant cellulose is impure due to the presence of lignin, hemicellulose and pectin. Apart from that, plant cellulose is slightly cytotoxic, less malleable and has small pore sizes due to less space between fibrils [72].

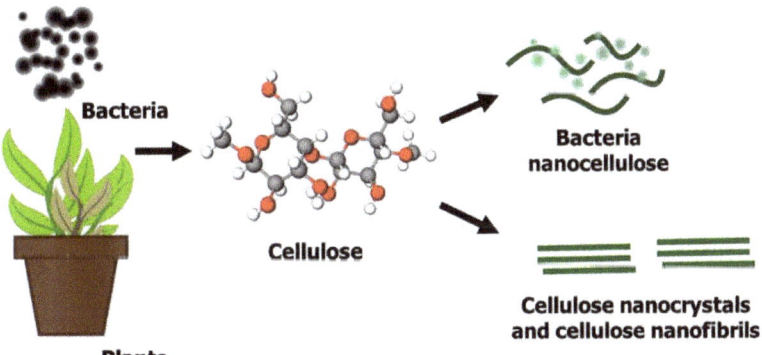

Figure 1. Formation of nanocellulose from cellulose-containing materials.

Nanocellulose can be prepared via mechanical or chemical methods. The most commonly used mechanical methods include high intensity ultrasonication, high-pressure homogenisation, micro-grinding and PFI milling [62]. However, these processes require high-energy consumption [73,74]. Chemical methods of nanocellulose preparation, such as acid hydrolysis, enzyme hydrolysis and (2,2,6,6-tetramethylpiperidin-1-yl) oxidanyl (TEMPO) oxidation, have also been rigorously employed. Yet, the drawbacks of the chemical methods are that they are time-consuming and yield low results [75]. Therefore, chemical and mechanical methods have been combined to overcome these issues, as well as reduce energy consumption. Acid hydrolysis is the most common type of this method used to date. The hydrolysis method reduces the sizes of nanofibres from microns to nanometres [76–79]. However, this method is not without drawbacks, such as the use of high concentrations of acid, which leads to acid waste and considerably adverse effects on the environment.

2. Preparation of Graphene Quantum Dots–Nanocellulose Composites

Over the past few years, GQDs–nanocellulose composites have been used in a variety of applications that have since generated a lot of interest in their synthesis and composite processing. Multiple studies have proven that GQDs–nanocellulose composites perform better than both GQDs and nanocellulose on their own. Different preparation techniques could impart different abilities to a composite as the preparation technique can affect the particle size, as well as the mechanical properties [80]. However, the major issues encountered when preparing nanocellulose composites include enhancing compatibility with hydrophobic polymers, uniform dispersion within the matrix and large-scale production [81]. Although melt-blending and in situ techniques can be employed during composite preparation, solution blending is the most common technique of GQDs–nanocellulose composite preparation. However, specific conditions, such as maintaining the temperature of the substrate, solution and solvent, are required during solution blending. Furthermore, material concentration and pH also affect the efficacy of the final composite.

Solution blending facilitates molecular mixing, which ensures that a composite is soluble in the appropriate solution. According to Ruiz-Palomero, Benítez-Martínez et al., GQDs–nanocellulose hydrogel particles can be prepared by combining carboxylated nanocellulose into sulphur–nitrogen co-doped GQDs (S, N-GQDs) [30]. Another study found that immersing S, N-GQDs synthesized via a hydrothermal process after using thiourea as the S and N source and citric acid as the source of carbon into the nanocellulose improved the penetration of the excitation light in the hydrogel composite and maintained higher fluorescence energy [82]. Carboxylated nanocellulose can be prepared using (2,2,6,6-tetramethylpiperidine-1-oxyl) TEMPO-radical catalysed oxidation. TEMPO was found to selectively oxidise the C6 primary alcohol moiety that is present on the surface of the nanocellulose particles and formed carboxylated groups [83], as presented in Figure 2.

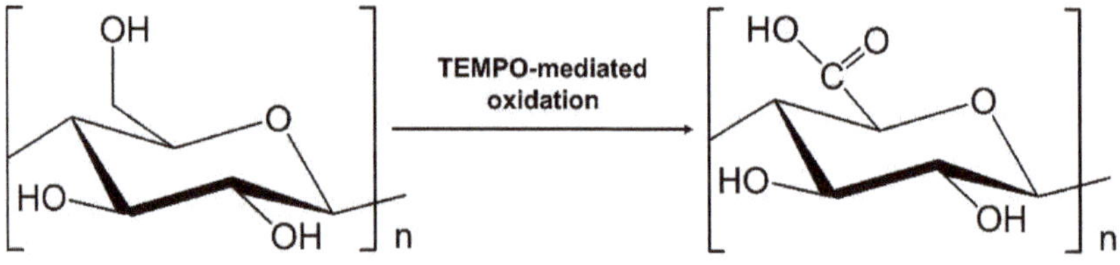

Figure 2. Oxidation of C6 hydroxyl group of cellulose in a TEMPO-system.

The regioselective conversion of the primary hydroxyl group to carboxylate weakens the adhesion between nanocellulose fibrils, as it prevents the formation of interfibrillar hydrogen bonds [84]. The TEMPO-mediated oxidation of the cellulose slurry starts to form when the NaOCl solution is added at room temperature and the pH is maintained at 10. The temperature is maintained at 25 °C, as the contents of carboxyl groups increases as the temperature increases up to only 25 °C and subsequently decreases at 35 °C. As such, optimum temperature is critical to form complete reactions, as increases in temperature decrease carboxyl group content, leading to de-polymerisation and losses in total yield. Meanwhile, lower temperatures may cause incomplete reactions to occur, since unreacted NaOCl forms and NaOH needs to be added to the solution to neutralise the generated carboxylic acid groups and maintain a pH of 10. The amount of NaOH added is gradually increased and remains constant after a certain time, once the limiting reactant, NaOCl, is fully consumed [82]. This technique provides a more uniform dispersion of nanocellulose in an aqueous phase. After conducting an inversion test on the composite hydrogels, Ruiz-Palomero, Benítez-Martínez et al. found that the hydrogels could be reformed several times as non-covalent interactions were involved [30].

The same researchers also reported a similar GQDs–nanocellulose hydrogel preparation procedure for sensing 2,4,5-trichlorophenol [31]. In short, the carboxylated nanocellulose was mixed with an aqueous solvent that contained S, N-GQDs, where the nanocellulose acted as a gelator. The sample was mixed using a vortex and sonication, then centrifuged for 0.5 min at 1300 rpm and heated for 20 s in a vial. The transparent hydrogel was observed following cooling to room temperature after the last centrifugation stage. An inversion test was used to assess gel formation. Due to the interacting surface hydroxyl and carboxyl groups, the carboxylated nanocellulose was found to be an effective gelator. This caused a significant self-association that resulted in nano-fibre entanglement due to hydrogen bonding. Therefore, nanocellulose is a suitable gel matrix for hosting GQDs. The study also found that 10 wt% of carboxylated nanocellulose provided the most stable hydrogel when mixed with the S, N-GQDs. This successful gel formation and stability confirmed the compatibility and suitability of the nanocellulose in hosting the GQDs and that it was potentially capable of detecting various analytes. This was due to its exceptional optical features which were accentuated by the network formation combined with the photoluminescence behaviour of the GQDs, which rendered them suitable for analytical purposes.

Apart from nanocellulose and GQDs concentration, the pH of the solvent containing GQDs and the doping characteristics of the GQDs play a significant role in the successful formation of a GQDs–nanocellulose hydrogel composite. The integrity of a composite structure can be weak in some cases due to the pH of the media or material concentration. For instance, Ruiz-Palomero, Soriano et al. [31] reported that only a solution of dialysed GQDs that had a neutral pH could support gel formation. Moreover, pure undoped GQDs produced weak gels, while N-doped GQDs did not form hydrogels even after purification. However, the S-GQDs produced low photoluminescence albeit gels. Doping GQDs with a combination of S and N heteroatoms (S, N-GQDs) produced the most stable gels with strong fluorescence features. On the other hand, the concentration of GQDs needed to be optimised as a very high concentration may de-stabilise the hydrogen bonding network via additional π–π stacking interactions. The study revealed that an 8 mg mL^{-1} concentration of S, N-GQDs and 10 wt% carboxylated nanocellulose produced the best and strongest hydrogel with the highest photoluminescent features.

Tetsuka et al. [85] was one of the earliest attempts at producing a GQDs–nanocellulose composite via solvent blending by producing a transparent clay film comprising of amino-functionalised GQDs (af-GQDs) and cellulose nanofibrils (CNF) as a colour converting material for blue light-emitting diode (LED) [85]. Treating the heavy oxidised graphene sheets (OGSs) with a mild amino-hydrothermal method produced uniform-sized af-GQDs with a tuneable photoluminescence. The amino groups were found to alter the electronic structures and shift the HOMO levels to a higher energy with a maximal photoluminescence at the long wavelength [85]. The af-GQDs–CNF composite was prepared by mixing the CNF suspension with aqueous solution containing af-GQDs. A clay suspension was prepared by using a high shear mixer to disperse clay in a solution containing 15 wt% of the af-GQDs–CNF. The mixture was then degassed and centrifuged to remove any flocculated clay impurities before it was poured on a glass mould. It was then heated at 60 °C overnight and detached from the glass mould to produce af-GQDs–CNF clay film. Unlike Ruiz-Palomero, Soriano et al. [31], Tetsuka et al. [85] introduced the incorporation of clay in the composite matrix via electrostatic interactions between the clay and af-GQDs–CNF to form a flexible and transparent film. The resultant film exhibited bright colourful photoluminescence and is a promising future light emitting diode application.

Khabibullin et al. [86] produced injectable fluorescent hydrogels composed of GQDs and cellulose nanocrystals (CNCs). Unlike Ruiz-Palomero, Soriano, et al. [31] and Tetsuka et al. [85], who used the hydrothermal method, Khabibullin et al. [86] employed the Hummers method to prepare the GQDs. The CNCs were functionalised with amino groups before being subjected to the composite hydrogel preparation. The hydrogel was prepared by dispersing the powdered GQDs in aqueous solution containing CNCs. The sample was mixed via vortex, then left to rest. The time of hydrogel formation varied depending on the concentration of GQDs and CNCs. The sample with 50 mg/mL of CNCs and 7 mg/mL of GQDs yielded a strong hydrogel formation (gelation within 30 min), while the sample with lower concentrations of CNCs and GQDs yielded a weak hydrogel formation (gelation within four hours) and the sample with <20 mg/mL CNCs and <5 mg/mL GQDs yielded no hydrogel formation. Similar to the neutral pH that facilitated hydrogel formation in Ruiz-Palomero, Soriano, et al.'s [31] study, the pH used for the preparation of hydrogel was maintained at seven. As the produced GQDs–CNCs composite possessed a shear thinning behaviour, it was a suitable injectable material for 3D printing with additional fluorescence features.

Alizadehgiashi et al. [8] also employed a solution mixing procedure for the preparation of GQDs–nanocellulose composites. Similar to Tetsuka et al. [85], Alizadehgiashi et al. [8] utilised amino-functionalised GQDs (af-GQDs) that had been prepared using the hydrothermal method, except that an aldehyde-modified CNCs was used for the composite hydrogel formation instead. The composite hydrogel was prepared by mixing various concentrations of aldehyde-functionalised CNCs (from 10 to 60 mg/mL) and af-GQDs (from 2.5 to 60 mg/mL) suspensions in various volumetric ratios, which may change the structure of the composite from lamellar to nanofibrillar, as well as enabling the controlling of the permeability of the hydrogel. A thick lamellar structure (large pores) was observed at a lower CNC concentration ratio, while a nanofibrillar structure (small pores) was observed at a high CNC concentration ratio (low af-GQDs content). The transition of the structure can be attributed to the number of crosslinking points available (higher CNC content imparts fewer crosslinking points), which, in turn, determines the thickness of the wall and size of the pores. In terms of permeability, the hydrogel composite with lower CNC content (lamellar structure) had higher permeability than the composite with higher CNC content (nanofibrillar structure) due to the larger pore size [87].

On the other hand, Rosddi et al. [88] produced a GQDs–nanocellulose composite by dissolving carboxylated GQDs in an aqueous solution containing cationically modified CNCs. A thin film of the composite was then formed using the spin coating technique and the composite showed potential in analytical applications. The same researchers later reported the modification of surface plasmon resonance gold film with carboxylated GQDs–CNCs to enhance the detection sensitivity of glucose [89]. Similar to Ruiz-Palomero, Benítez-Martínez, et al. [30] and Ruiz-Palomero, Soriano, et al. [31], Mahmoud et al. [90] prepared GQDs–nanocellulose using S, N-GQDs, except that the nanocellulose used was not modified with the carboxyl group and no hydrogel formation was reported. The GQDs–nanocellulose was prepared by mixing of equal concentrations of S, N-GQDs and nanocellulose (3 mg/mL each). The composite was then sonicated and drop-casted onto glassy carbon electrode (GCE) to fabricate a modified GCE sensor. More recently, Xiong et al. [91] prepared a flexible GQDs–nanocellulose film via a combination of electrolysis and liquid dispersion for sensor and supercapacitor applications.

The average particle size of GQDs–nanocellulose composites can be controlled by altering the reaction conditions. The reactive chemical groups and its distinctive morphology make nanocellulose a good biological template for GQDs–nanocellulose synthesis. Solution mixing is the most common method of preparing nanocomposite films by far. This method is suitable for obtaining good dispersion of nanocellulose with a polymer solution due to the good dispersion of these nanoparticles in water. Figure 3 presents a schematic illustration of GQDs–nanocellulose hydrogel formation.

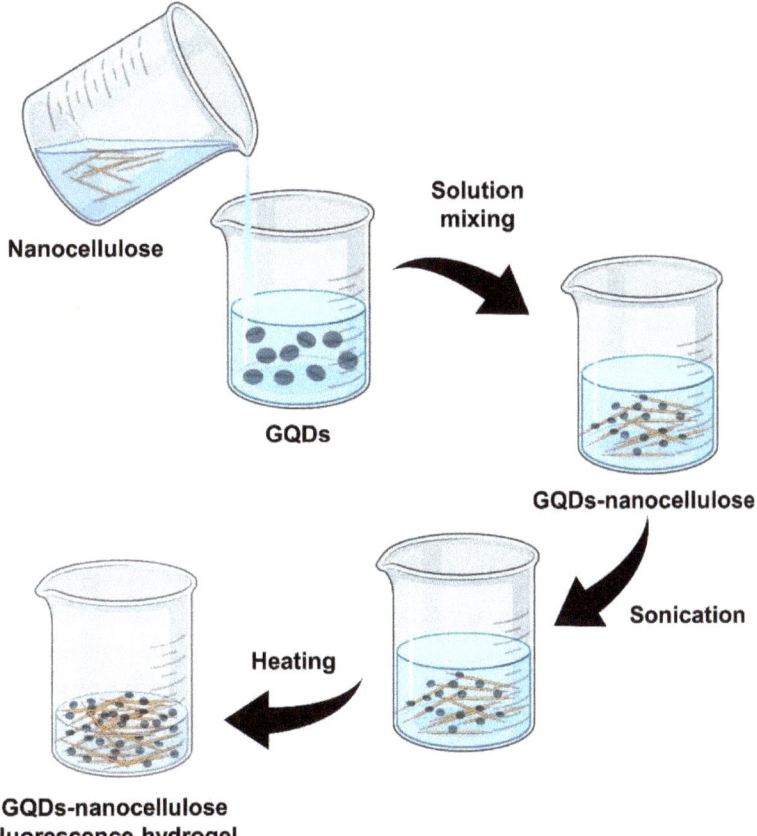

Figure 3. Preparation of GQDs–nanocellulose fluorescence hydrogel via solution mixing.

3. Marriage Chemistry of GQDs and Nanocellulose

As GQDs can significantly affect the properties of a composite, determining the GQD content threshold of a nanocomposite depends on its specific application, as different concentrations are required for different applications. The GQD, a functional derivative of graphene, consists of many oxygen-containing functional groups, such as ether, hydroxyl, carbonyl and carboxyl, on the edges and basal planes [92,93]. Despite the smaller size of GQDs, their edge sites are more reactive than the carbon matrix [94]. Furthermore, the presence of functional groups not only facilitates the easy blending of GQDs into a polymer matrix [95,96] but also improves its electrical, mechanical and dielectric properties. The equatorial direction of the glucopyranose ring present in the nanocellulose exhibited a hydrophilic nature, as all three hydroxyl groups were placed at an equatorial position on this ring, while the hydrogen atoms in the carbon and hydrogen bonds were located at the axial positions of the rings, which is responsible for the hydrophobic nature of the axial direction of the glucopyranose ring [86].

As illustrated in Figure 4, the attractive forces from the hydrogen bonds between the hydroxyl groups on the surface of nanocellulose particles and the oxygen-rich carboxyl groups present on the GQD edges produce a strong and stable GQDs–nanocellulose composite. Hydrophobic interactions are noted between the basal planes of GQDs and hydrophobic faces of the nanocellulose [86]. Although repulsive forces may occur between the negatively charged sulphate half-ester groups present on the surface of the nanocellulose surface and the carboxylic groups present on GQDs due to electrostatic interactions, the hydrogen bonds and the hydrophobic forces can overcome the repulsive forces leading to the crosslinking of the GQDs and the nanocellulose. The functional groups containing the oxygen atom in GQDs and the aromatic sp^2 domains can generate an interfacial bonding in the GQDs and nanocellulose composites [31]. The network noted in nanocellulose composites was seen to alter the structural region and form amorphous and crystalline regions, thereby increasing the mechanical strength of the composites [97].

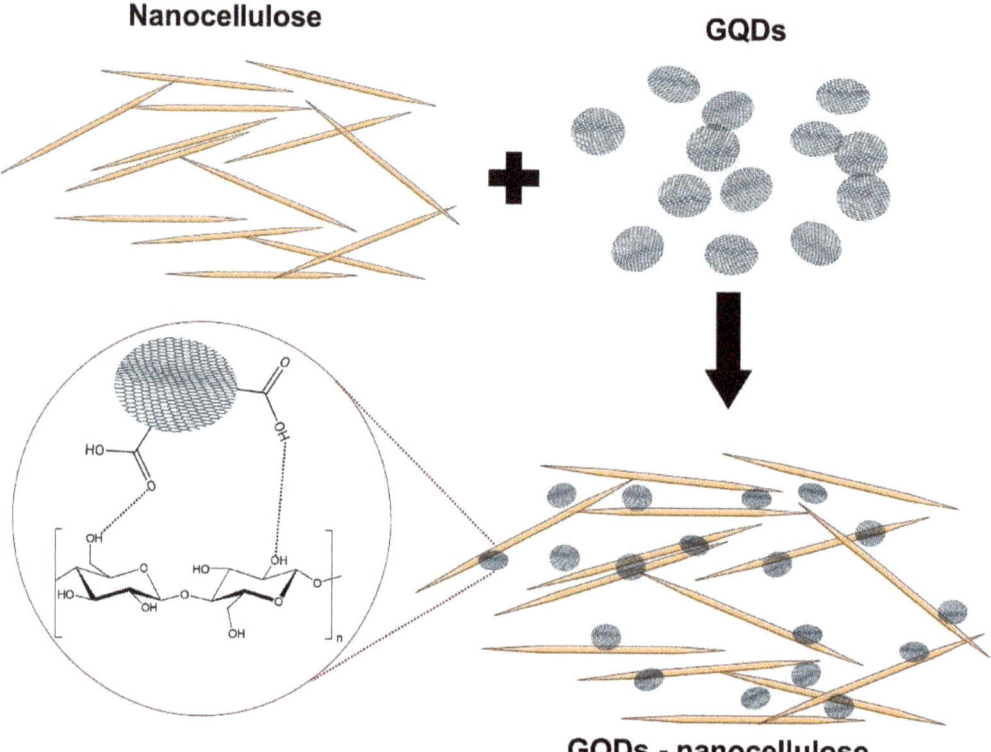

Figure 4. A schematic illustrating the interaction and formation of the hydrogen bonding between the carboxyl groups of graphene quantum dots (GQDs) and the hydroxyl groups of nanocellulose.

Therefore, the addition of GQDs improves the tensile strength, toughness and stiffness of nano-cellulosic composites due to the hydrogen bonds between the functional groups of both GQDs and nanocellulose. The intralayer of the hydrogen bonds present in the nanocellulose is constant, while the hydrogen bonding in the nanocellulose interlayers can be significantly improved. Additionally, the interaction between GQDs and nanocellulose induces a locally stable energy state at the interfacial plane of nanocellulose. The marriage chemistry of GQDs-nanocellulose composites is summarised in Table 1.

Table 1. A summary of marriage chemistry of GQDs–nanocellulose composites.

Composite	Marriage Chemistry	Ref.
Amino-functionalised GQDs + aldehyde-modified CNCs	• Chemical crosslinking due to hydrogen bonding between amino groups of the amino-functionalised GQDs and hydroxyl groups of aldehyde-modified CNCs. • Ionic interactions between negatively charged sulphate half-ester groups of CNCs and protonated amines of GQDs. • Hydrophobic interactions between basal planes of GQDs and hydrophobic backbone of CNCs.	[8]
S, N-GQDs + carboxylated nanocellulose	• Electrostatic interactions between superficial groups of GQDs and oxygen-containing nanocellulose.	[30]
S, N-GQD + carboxylated nanocellulose	• Hydrogen bonding interactions between carboxyl and hydroxyl groups of nanocellulose and oxygen-containing groups of GQDs. • Hydrogen bonding network de-stabilised with higher GQDs content due π–π stacking-induced GQD aggregation.	[31]
GQDs + amino-modified CNCs	• Hydrophobic interactions between basal plane of GQDs and C–H grouped located on axial positions of glucopyranose ring of CNCs. • Hydrogen bonding interactions between amino or hydroxyl groups of CNCs and oxygen-rich groups on GQD edges.	[86]
Carboxylated GQDs + cationically modified CNCs	• Hydrogen bonding interactions between oxygen-containing groups of GQDs and hydroxyl and oxygen atoms in CNCs.	[88]
Carboxylated GQDs + cationically modified CNCs	• Hydrogen bonding interaction between carboxylated GQDs, and hydroxyl groups and oxygen atoms in CNCs	[89]
S, N-GQDs + nanocellulose	• Hydrogen bonding formation between polar groups of S, N-GQDs and nanocellulose. • π–π stacking between GQDs	[90]

The concentration of GQDs and nanocellulose may affect the interactions within a composite. For instance, a higher concentration of GQDs may induce the formation of π–π stacking between GQDs, thus affecting the hydrogen bonding network of the nanocellulose and hydrogel formation. Although such an interaction (π–π stacking) and graphitic aggregation are accounted for and may be accentuated, the surface modification and functionalisation of a nanomaterial also affect the inter-nanomaterial interactions. For example, the carboxyl and hydroxyl groups of carboxylated nanocellulose interact with the oxygen-containing groups of GQDs and these interactions may facilitate significant improvements and provide a synergistic enhancement of the composite's properties. Therefore, this excellent combination of GQDs and nanocellulose appears to be an outstanding strategy because it results in the improvement of the properties of each individual component, including the mechanical strength of the composite or hydrogel, as well as the enhancement of optical or fluorescent features [31,85]. GQDs–nanocellulose composites are promising for sensing, printing materials and analytical applications by virtue of their unique optical, structural and mechanical properties. The synergistic manner and improvement of the fluorescent properties of a composite can be indicative of the feasible interaction and marriage chemistry between the superficial groups of GQDs and the oxygen-containing groups of nanocellulose, thus reducing π–π stacking-induced GQD aggregation.

Although cellulose is usually described as a hydrophilic material due to the presence of hydroxyl groups, it is noteworthy that the significant amphiphilicity of cellulose is due to the hydrophobic effects of C–H grouped in the glucopyranose unit backbone of the cellulose. This cellulose behaviour, as dictated by the hydrophobic effect, was conceptualised by Lindman et al. [98] and subsequently prompted a unique and intriguing debate among cellulose experts [99]. The hydrophobicity of cellulose is also corroborated by Yamane et al. [100], based on the anisotropy of cellulose's structure. Despite its hydrophilicity, cellulose has been found to interact strongly with non-polar (hydrophobic) organic solvents, such as toluene, dichloromethane and hexane [100]. An investigation of its structural anisotropy proposed that the hydrophobic behaviour of cellulose stems from the C–H bonds of the glucopyranose rings located on the axial position of the rings. On the other hand, it is well known that graphene materials, including GQDs, have hydrophobic properties. Owing to the hydrophobicity of graphene and amphiphilicity of cellulose, the marriage chemistry and interaction between these two materials can be elucidated. Moreover, Alqus et al.'s [101] theoretical modelling and molecular dynamic simulation investigated the interactions between cellulose and graphene. The study reported stable complex formation of graphene-cellulose through a hydrophobic interface which was primarily formed by CH–π interactions. Therefore, this further emphasises that the amphiphilic nature of cellulose plays an important role in favoured interactions when fabricating a GQD–cellulose composite.

4. Applications of GQDs–Nanocellulose Composite

Nanocellulose and GQDs have garnered a lot of research interest in different fields, such as energy storage devices, electronics, photovoltaic devices and biosensors [102–104]. The various functional groups present on the surface of GQDs, such as -OH, -COOH and -NH$_2$, act as active coordination sites for the transition metal ions. Nanocellulose is biocompatible, environmentally friendly, flexible and thermally stable [105–107]. Nanocellulose can improve mechanically flexible materials, as they are receptive to light, heat, chemicals and magnetic fields which can connect stimulus responses and allow the composites to work optimally [108]. These intriguing properties of GQDs–nanocellulose composites have encouraged researchers to deeply explore and unlock their potential for use in a multitude of applications, including sensors, light-emitting diodes, 3D printing materials and supercapacitors, as summarised in Figure 5.

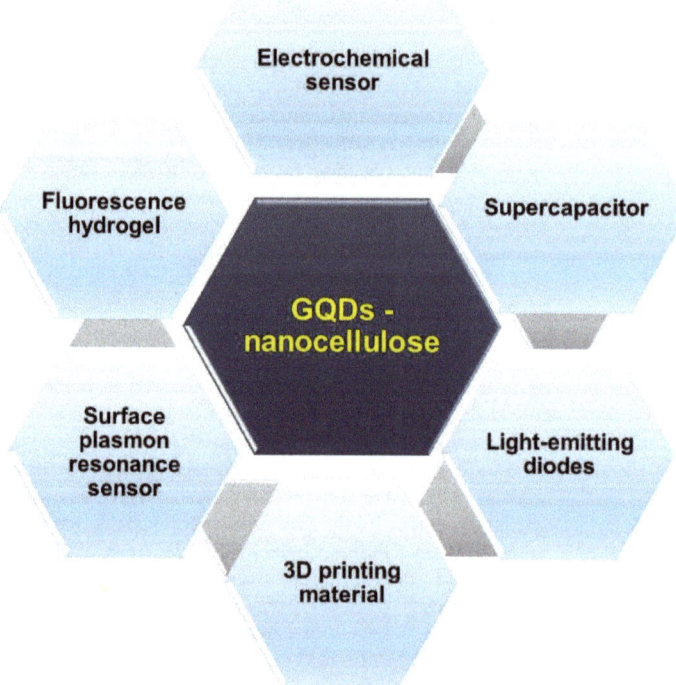

Figure 5. The applications of GQDs–nanocellulose composites.

4.1. Sensors

As seen in Table 2, GQDs–nanocellulose composites can be used to produce sensors with which to detect various analytes. Multiple types of sensors that were developed from GQDs–nanocellulose, such as drug sensors, humidity sensors, laccase monitoring sensors, glucose sensors and metal ion sensors, were investigated. The functional capabilities of each sensor improved when GQDs were added into the nanocellulose matrix. These sensors are materials or electronic devices that convert one form of energy to another [102]. The conditions required to develop the best sensor include stability and sensitivity, so that they can detect trace quantities of molecules in various applications. Apart from that, the sensors also need to provide accurate results with rapid performance [108].

Table 2. A summary of various GQDs–nanocellulose composite-based platforms used for sensing applications.

Sensor	Prior to GQDs–Nanocellulose Composite Formation	Post GQDs–Nanocellulose Composite Addition
Drug	Electrochemical sensor with poor electro-catalytic performance and low active surface area for drug detection.	Good conductivity, increased active surface area and good reproducibility, repeatability and stability for drug detection.
Metal ions	Low detection limit and unable to measure many metal ions.	Large surface area and higher number of active sites on the membrane increased the efficiency of scavenging metal ions from samples.
Laccase	Based on catalytic activity which is dependent on environmental parameters and the presence of inhibitors and inducers.	Based on fluorescence response of hydrogels containing GQDs acting as luminophore towards laccase.
Humidity	Less sensitive and produce less accurate result.	Sensitivity is expected to increase due to the abundance of hydrophilic functional groups in both GQDs and nanocellulose.
Glucose	Low sensitivity and binding affinity.	More sensitive and higher binding affinity constant.

4.1.1. Modified Electrochemical Sensor for Drug Detection

An electrochemical sensor functions when gases react or generate electrical signals according to a concentration of gas. These sensors consist of a counter electrode and electrodes separated by an electrolyte layer [109]. The gas concentration is proportional to the electrical signal that is generated. The mechanism of action of an electrochemical sensor is that the gas which interacts with the sensor passes through the capillary opening and diffuses through the hydrophobic barrier before it reaches the surface of the electrode [110]. This ensures a proper gas flow, that can react with the sensing electrode. An appropriate amount of electrical signal is generated in this procedure, without causing any leakage [111]. The larger number of active sites, higher conductivity for charge transfer to the electrodes and the stable mechanical strength of GQDs–nanocellulose composites help electrochemical sensing, as the structure of the electrode facilitates high electrical conductivity and rapid electron transfer at the surface of the electrodes. Electrochemical methods have garnered considerable interest in recent times due to their high sensitivity, simplicity and rapid analysis. Mahmoud et al. [90] recently used a S, N-GQDs–nanocellulose composite to modify electrochemical sensors for drug analysis. The nanomaterial was drop-casted onto the surface of a bare electrode and dried prior to analysis. It was revealed that the modified electrode sensor provided good electro-catalytic performance for drug detection. Furthermore, the active surface area increased two-fold, in comparison to the bare electrode. The modified electrode also showed good reproducibility and repeatability with relative standard deviations of less than 2.4% and 2.1%, respectively. The electrode maintained its original performance of 95% within 28 days of storage, indicating good stability. This can be explained by the intermolecular interactions of various functional groups, unique structure and high active surface area due to the addition of S, N-GQDs and nanocellulose, which enhanced the analytical performance of the electrode in a synergistic manner.

An earlier study developed an electrochemical sensor using only GQDs and found poor electrical conductivity. The performance of the sensor was improved by adding heteroatoms to the core of the GQDs. The oxygen-rich functional groups of the GQDs could improve the properties of the electrochemical sensor, wherein the sensors showed good quantum confinement, edge effect and a larger surface area [90]. Furthermore, the addition of nanocellulose particles to the composites improved the sensitivity of the electrochemical sensors in detecting many analytes. The nanocellulose particles contained many OH groups which were modified to ensure the selective binding of the analytes. The GQDs–nanocellulose composites improved ionic conductivity as they increased ion transportation by introducing many ion conduction pathways. The addition of heteroatoms, such as nitrogen, sulphur, boron and phosphorous, to the composite helped form strong bonds due to hydrogen bonding or π–π stacking. Therefore, GQDs–nanocellulose composites facilitate the adsorption and electrochemical oxidation of drugs onto the surfaces of modified electrodes due to the presence of π–π stacking or hydrogen bonds.

4.1.2. Hydrogel Sensor for Enzyme Detection

Combining GQDs and nanocellulose to develop fluorescent hydrogels has quickly become a new strategy of detecting laccase enzymes over the last few years [30]. The unique optical features provided by GQDs, as well as the three-dimensional framework provided by nanocellulose, enable significant detection of laccase through fluorescent quenching and ideal enzyme immobilisation, respectively. In short, the fluorescent intensity decreased as the concentration of laccase enzymes increased. The hydrogels also possessed remarkably better signal stability than non-nanocellulose framework platforms due to the enzyme immobilisation behaviour imparted by the nanocellulose framework platform. These hydrogels are described as a heterogeneous mixture of multiple phases. The dispersed phase in the hydrogel is water, whereas the solid 3D network forms a solid phase. The hydrogel consists of a 3D polymeric network which is filled with water and displays a gel-like behaviour due to the hydrogen bonding between the nanofibrils in the aqueous medium. Water retention occurs due to the formation of hydrogen bonds between the hydrophilic groups in these composites. As such, hydration can cause the deformation of the polymeric chains when compensating the stresses within this structure [112]. Fluorescent hydrogels that have been developed using GQDs–nanocellulose composites are reversible and can easily convert to form a fluid when heated or stirred at a higher shear rate. This process can be repeated up to 10 times without issues owing to the presence of non-covalent interactions between the molecules [30].

Hydrogel fluorescent sensors are cost-effective, simple and eco-friendly. They can also detect and stabilise laccase molecules, as well as storing or recycling the enzymes. It is also serves as a stabilising matrix for trapping enzyme molecules from the complex matrix and reusing it after being stored. This helps protect the enzyme molecules from extreme environmental conditions. Therefore, the enzyme molecules can be stored in a nano-cellulosic hydrogel and recovered for use at a later stage [30]. Laccase detection was attributed to the quenching of fluorescent molecules in the GQDs–nanocellulose hydrogels. This was caused by the interactions between the laccase enzymes and the graphitic layers which are stabilised by the nanocellulose. Another advantage of using a fluorescent GQDs–nanocellulose hydrogel is that it follows a simple mechanism, as the enzymatic activity does not have to be measured. Furthermore, it is eco-friendly and can be easily prepared using biocompatible nanoparticles which are self-organised due to electrostatic interactions [113].

4.1.3. Hydrogel Sensor for Pesticide Detection

The use of GQDs–nanocellulose composites for developing a hydrogel sensor to detect 2,4,5-trichlorophenol (TCP) has also been investigated [31]. An organic compound belonging to the chlorophenols family, TCP is carcinogenic to humans [114]. As such, it is commonly used in fungicides and herbicides or as an intermediate in the production of other pesticides [115]. As TCP has been detected in soil, groundwater and drinking water, it poses a health and environmental risk. Therefore, it is imperative to detect and remove these hazardous compounds from the environment. Consequently, the development of a GQDs–nanocellulose hydrogel as sensing platform is as strategic endeavour that is convenient, rapid and simple with a selective sensing ability for TCP determination. Such fluorometric hydrogel sensor provided remarkable performance, including high sensitivity, good repeatability and reproducibility, indicating the immense potential of composites as sensing analytical tools. Although this hydrogel, in particular, is selective towards TCP, its potential to detect various other types of pesticides should be further explored. By virtue of its versatility, unique properties and further surface function modification, GQDs–nanocellulose may serve as a promising sensing candidate for pesticide detection.

4.1.4. Hydrogel Sensor for Metal Ion Detection

In recent years, Alizadehgiashi et al. [8] produced a nano-colloidal heavy metal ion scavenger hydrogel composed of GQDs and CNCs. This hydrogel possessed more active surface sites as well as a larger surface area and metal ion sensing capacity. The presence of carboxylic acid groups on the edges of the GQDs allowed them to serve as metal ion scavengers, while nanocellulose or CNCs provided the hydrogel with immobilisation capabilities that facilitate easy water separation. Therefore, the GQD and CNC content determined the number of active sites and the pore size of the hydrogel composite, respectively.

4.1.5. Humidity Sensor

Earlier studies describe two types of humidity sensors, capacitive and resistive [116,117]. Humidity sensors are used in the window defogger system of the automobile industry, in respiratory equipment and for medicine processing in the medical field, as well as controlling greenhouse air and monitoring soil conditions in the agricultural field. Existing studies have investigated the synergistic benefits of graphene oxide–nanocellulose composite films and their application while designing flexible, cheap and renewable humidity sensors.

Kafy et al. [95] found that graphene oxide–CNCs composite films had higher sensitivity than CNC films while developing a humidity sensor. In general, the inclusion of the graphene oxide into the CNC matrix increased the number of hydrophilic functional groups, such as carboxyl and hydroxyl. It was observed that the sensing capability of CNCs was dependent on the hydrophilic functional groups that attracted water molecules and improved the capacitance. Although the efficacy of using a GQDs–nanocellulose composite as a humidity sensor has yet to be investigated, the presence of carboxyl and hydroxyl groups in GQDs, both of which supported the chemical surface membrane of CNCs, may enable it to outperform the GO–nanocellulose humidity sensor. Furthermore, GQDs–nanocellulose composite films, which have a higher rigidity and mechanical strength, could be used for long-lasting applications.

4.1.6. Surface Plasmon Resonance Sensor for Glucose Detection

Rosddi et al. [88] recently fabricated a thin GQDs–nanocellulose composite film using a combination of carboxylated GQDs and cationically functionalised nanocellulose. The same researchers later used GQDs–nanocellulose composites to modify a surface plasmon resonance gold film sensor to enhance the detection sensitivity to glucose [89]. Their results indicated that the thin GQDs–nanocellulose film had better sensitivity and glucose binding affinity than an unmodified sensor, which can detect various concentrations of glucose with the lowest detection of 5 nM.

4.2. Light Emitting Diode

One of the earliest studies on GQDs–nanocellulose composites showed its significant potential when incorporated into a flexible film as a colour-converting material in light-emitting diode (LED) [85]. The film was prepared by mixing amino-functionalised GQDs and cellulose nanofibres into the clay matrix. The resultant film exhibited excellent physical stability, as well as bright photoluminescence, and emitted a white light when casted onto a blue LED. Therefore, the incorporation of GQDs–nanocellulose into a composite film is promising for the future of light emitting devices, such as bio-imaging, photovoltaics and colour-converting material.

4.3. Three-Dimensional Printing Material

GQDs–nanocellulose has also been investigated for its potential as an injectable 3D printing material [86] by controlling its gelling properties and shear-thinning behaviour. The performance of the injectable material was investigated using a 3D printer. The composite was able to form a pre-designed pattern and retain its thread-like shape. Furthermore, the extruded thread exhibited photoluminescent properties due to the fluorescence characteristics of GQDs. The birefringent properties observed on the hydrogel thread under polarised optical microscopy stemmed from the shear-induced alignment of CNCs within the composite. In light of all these advantages, GQDs–nanocellulose is a useful injectable material and intriguing candidate for a wide range of applications, including tissue engineering, bio-imaging, drug delivery, biomedical and 3D printing.

4.4. Supercapacitor

Xiong et al. recently combined electrolysis and liquid dispersion to produce a composite film comprising of cellulose nanofibres and GQDs for use as a supercapacitor [91]. The produced hybrid film had remarkable electrochemical storage performance and mechanical properties. It also exhibited a specific capacitance of 118 mF cm^{-2} even at an ultrahigh scan rate of 1000 mV s^{-1} and a high capacitance retention of more than 93% after 5000 cycles at various current densities. Moreover, the supercapacitor constructed based on the GQDs–nanocellulose presented high energy densities and power indicating exceptional performance rate and cycle stability. Therefore, combining GQDs and nanocellulose not only overcame the drawbacks of individual materials, such as low efficiency and poor conductivity, but enhanced performance in a synergistic manner. This indicates that the scientific community should further investigate the suitability of GQDs–nanocellulose for fundamental studies as well as advanced applications.

5. Conclusions

Owing to unique characteristics, such as inherent luminescence, biocompatibility, high surface area, high adsorption, good surface functionality, significant strides have been made in the preparation and application of GQDs–nanocellulose over the past several years. Studies have shown that the hybridisation of these novel materials not only improves existing applications but also provides additional advantages, as well as further improvement of desirable features, all of which are unattainable if GQDs and nanocellulose are used individually. Therefore, this advantageous composite material warrants remarkable applications. This review provided a brief overview of this evolving body of research which will unlock windows of opportunities for future research and multifunctional applications. GQDs–nanocellulose can be produced via rapid and simple solution blending or drop-casting onto a selected matrix depending on the targeted application. Furthermore, nanomaterial concentration, composition ratio, pH of the media, heteroatoms doping and surface functionalisation determine the properties, marriage chemistry and final GQDs–nanocellulose composites. The remarkable and synergistic properties of GQDs–nanocellulose certainly unlock their potential for use in a multitude of applications, including sensors, light-emitting diodes, injectable shear-thinning materials and supercapacitors. Although the development of GQDs–nanocellulose composites is limited and still at its infancy, this hybrid material is anticipated to be commercially accessible and more practical in the future. This review disclosed a range of opportunities to apply nanomaterials derived from naturally occurring cellulose polymers or renewable biomass resources and assisted in more innovative nanomaterial production methods and developments in the future.

Author Contributions: Conceptualization, W.H.D. and Z.A.M.; writing—original draft preparation, W.H.D. and N.F.M.B.; writing—review and editing, W.H.D. and N.F.M.B.; supervision, Z.A.M.; project administration, W.H.D.; funding acquisition, W.H.D. All authors have read and agreed to the published version of the manuscript.

Funding: The authors are grateful for funding received from the Fundamental Research Grant Scheme (FRGS19-015-0623), Ministry of Higher Education (MOHE), Malaysia.

Data Availability Statement: Data supporting reported results can be found in publicly archived datasets such as ScienceDirect and Scopus.

Acknowledgments: This work was supported by the Fundamental Research Grant Scheme (FRGS/1/2018/STG01/UIAM/03/2) (FRGS19-015-0623), Ministry of Higher Education (MOHE), Malaysia, and Department of Chemistry, Kulliyyah of Science, International Islamic University, Malaysia.

Conflicts of Interest: The authors declare no conflict of interest.

References

1. Biswas, M.C.; Islam, M.T.; Nandy, P.K.; Hossain, M.M. Graphene Quantum Dots (GQDs) for Bioimaging and Drug Delivery Applications: A Review. *ACS Mater. Lett.* **2021**, *3*, 889–911. [CrossRef]
2. Maio, A.; Pibiri, I.; Morreale, M.; La Mantia, F.P.; Scaffaro, R. An overview of functionalized graphene nanomaterials for advanced applications. *Nanomaterials* **2021**, *11*, 1717. [CrossRef] [PubMed]
3. Danial, W.H.; Norhisham, N.A.; Ahmad Noorden, A.F.; Abdul Majid, Z.; Matsumura, K.; Iqbal, A. A short review on electrochemical exfoliation of graphene and graphene quantum dots. *Carbon Lett.* **2021**, *31*, 371–388. [CrossRef]
4. Bressi, V.; Ferlazzo, A.; Iannazzo, D.; Espro, C. Graphene quantum dots by eco-friendly green synthesis for electrochemical sensing: Recent advances and future perspectives. *Nanomaterials* **2021**, *11*, 1120. [CrossRef] [PubMed]
5. Armano, A.; Agnello, S. Two-Dimensional Carbon: A Review of Synthesis Methods, and Electronic, Optical, and Vibrational Properties of Single-Layer Graphene. *C J. Carbon Res.* **2019**, *5*, 67. [CrossRef]
6. Li, M.; Chen, T.; Gooding, J.J.; Liu, J. Review of carbon and graphene quantum dots for sensing. *ACS Sensors* **2019**, *4*, 1732–1748. [CrossRef]
7. Xu, L.; Zhang, Y.; Pan, H.; Xu, N.; Mei, C.; Mao, H.; Zhang, W.; Cai, J.; Xu, C. Preparation and performance of radiata-pine-derived polyvinyl alcohol/carbon quantum dots fluorescent films. *Materials* **2020**, *13*, 67. [CrossRef] [PubMed]
8. Alizadehgiashi, M.; Khuu, N.; Khabibullin, A.; Henry, A.; Tebbe, M.; Suzuki, T.; Kumacheva, E. Nanocolloidal hydrogel for heavy metal scavenging. *ACS Nano* **2018**, *12*, 8160–8168. [CrossRef] [PubMed]
9. Chen, W.; Lv, G.; Hu, W.; Li, D.; Chen, S.; Dai, Z. Synthesis and applications of graphene quantum dots: A review. *Nanotechnol. Rev.* **2018**, *7*, 157–185. [CrossRef]
10. Kim, D.J.; Yoo, J.M.; Suh, Y.; Kim, D.; Kang, I.; Moon, J.; Park, M.; Kim, J.; Kang, K.S.; Hong, B.H. Graphene quantum dots from carbonized coffee bean wastes for biomedical applications. *Nanomaterials* **2021**, *11*, 1423. [CrossRef] [PubMed]
11. Cui, Y.; Wang, T.; Liu, J.; Hu, L.; Nie, Q.; Tan, Z.; Yu, H. Enhanced solar photocatalytic degradation of nitric oxide using graphene quantum dots/bismuth tungstate composite catalysts. *Chem. Eng. J.* **2021**, *420*, 129595. [CrossRef]
12. Prabhu, S.A.; Kavithayeni, V.; Suganthy, R.; Geetha, K. Graphene quantum dots synthesis and energy application: A review. *Carbon Lett.* **2021**, *31*, 1–12. [CrossRef]
13. Lee, S.H.; Kim, D.Y.; Lee, J.; Lee, S.B.; Han, H.; Kim, Y.Y.; Mun, S.C.; Im, S.H.; Kim, T.H.; Park, O.O. Synthesis of Single-Crystalline Hexagonal Graphene Quantum Dots from Solution Chemistry. *Nano Lett.* **2019**, *19*, 5437–5442. [CrossRef] [PubMed]
14. Yang, Y.; Xiao, X.; Xing, X.; Wang, Z.; Zou, T.; Wang, Z.; Zhao, R.; Wang, Y. One-pot synthesis of N-doped graphene quantum dots as highly sensitive fluorescent sensor for detection of mercury ions water solutions. *Mater. Res. Express* **2019**, *6*, 095615. [CrossRef]
15. Nair, R.V.; Thomas, R.T.; Sankar, V.; Muhammad, H.; Dong, M.; Pillai, S. Rapid, Acid-Free Synthesis of High-Quality Graphene Quantum Dots for Aggregation Induced Sensing of Metal Ions and Bioimaging. *ACS Omega* **2017**, *2*, 8051–8061. [CrossRef] [PubMed]
16. Gu, S.; Hsieh, C.T.; Chiang, Y.M.; Tzou, D.Y.; Chen, Y.F.; Gandomi, Y.A. Optimization of graphene quantum dots by chemical exfoliation from graphite powders and carbon nanotubes. *Mater. Chem. Phys.* **2018**, *215*, 104–111. [CrossRef]
17. Deng, J.; Lu, Q.; Mi, N.; Li, H.; Liu, M.; Xu, M.; Tan, L.; Xie, Q.; Zhang, Y.; Yao, S. Electrochemical synthesis of carbon nanodots directly from alcohols. *Chem. A Eur. J.* **2014**, *20*, 4993–4999. [CrossRef] [PubMed]
18. Moon, R.J.; Martini, A.; Nairn, J.; Simonsen, J.; Youngblood, J. Cellulose nanomaterials review: Structure, properties and nanocomposites. *Chem. Soc. Rev.* **2011**, *40*, 3941–3994. [CrossRef] [PubMed]
19. Lunardi, V.B.; Soetaredjo, F.E.; Putro, J.N.; Santoso, S.P.; Yuliana, M.; Sunarso, J.; Ju, Y.-H.; Ismadji, S. Nanocelluloses: Sources, Pretreatment, Isolations, Modification, and Its Application as the Drug Carriers. *Polymers* **2021**, *13*, 2052. [CrossRef]
20. Phanthong, P.; Reubroycharoen, P.; Hao, X.; Xu, G.; Abudula, A.; Guan, G. Nanocellulose: Extraction and application. *Carbon Resour. Convers.* **2018**, *1*, 32–43. [CrossRef]
21. Mishra, R.K.; Sabu, A.; Tiwari, S.K. Materials chemistry and the futurist eco-friendly applications of nanocellulose: Status and prospect. *J. Saudi Chem. Soc.* **2018**, *22*, 949–978. [CrossRef]

22. Dufresne, A. Nanocellulose: A new ageless bionanomaterial. *Mater. Today* **2013**, *16*, 220–227. [CrossRef]
23. Shojaeiarani, J.; Bajwa, D.S.; Chanda, S. Cellulose nanocrystal based composites: A review. *Compos. Part C Open Access* **2021**, *5*, 100164. [CrossRef]
24. Santos, R.F.; Ribeiro, J.C.L.; Franco de Carvalho, J.M.; Magalhães, W.L.E.; Pedroti, L.G.; Nalon, G.H.; de Lima, G.E.S. Nanofibrillated cellulose and its applications in cement-based composites: A review. *Constr. Build. Mater.* **2021**, *288*, 123122. [CrossRef]
25. Almeida, T.; Silvestre, A.J.D.; Vilela, C.; Freire, C.S.R. Bacterial nanocellulose toward green cosmetics: Recent progresses and challenges. *Int. J. Mol. Sci.* **2021**, *22*, 2836. [CrossRef] [PubMed]
26. Ferrer, A.; Pal, L.; Hubbe, M. Nanocellulose in packaging: Advances in barrier layer technologies. *Ind. Crops Prod.* **2017**, *95*, 574–582. [CrossRef]
27. Ni, Y.; Gu, Q.; Li, J.; Fan, L. Modulating in vitro gastrointestinal digestion of nanocellulose-stabilized pickering emulsions by altering cellulose lengths. *Food Hydrocoll.* **2021**, *118*, 106738. [CrossRef]
28. Li, K.; Jin, S.; Li, X.; Li, J.; Shi, S.Q.; Li, J. Bioinspired interface engineering of soybean meal-based adhesive incorporated with biomineralized cellulose nanofibrils and a functional aminoclay. *Chem. Eng. J.* **2021**, *421*, 129820. [CrossRef]
29. Wang, B.; Dai, L.; Hunter, L.A.; Zhang, L.; Yang, G.; Chen, J.; Zhang, X.; He, Z.; Ni, Y. A multifunctional nanocellulose-based hydrogel for strain sensing and self-powering applications. *Carbohydr. Polym.* **2021**, *268*, 118210. [CrossRef] [PubMed]
30. Ruiz-Palomero, C.; Benítez-Martínez, S.; Soriano, M.L.; Valcárcel, M. Fluorescent nanocellulosic hydrogels based on graphene quantum dots for sensing laccase. *Anal. Chim. Acta* **2017**, *974*, 93–99. [CrossRef]
31. Ruiz-Palomero, C.; Soriano, M.L.; Benítez-Martínez, S.; Valcárcel, M. Photoluminescent sensing hydrogel platform based on the combination of nanocellulose and S, N-codoped graphene quantum dots. *Sensors Actuators, B Chem.* **2017**, *245*, 946–953. [CrossRef]
32. Mao, Q. A Molecular Dynamics Study of the Cellulose-Graphene Oxide Nanocomposites: The Interface Effects. Ph.D. Thesis, Clemson University, Clemson, SC, USA, 2018.
33. Bacakova, L.; Pajorova, J.; Tomkova, M.; Matejka, R.; Broz, A.; Stepanovska, J.; Prazak, S.; Skogberg, A.; Siljander, S.; Kallio, P. Applications of Nanocellulose/Nanocarbon Composites: Focus on Biotechnology and Medicine. *Nanomaterials* **2020**, *10*, 196. [CrossRef] [PubMed]
34. Ates, B.; Koytepe, S.; Ulu, A.; Gurses, C.; Thakur, V.K. Chemistry, structures, and advanced applications of nanocomposites from biorenewable resources. *Chem. Rev.* **2020**, *120*, 9304–9362. [CrossRef] [PubMed]
35. Trache, D.; Thakur, V.K.; Boukherroub, R. Cellulose nanocrystals/graphene hybrids—A promising new class of materials for advanced applications. *Nanomaterials* **2020**, *10*, 1523. [CrossRef] [PubMed]
36. Zhu, S.; Song, Y.; Zhao, X.; Shao, J.; Zhang, J.; Yang, B. The photoluminescence mechanism in carbon dots (graphene quantum dots, carbon nanodots, and polymer dots): Current state and future perspective. *Nano Res.* **2015**, *8*, 355–381. [CrossRef]
37. Gu, B.; Liu, Z.; Chen, D.; Gao, B.; Yang, Y.; Guo, Q.; Wang, G. Solid-state fluorescent nitrogen doped graphene quantum dots with yellow emission for white light-emitting diodes. *Synth. Met.* **2021**, *277*, 116787. [CrossRef]
38. El-Shabasy, R.M.; Elsadek, M.F.; Ahmed, B.M.; Farahat, M.F.; Mosleh, K.M.; Taher, M.M. Recent developments in carbon quantum dots: Properties, fabrication techniques, and bio-applications. *Processes* **2021**, *9*, 388. [CrossRef]
39. Yuan, F.; Li, S.; Fan, Z.; Meng, X.; Fan, L.; Yang, S. Shining carbon dots: Synthesis and biomedical and optoelectronic applications. *Nano Today* **2016**, *11*, 565–586. [CrossRef]
40. Nie, H.; Li, M.; Li, Q.; Liang, S.; Tan, Y.; Sheng, L.; Shi, W.; Zhang, S.X.A. Carbon dots with continuously tunable full-color emission and their application in ratiometric pH sensing. *Chem. Mater.* **2014**, *26*, 3104–3112. [CrossRef]
41. Chung, S.; Revia, R.A.; Zhang, M. Graphene Quantum Dots and Their Applications in Bioimaging, Biosensing, and Therapy. *Adv. Mater.* **2021**, *33*, 1904362. [CrossRef]
42. Yan, Y.; Manickam, S.; Lester, E.; Wu, T.; Pang, C.H. Synthesis of graphene oxide and graphene quantum dots from miscanthus via ultrasound-assisted mechano-chemical cracking method. *Ultrason. Sonochem.* **2021**, *73*, 105519. [CrossRef]
43. Qi, B.P.; Zhang, X.; Shang, B.B.; Xiang, D.; Zhang, S. Solvothermal tuning of photoluminescent graphene quantum dots: From preparation to photoluminescence mechanism. *J. Nanoparticle Res.* **2018**, *20*, 20. [CrossRef]
44. Kapoor, S.; Jha, A.; Ahmad, H.; Islam, S.S. Avenue to Large-Scale Production of Graphene Quantum Dots from High-Purity Graphene Sheets Using Laboratory-Grade Graphite Electrodes. *ACS Omega* **2020**, *5*, 18831–18841. [CrossRef] [PubMed]
45. Danial, W.H.; Chutia, A.; Majid, Z.A.; Sahnoun, R.; Aziz, M. Electrochemical synthesis and characterization of stable colloidal suspension of graphene using two-electrode cell system. In Proceedings of the AIP Conference Proceedings, Tronoh, Malaysia, 22 July 2015; Volume 1669, p. 020020.
46. Li, H.; He, X.; Kang, Z.; Huang, H.; Liu, Y.; Liu, J.; Lian, S.; Tsang, C.H.A.; Yang, X.; Lee, S.-T. Water-Soluble Fluorescent Carbon Quantum Dots and Photocatalyst Design. *Angew. Chem.* **2010**, *122*, 4532–4536. [CrossRef]
47. Danial, W.H.; Farouzy, B.; Abdullah, M.; Majid, Z.A. Facile one-step preparation and characterization of graphene quantum dots suspension via electrochemical exfoliation. *Malays. J. Chem.* **2021**, *23*, 127–135.
48. Pan, D.; Zhang, J.; Li, Z.; Wu, M. Hydrothermal Route for Cutting Graphene Sheets into Blue-Luminescent Graphene Quantum Dots. *Adv. Mater.* **2010**, *22*, 734–738. [CrossRef] [PubMed]
49. Zhu, S.; Zhang, J.; Qiao, C.; Tang, S.; Li, Y.; Yuan, W.; Li, B.; Tian, L.; Liu, F.; Hu, R.; et al. Strongly green-photoluminescent graphene quantum dots for bioimaging applications. *Chem. Commun.* **2011**, *47*, 6858. [CrossRef]

50. Zheng, L.; Chi, Y.; Dong, Y.; Lin, J.; Wang, B. Electrochemiluminescence of water-soluble carbon nanocrystals released electrochemically from graphite. *J. Am. Chem. Soc.* **2009**, *131*, 4564–4565. [CrossRef]
51. Valappil, M.O.; Pillai, V.K.; Alwarappan, S. Spotlighting graphene quantum dots and beyond: Synthesis, properties and sensing applications. *Appl. Mater. Today* **2017**, *9*, 350–371. [CrossRef]
52. Wang, D.; Chen, J.F.; Dai, L. Recent advances in graphene quantum dots for fluorescence bioimaging from cells through tissues to animals. *Part. Part. Syst. Charact.* **2015**, *32*, 515–523. [CrossRef]
53. Hong, G.L.; Zhao, H.L.; Deng, H.H.; Yang, H.J.; Peng, H.P.; Liu, Y.H.; Chen, W. Fabrication of ultra-small monolayer graphene quantum dots by pyrolysis of trisodium citrate for fluorescent cell imaging. *Int. J. Nanomed.* **2018**, *13*, 4807–4815. [CrossRef] [PubMed]
54. Xu, M.; Zaohui, L.; Zu, X.; Hu, N.; Wei, H.; Yang, Z.; Zang, Y. Hydrothermal/Solvothermal Synthesis of Graphene Quantum Dots and Their Biological Applications. *Nano Biomed. Eng.* **2013**, *5*, 65–71. [CrossRef]
55. Tetsuka, H.; Asahi, R.; Nagoya, A.; Okamoto, K.; Tajima, I.; Ohta, R.; Okamoto, A. Optically Tunable Amino-Functionalized Graphene Quantum Dots. *Adv. Mater.* **2012**, *24*, 5333–5338. [CrossRef] [PubMed]
56. Peng, J.; Gao, W.; Gupta, B.K.; Liu, Z.; Romero-Aburto, R.; Ge, L.; Song, L.; Alemany, L.B.; Zhan, X.; Gao, G.; et al. Graphene quantum dots derived from carbon fibers. *Nano Lett.* **2012**, *12*, 844–849. [CrossRef]
57. Sun, X.; Liu, Z.; Welsher, K.; Robinson, J.T.; Goodwin, A.; Zaric, S.; Dai, H. Nano-graphene oxide for cellular imaging and drug delivery. *Nano Res.* **2008**, *1*, 203–212. [CrossRef] [PubMed]
58. Zhao, C.; Song, X.; Liu, Y.; Fu, Y.; Ye, L.; Wang, N.; Wang, F.; Li, L.; Mohammadniaei, M.; Zhang, M.; et al. Synthesis of graphene quantum dots and their applications in drug delivery. *J. Nanobiotechnology* **2020**, *18*, 142. [CrossRef] [PubMed]
59. Garg, M.; Vishwakarma, N.; Sharma, A.L.; Singh, S. Amine-Functionalized Graphene Quantum Dots for Fluorescence-Based Immunosensing of Ferritin. *ACS Appl. Nano Mater.* **2021**, *4*, 7416–7425. [CrossRef]
60. Liu, Y.; Tang, X.; Deng, M.; Cao, Y.; Li, Y.; Zheng, H.; Li, F.; Yan, F.; Lan, T.; Shi, L.; et al. Nitrogen doped graphene quantum dots as a fluorescent probe for mercury(II) ions. *Microchim. Acta* **2019**, *186*. [CrossRef] [PubMed]
61. Qu, C.; Zhang, D.; Yang, R.; Hu, J.; Qu, L. Nitrogen and sulfur co-doped graphene quantum dots for the highly sensitive and selective detection of mercury ion in living cells. *Spectrochim. Acta Part A Mol. Biomol. Spectrosc.* **2019**, *206*, 588–596. [CrossRef] [PubMed]
62. Luo, X.; Wang, X. Preparation and characterization of nanocellulose fibers from NaOH/urea pretreatment of oil palm fibers. *BioResources* **2017**, *12*, 5826–5837. [CrossRef]
63. Beluns, S.; Gaidukovs, S.; Platnieks, O.; Gaidukova, G.; Mierina, I.; Grase, L.; Starkova, O.; Brazdausks, P.; Thakur, V.K. From Wood and Hemp Biomass Wastes to Sustainable Nanocellulose Foams. *Ind. Crops Prod.* **2021**, *170*, 113780. [CrossRef]
64. Malainine, M.E.; Mahrouz, M.; Dufresne, A. Thermoplastic nanocomposites based on cellulose microfibrils from Opuntia ficus-indica parenchyma cell. *Compos. Sci. Technol.* **2005**, *65*, 1520–1526. [CrossRef]
65. Farooq, A.; Patoary, M.K.; Zhang, M.; Mussana, H.; Li, M.; Naeem, M.A.; Mushtaq, M.; Farooq, A.; Liu, L. Cellulose from sources to nanocellulose and an overview of synthesis and properties of nanocellulose/zinc oxide nanocomposite materials. *Int. J. Biol. Macromol.* **2020**, *154*, 1050–1073. [CrossRef]
66. Gao, K.; Shao, Z.; Wang, X.; Zhang, Y.; Wang, W.; Wang, F. Cellulose nanofibers/multi-walled carbon nanotube nanohybrid aerogel for all-solid-state flexible supercapacitors. *RSC Adv.* **2013**, *3*, 15058–15064. [CrossRef]
67. Rangaswamy, B.E.; Vanitha, K.P.; Hungund, B.S. Microbial Cellulose Production from Bacteria Isolated from Rotten Fruit. *Int. J. Polym. Sci.* **2015**, *2015*, 1–8. [CrossRef]
68. Aswini, K.; Gopal, N.O.; Uthandi, S. Optimized culture conditions for bacterial cellulose production by Acetobacter senegalensis MA1. *BMC Biotechnol.* **2020**, *20*, 46. [CrossRef]
69. Costa, A.F.S.; Almeida, F.C.G.; Vinhas, G.M.; Sarubbo, L.A. Production of bacterial cellulose by Gluconacetobacter hansenii using corn steep liquor as nutrient sources. *Front. Microbiol.* **2017**, *8*. [CrossRef] [PubMed]
70. Portela, R.; Leal, C.R.; Almeida, P.L.; Sobral, R.G. Bacterial cellulose: A versatile biopolymer for wound dressing applications. *Microb. Biotechnol.* **2019**, *12*, 586 610. [CrossRef] [PubMed]
71. Abdul Khalil, H.P.S.; Jummaat, F.; Yahya, E.B.; Olaiya, N.G.; Adnan, A.S.; Abdat, M.; Nasir, N.A.M.; Halim, A.S.; Seeta Uthaya Kumar, U.; Bairwan, R.; et al. A review on micro- to nanocellulose biopolymer scaffold forming for tissue engineering applications. *Polymers* **2020**, *12*, 2043. [CrossRef] [PubMed]
72. Naomi, R.; Idrus, R.B.H.; Fauzi, M.B. Plant-vs. Bacterial-derived cellulose for wound healing: A review. *Int. J. Environ. Res. Public Health* **2020**, *17*, 6803. [CrossRef] [PubMed]
73. Spence, K.L.; Venditti, R.A.; Rojas, O.J.; Habibi, Y.; Pawlak, J.J. A comparative study of energy consumption and physical properties of microfibrillated cellulose produced by different processing methods. *Cellulose* **2011**, *18*, 1097–1111. [CrossRef]
74. Malucelli, L.C.; Matos, M.; Jordão, C.; Lomonaco, D.; Lacerda, L.G.; Carvalho Filho, M.A.S.; Magalhães, W.L.E. Influence of cellulose chemical pretreatment on energy consumption and viscosity of produced cellulose nanofibers (CNF) and mechanical properties of nanopaper. *Cellulose* **2019**, *26*, 1667–1681. [CrossRef]
75. Tang, L.R.; Huang, B.; Ou, W.; Chen, X.R.; Chen, Y.D. Manufacture of cellulose nanocrystals by cation exchange resin-catalyzed hydrolysis of cellulose. *Bioresour. Technol.* **2011**, *102*, 10973–10977. [CrossRef] [PubMed]
76. Azizi Samir, M.A.S.; Alloin, F.; Dufresne, A. Review of recent research into cellulosic whiskers, their properties and their application in nanocomposite field. *Biomacromolecules* **2005**, *6*, 612–626. [CrossRef] [PubMed]

77. Danial, W.H.; Mohd Taib, R.; Abu Samah, M.A.; Mohd Salim, R.; Abdul Majid, Z. The valorization of municipal grass waste for the extraction of cellulose nanocrystals. *RSC Adv.* **2020**, *10*, 42400–42407. [CrossRef]
78. Danial, W.H.; Abdul Majid, Z.; Mohd Muhid, M.N.; Triwahyono, S.; Bakar, M.B.; Ramli, Z. The reuse of wastepaper for the extraction of cellulose nanocrystals. *Carbohydr. Polym.* **2015**, *118*, 165–169. [CrossRef]
79. Hanafiah, S.F.M.; Danial, W.H.; Samah, M.A.A.; Samad, W.Z.; Susanti, D.; Salim, R.M.; Majid, Z.A. Extraction and characterization of microfibrillated and nanofibrillated cellulose from office paper waste. *Malaysian J. Anal. Sci.* **2019**, *23*, 901–913. [CrossRef]
80. Zhang, Q.; Zhang, L.; Wu, W.; Xiao, H. Methods and applications of nanocellulose loaded with inorganic nanomaterials: A review. *Carbohydr. Polym.* **2020**, *229*, 115454. [CrossRef]
81. Ray, D.; Sain, S. In situ processing of cellulose nanocomposites. *Compos. Part A Appl. Sci. Manuf.* **2016**, *83*, 19–37. [CrossRef]
82. Qu, D.; Zheng, M.; Du, P.; Zhou, Y.; Zhang, L.; Li, D.; Tan, H.; Zhao, Z.; Xie, Z.; Sun, Z. Highly luminescent S, N co-doped graphene quantum dots with broad visible absorption bands for visible light photocatalysts. *Nanoscale* **2013**, *5*, 12272–12277. [CrossRef]
83. Leung, A.C.W.; Hrapovic, S.; Lam, E.; Liu, Y.; Male, K.B.; Mahmoud, K.A.; Luong, J.H.T. Characteristics and properties of carboxylated cellulose nanocrystals prepared from a novel one-step procedure. *Small* **2011**, *7*, 302–305. [CrossRef] [PubMed]
84. Kumar, Y.R.; Deshmukh, K.; Sadasivuni, K.K.; Pasha, S.K.K. Graphene quantum dot based materials for sensing, bio-imaging and energy storage applications: A review. *RSC Adv.* **2020**, *10*, 23861–23898. [CrossRef]
85. Tetsuka, H.; Nagoya, A.; Asahi, R. Highly luminescent flexible amino-functionalized graphene quantum dots@cellulose nanofiber-clay hybrids for white-light emitting diodes. *J. Mater. Chem. C* **2015**, *3*, 3536–3541. [CrossRef]
86. Khabibullin, A.; Alizadehgiashi, M.; Khuu, N.; Prince, E.; Tebbe, M.; Kumacheva, E. Injectable Shear-Thinning Fluorescent Hydrogel Formed by Cellulose Nanocrystals and Graphene Quantum Dots. *Langmuir* **2017**, *33*, 12344–12350. [CrossRef] [PubMed]
87. Prince, E.; Alizadehgiashi, M.; Campbell, M.; Khuu, N.; Albulescu, A.; De France, K.; Ratkov, D.; Li, Y.; Hoare, T.; Kumacheva, E. Patterning of Structurally Anisotropic Composite Hydrogel Sheets. *Biomacromolecules* **2018**, *19*, 1276–1284. [CrossRef]
88. Rosddi, N.N.M.; Fen, Y.W.; Anas, N.A.A.; Omar, N.A.S.; Ramdzan, N.S.M.; Mohd Daniyal, W.M.E.M. Cationically modified nanocrystalline cellulose/carboxyl-functionalized graphene quantum dots nanocomposite thin film: Characterization and potential sensing application. *Crystals* **2020**, *10*, 875. [CrossRef]
89. Rosddi, N.N.M.; Fen, Y.W.; Omar, N.A.S.; Anas, N.A.A.; Hashim, H.S.; Ramdzan, N.S.M.; Fauzi, N.I.M.; Anuar, M.F.; Daniyal, W.M.E.M.M. Glucose detection by gold modified carboxyl-functionalized graphene quantum dots-based surface plasmon resonance. *Optik* **2021**, *239*, 166779. [CrossRef]
90. Mahmoud, A.M.; Mahnashi, M.H.; Alkahtani, S.A.; El-Wekil, M.M. Nitrogen and sulfur co-doped graphene quantum dots/nanocellulose nanohybrid for electrochemical sensing of anti-schizophrenic drug olanzapine in pharmaceuticals and human biological fluids. *Int. J. Biol. Macromol.* **2020**, *165*, 2030–2037. [CrossRef] [PubMed]
91. Xiong, C.; Xu, J.; Han, Q.; Qin, C.; Dai, L.; Ni, Y. Construction of flexible cellulose nanofiber fiber@graphene quantum dots hybrid film applied in supercapacitor and sensor. *Cellulose* **2021**, *28*, 10359–10372. [CrossRef]
92. Facure, M.H.M.; Schneider, R.; Mercante, L.A.; Correa, D.S. A review on graphene quantum dots and their nanocomposites: From laboratory synthesis towards agricultural and environmental applications. *Environ. Sci. Nano* **2020**, *7*, 3710–3734. [CrossRef]
93. Milenković, M.; Mišović, A.; Jovanović, D.; Popović Bijelić, A.; Ciasca, G.; Romanò, S.; Bonasera, A.; Mojsin, M.; Pejić, J.; Stevanović, M.; et al. Facile Synthesis of L-Cysteine Functionalized Graphene Quantum Dots as a Bioimaging and Photosensitive Agent. *Nanomaterials* **2021**, *11*, 1879. [CrossRef] [PubMed]
94. Elvati, P.; Baumeister, E.; Violi, A. Graphene quantum dots: Effect of size, composition and curvature on their assembly. *RSC Adv.* **2017**, *7*, 17704–17710. [CrossRef]
95. Kafy, A.; Akther, A.; Shishir, M.I.R.; Kim, H.C.; Yun, Y.; Kim, J. Cellulose nanocrystal/graphene oxide composite film as humidity sensor. *Sens. Actuators A Phys.* **2016**, *247*, 221–226. [CrossRef]
96. Smith, A.T.; LaChance, A.M.; Zeng, S.; Liu, B.; Sun, L. Synthesis, properties, and applications of graphene oxide/reduced graphene oxide and their nanocomposites. *Nano Mater. Sci.* **2019**, *1*, 31–47. [CrossRef]
97. Pottathara, Y.B.; Bobnar, V.; Gorgieva, S.; Grohens, Y.; Finšgar, M.; Thomas, S.; Kokol, V. Mechanically strong, flexible and thermally stable graphene oxide/nanocellulosic films with enhanced dielectric properties. *RSC Adv.* **2016**, *6*, 49138–49149. [CrossRef]
98. Lindman, B.; Karlström, G.; Stigsson, L. On the mechanism of dissolution of cellulose. *J. Mol. Liq.* **2010**, *156*, 76–81. [CrossRef]
99. Glasser, W.G.; Atalla, R.H.; Blackwell, J.; Brown, M.M.; Burchard, W.; French, A.D.; Klemm, D.O.; Nishiyama, Y. About the structure of cellulose: Debating the Lindman hypothesis. *Cellulose* **2012**, *19*, 589–598. [CrossRef]
100. Yamane, C.; Aoyagi, T.; Ago, M.; Sato, K.; Okajima, K.; Takahashi, T. Two different surface properties of regenerated cellulose due to structural anisotropy. *Polym. J.* **2006**, *38*, 819–826. [CrossRef]
101. Alqus, R.; Eichhorn, S.J.; Bryce, R.A. Molecular dynamics of cellulose amphiphilicity at the graphene-water interface. *Biomacromolecules* **2015**, *16*, 1771–1783. [CrossRef] [PubMed]
102. Sabo, R.; Yermakov, A.; Law, C.T.; Elhajjar, R. Nanocellulose-enabled electronics, energy harvesting devices, smart materials and sensors: A review. *J. Renew. Mater.* **2016**, *4*, 297–312. [CrossRef]
103. Golmohammadi, H.; Morales-Narváez, E.; Naghdi, T.; Merkoçi, A. Nanocellulose in Sensing and Biosensing. *Chem. Mater.* **2017**, *29*, 5426–5446. [CrossRef]

104. Danial, W.H.; Mohamed, N.A.S.; Majid, Z.A. Recent advances on the preparation and application of graphene quantum dots for mercury detection: A systematic review. *Carbon Lett.* **2021**. [CrossRef]
105. Curvello, R.; Raghuwanshi, V.S.; Garnier, G. Engineering nanocellulose hydrogels for biomedical applications. *Adv. Colloid Interface Sci.* **2019**, *267*, 47–61. [CrossRef] [PubMed]
106. Norrrahim, M.N.F.; Mohd Kasim, N.A.; Knight, V.F.; Ujang, F.A.; Janudin, N.; Abdul Razak, M.A.I.; Shah, N.A.A.; Noor, S.A.M.; Jamal, S.H.; Ong, K.K.; et al. Nanocellulose: The next super versatile material for the military. *Mater. Adv.* **2021**, *2*, 1485–1506. [CrossRef]
107. Dias, O.A.T.; Konar, S.; Leão, A.L.; Yang, W.; Tjong, J.; Sain, M. Current State of Applications of Nanocellulose in Flexible Energy and Electronic Devices. *Front. Chem.* **2020**, *8*, 420. [CrossRef]
108. Nizam, P.A.; Gopakumar, D.A.; Pottathara, Y.B.; Pasquini, D.; Nzihou, A.; Thomas, S. Nanocellulose-based composites. In *Nanocellulose Based Composites for Electronics*; Elsevier: Amsterdam, The Netherlands, 2021; pp. 15–29.
109. Ansari, J.R.; Hegazy, S.M.; Houkan, M.T.; Kannan, K.; Aly, A.; Sadasivuni, K.K. Nanocellulose-based materials/composites for sensors. In *Nanocellulose Based Composites for Electronics*; Elsevier: Amsterdam, The Netherlands, 2021; pp. 185–214.
110. Habibi, Y.; Lucia, L.A.; Rojas, O.J. Cellulose nanocrystals: Chemistry, self-assembly, and applications. *Chem. Rev.* **2010**, *110*, 3479–3500. [CrossRef] [PubMed]
111. Wang, J. Carbon-nanotube based electrochemical biosensors: A review. *Electroanalysis* **2005**, *17*, 7–14. [CrossRef]
112. Nascimento, D.M.; Nunes, Y.L.; Figueirêdo, M.C.B.; De Azeredo, H.M.C.; Aouada, F.A.; Feitosa, J.P.A.; Rosa, M.F.; Dufresne, A. Nanocellulose nanocomposite hydrogels: Technological and environmental issues. *Green Chem.* **2018**, *20*, 2428–2448. [CrossRef]
113. Palanisamy, S.; Ramaraj, S.K.; Chen, S.M.; Yang, T.C.K.; Pan, Y.F.; Chen, T.W.; Velusamy, V.; Selvam, S. A novel Laccase biosensor based on laccase immobilized graphene-cellulose microfiber composite modified screen-printed carbon electrode for sensitive determination of catechol. *Sci. Rep.* **2017**, *7*, 41214. [CrossRef]
114. Oluwasanu, A.A. Fate and Toxicity of Chlorinated Phenols of Environmental Implications: A Review. *Med. Anal. Chem. Int. J.* **2018**, *2*, 000126. [CrossRef]
115. Olaniran, A.O.; Igbinosa, E.O. Chlorophenols and other related derivatives of environmental concern: Properties, distribution and microbial degradation processes. *Chemosphere* **2011**, *83*, 1297–1306. [CrossRef] [PubMed]
116. Sakai, Y.; Sadaoka, Y.; Matsuguchi, M. Humidity sensors based on polymer thin films. *Sens. Actuators B Chem.* **1996**, *35*, 85–90. [CrossRef]
117. Sakai, Y.; Matsuguchi, M.; Yonesato, N. Humidity sensor based on alkali salts of poly(2-acrylamido-2-methylpropane sulfonic acid). *Electrochim. Acta* **2001**, *46*, 1509–1514. [CrossRef]

Review

Locust Bean Gum, a Vegetable Hydrocolloid with Industrial and Biopharmaceutical Applications

Max Petitjean and José Ramón Isasi *

Department of Chemistry, University of Navarra, 31080 Pamplona, Spain
* Correspondence: jrisasi@unav.es; Tel.: +34-948425600

Abstract: Locust bean gum (LBG), a vegetable galactomannan extracted from carob tree seeds, is extensively used in the food industry as a thickening agent (E410). Its molecular conformation in aqueous solutions determines its solubility and rheological performance. LBG is an interesting polysaccharide also because of its synergistic behavior with other biopolymers (xanthan gum, carrageenan, etc.). In addition, this hydrocolloid is easily modified by derivatization or crosslinking. These LBG-related products, besides their applications in the food industry, can be used as encapsulation and drug delivery devices, packaging materials, batteries, and catalyst supports, among other biopharmaceutical and industrial uses. As the new derivatized or crosslinked polymers based on LBG are mainly biodegradable and non-toxic, the use of this polysaccharide (by itself or combined with other biopolymers) will contribute to generating greener products, considering the origin of raw materials used, the modification procedures selected and the final destination of the products.

Keywords: locust bean gum; galactomannans; carob; thickening agents

1. Introduction

Locust bean gum (LBG) is a high molecular weight non-ionic galactomannan polysaccharide, extracted from the seeds of *Ceratonia Siliqua* (carob tree, or locust bean tree, mainly found in the Mediterranean region). Although both structurally and chemically similar to guar gum, it shows important differences. Soluble in water with the addition of heat, LBG solutions do not form gels by themselves but enhance those produced by other types of hydrocolloids such as xanthan and carrageenan.

Among the many constituents of the carob fruit, including sugars and bioactive compounds [1–5], their polysaccharides are in both the carob fiber and the carob bean gum [6,7]. Recently, potential health benefits of carob products have been reported because the polyphenols in the fruit are powerful antioxidants [2,3], and the dietary fibers and sugars prevent diabetes, heart diseases and gastrointestinal disturbances [6]. The carob mucilage, also known as locust bean gum, was critically described at the end of the 19th century, and its colloidal properties were well-studied many years ago [8].

As mentioned by Gioxary et al. in their recent review, the development of carob tree cultivation can be useful for the environment thanks to the capacity of this species to prevent soil degradation, besides its high CO_2 absorption ratio and its potential usage not only to produce animal feed but also for the human diet in the Mediterranean regions [1].

LBG is mainly used as an additive (E410) in the food and beverage industry, most often as a thickening, stabilizing and gelling agent, or emulsifier, the texture being an intangible property of food of great importance [9]. This polysaccharide has found applications in other sectors such as pharmaceuticals, cosmetics, textiles, paper, or the petroleum industry [10–12]. In fact, various non-starch polysaccharides isolated from plants, including LBG, show a considerable potential to prepare drug delivery systems to achieve tailored and/or site-specific drug release [10]. In addition, natural gum-based hydrogels can be used in tissue engineering, wound dressing, hygienic products, agriculture, and water purification [11].

In this review, after the description of the structural characteristics of locust bean gum, its biosynthetic origin and its chemical isolation will be also accounted. Its chemical composition originates a particular conformation in aqueous solutions, responsible for its rheological properties, also presented here. Besides the applications of these solutions due to their own viscosity, an essential aspect is their synergistic behavior when mixed with other polysaccharides. The last section of this review deals with the derivatives of locust bean gum, either by functionalization or cross-linking.

2. Structure, Processing and Properties

2.1. Composition

Locust bean gum (LBG), a polysaccharide of vegetal origin that belongs to the galactomannan family [13], is composed of β-(1-4)-mannose backbones randomly branched by α-(1-6)-galactose (Figure 1) [14]. The mannose:galactose ratios, usually determined by the Blakeney method [15], are different for each type of galactomannan, depending on their origin [16]. Thus, the ratios found are between 1 to 1 and 10 to 1 [14,17] for different gums: fenugreek gum 1:1; guar gum 2:1; tara gum 3:1; LBG 4:1; cassia gum 5:1, etc. [18]. The first elucidation of the fine structure of LBG was proposed by Baker et al. in 1975, using an alkaline degradation method [19], but its discovery and analysis come from much earlier, as explained by Dea et al. [14]. There are other characterization techniques available, such as those involving hydrolysis, periodate oxidation [20], 13C NMR [21], methylation [22,23], partial or enzymatic hydrolysis [17] and the development of sulfonyl derivatives [24].

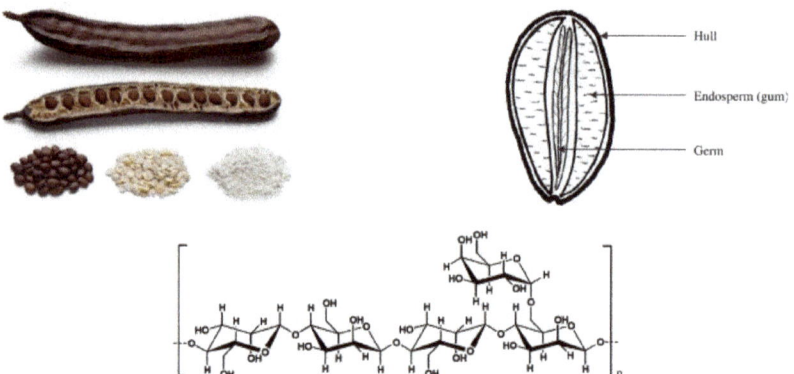

Figure 1. Carob fruit and locust bean gum.

LBG and other galactomannans can be obtained from the Leguminosae plant family. The carob tree, a Mediterranean plant also known as *Ceratonia siliqua*, can be found in Portugal, Spain, Italy, Cyprus, Greece, Morocco, and the rest of northern Africa, but it is also grown in Asia, Australia and South America. A pod of the carob bean will be biologically composed of a seed coat (≈30%), germ (≈25%) and endosperm (≈42%). The rest of the mass will be moisture (≈8%) [25]. Chemically, a pod is a mixture of galactomannan (≈85%), water (≈8%), protein (≈5%), ash, fibers and fat, each one around 1% [26].

2.2. Biosynthesis

The biosynthesis path takes place in the lumen of the Golgi apparatus. The product is then transported to the surface cell by secretory vesicles and introduced into its wall matrix [27]. The biosynthesis of galactomannan has been well described by Sharma et al. [28]. Briefly, it begins with the transformation of sucrose in uridine diphosphate (UDP)-glucose and UDP-fructose by a synthase, and in glucose and fructose by an invertase. Fructose is then phosphorylated and isomerized to produce mannose-6-phosphate. The phosphate group is delocalized into position 1 by phosphomannomutase. This mannose-1-phosphate is then transformed into GDP-mannose using GDP-mannose pyrophosphorylase. UDP-glucose, previously formed, is then converted into UDP-galactose by UDP-galactose 4-epimerase. The synthesis of both substrates, GDP-mannose and UDP-galactose, is also an enzymatic process. The mannan backbone is formed by using the GDP-mannose thanks to the mannan synthase, and the branched galactosyl units on the backbone come from UDP-galactose by the galactosyltransferase enzyme [29,30]. The M:G ratio can be modified in vitro by changing the GDP-mannose concentration [31] along the synthesis process, or by removing galactosyl units during hydrolysis thanks to α-galactosidase [32]. It is also modified by the culture conditions and depends on the seed origin and the gum fabrication process [14,33,34].

2.3. Extraction

The extraction of LBG can proceed following different methods. First, the removal of the seeds from the pod must be performed mechanically. After that, to eliminate the hull, diverse procedures are available, such as roasting [35], acid extraction [36,37], water extraction [36], mechanical processes, or by swelling and freezing [12]. The endosperm is then milled and pulverized under different conditions to remove the remaining husk. The endosperm is a mixture of polysaccharides, proteins, and other impurities, which necessitates a purification step, and this can be performed by precipitation coupled with dialysis. After dissolving the powder in water, the addition of an alcohol such as ethanol [38,39], methanol [40], or isopropanol [14]; a copper complex [41] or a barium-complex [42] produces the precipitation of the galactomannan [28]. Azero et al. studied different purification techniques and their impact on the physicochemical properties of the formed gum, and they showed better inter- and intramolecular associations for LBG for the one filtered over the centrifuged product [43]. Isopropanol decreases the content of ashes and proteins and produces a more stable solution due to the elimination of enzymes and impurities [33]. Dakia et al. compared two types of processes: the first one using water, removing the different seed layers by letting the seed swell in boiling water, and the germ removed after drying the seed; the second one by an acidic extraction. The seed is macerated in H_2SO_4/H_2O 60/40 (v/v) at 60 °C for 1 h. The carbonized hull is removed by washing for 2 min with a metallic sieve. After drying the seeds, they are crushed to release and remove the germ. Both procedures mill and sift the endosperm using the same conditions [36]. Some physicochemical differences arise between the LBGs coming from the two processes: acid extraction produces better thickening properties, while water extraction is responsible for a higher solubility at high temperatures, for example.

2.4. Conformation

X-ray analysis shows that the LBG powder is mainly amorphous [21]. As shown by Grimaud et al., ordered conformations similar to that of the LBG backbone (i.e., a chain of α-(1-4)-mannose) favor the presence of crystalline structures and then an inter- or intramolecular complexation, which creates hydrophobic regions preventing good solubilization in water [44]. The galactose-branched units facilitate the solubilization of the backbone, and this property increases with the degree of substitution [45]. The conformation of the galactomannan depends on the inter- and intra-molecular interactions and on their hydrophobic interactions [46], passing from elongated ribbon-like forms [12] to aggregates, and, therefore, forming hydrophobic microdomains. These hydrophobic microdomains also depend on the distribution of galactose, the mannose:galactose ratio, and the solution temperature [47]. These characteristics also influence its critical association concentration. Molecular modeling showed the influence of the galactose-branched chains on the flexibility of the mannose backbone [48]. Then, a method to measure this flexibility by using the persistent length was found, which produces, for a 1:1 M:G, an error of 3 Å [45]. Other techniques have facilitated the study of the fine structure of galactomannans, such as X-ray scattering [49], one- or two-dimensional NMR [50], size exclusion chromatography coupled with multi-angle static light scattering [51] and fluorescence spectroscopy [47].

2.5. Physico-Chemical Properties

The galactomannan aqueous solubility depends on the temperature but it is also associated with the M:G ratio. Thus, the higher the amount of branched galactose units present in the polysaccharide, the higher its solubility at a low temperature [52,53]. In that sense, for LBG, the solubility value originates from a thermodynamic equilibrium between the amorphous solid phase swollen by the solvent and the pure solvent phase [54]. Because LBG does not possess ionizable functions, solubilization depends on the amount of hydrogen bonds and the quality of the solvent [55]. As mentioned above, a pure mannose backbone possesses a high level of intra- and intermolecular interactions via hydrogen bonding, permitting the aggregation that leads to precipitation [44,56,57]. The galactose branched units have two functions: solubilization and anti-aggregation of the polysaccharide. The solubility temperature depends also on the distribution of galactosyl-branched units along the backbone (Figure 2). A high-temperature solubility means compact galactosyl branched units, so large smooth regions [58].

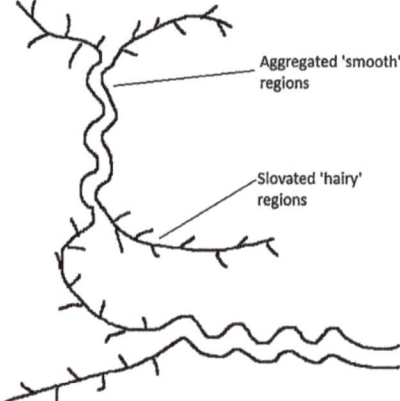

Figure 2. Entanglement of galactomannan polymers; the smooth regions form the insoluble part of the gel while the hairy regions permit the solubility of the polysaccharide [12].

For dilute galactomannan solutions, the viscosity, as the solubility, will depend on the molecular mass, the M:G ratio and the distribution of the galactosyl branched units, as expressed by the Mark–Houwink equation [59]. A higher molar mass yields a higher intrinsic viscosity, as occurs for a large number of galactosyl-branched units [14,60]. Therefore, Morris et al. proposed an equation introducing the gyration radius related to the distribution of galactosyl-branched units [61]. For a higher concentration, the polymer interpenetration phenomenon increases the viscosity by creating a physical covering [62]. If this phenomenon is the only one taking place in the solution, the theoretical viscosity will depend merely on the concentration and molar mass [63]. However, LBG solutions have a higher viscosity than the theoretical ones, meaning that another phenomenon is also occurring. This, as explained by Sittikijyothin et al., consists of the 'hyperentanglements' [64], an intermolecular aggregation influenced by the M:G ratio and galactose distribution [65]. A higher amount of smooth regions or a lower quantity of galactose branched units permits the formation of these additional entanglements [56,66,67].

The dynamic viscosity of galactomannans and, more specifically, that of LBG, has been well studied. As for many polysaccharides, this hydrocolloid shows a pseudoplastic behavior in solution [64,68]. The galactomannan concentration and its microstructure are two factors influencing the dynamic viscosity of the solutions; both of these characteristics are directly related to the intra- and intermolecular associations and, consequently, depend on the smooth locust bean gum region [66,69]. The ionic strength, temperature and pH have a small influence on the viscosity [14,17]. Temperature and pH can break the polysaccharides, modifying the final viscosity due to the changes in the molecular mass of the polymer chains. The solution temperature has an impact on its viscosity since a higher temperature permits an efficient solubilization process and also promotes a higher solubility [70]. When this solution is cooling down, the observed viscosity will be higher than that of the one that was first dissolved at a lower temperature. This can be explained by a higher entanglement probability when we have produced better solubilization of the hydrocolloid [60].

Galactomannans are mainly incorporated as a powder, solubilized in the desired solution and used, due to their rheological properties, for food improvement purposes in sauces [71], beverages [72], ice creams [73], low-fat [72] or bakery products [74]. Mixed with other natural compounds, it is possible to produce edible films [75]. Carob bean is used in food recipes that can benefit health [6], as antioxidants [76], because of its polyphenols and flavonoids contents; anti-diabetic effect [77], also because of its LBG, flavonoids and phenolic acids; anti-hyperlipidemia properties [78], conferred by the fibers and gastrointestinal benefits [79], thanks to the locust bean gum. It can also be used for pharmaceutical/medical purposes [80,81], in buccal [82], oral [83], gastric [84], colon [85], ocular [86], or topical [87] drug delivery [88] formulations. Recently, with the necessity of finding greener energy devices, LBG has been proposed as a component of bio-batteries [89], as the binder of $ZnSO_4$ and MnO_2, in order to form a "quasi-solid-state" LBG electrolyte. Its high specific capacity, rate performance and capacity retention, make LBG a viable ingredient with a high potential for use as a binder for green batteries.

2.6. Synergistic Behaviors of LBG Mixtures

Locust bean gum shows synergistic behaviors with different polysaccharides, such as xanthan gum, carrageenan and alginate, for example. As reported by Dionísio and Grenha [90], and by Verma et al. [81], the rheological synergy can be of interest for pharmaceutical applications because of the non-toxicity of the products and the different entanglement levels feasible.

The synergies between xanthan gum (XG) and galactomannans (GM) have been well studied since they were discovered. Nevertheless, the molecular mechanisms of such synergies continue to be debated. Historically, the synergistic behavior was first explained by poor or inexistent interactions because of gum incompatibility [91], volume exclusion [92], or because of weak connections other than specific intermolecular interactions [93]. Another explanation of such synergy is the existence of cooperative interactions between both polysaccharides [94]. The galactomannan branches are not regularly placed along the backbone: some parts are more branched than other sections, which are considered 'smooth' regions [19]. Different authors tried to explain where these interactions take place: between side chains of xanthan helices and smooth regions of the galactomannan backbones [95], between the xanthan helix and those smooth regions [96], or between the disordered xanthan and galactomannan structures [97]. Respectively, those possible scenarios are called the "Tako model", the "Unilever model" and the "Norwich model", as described by Takemasa and Nishinari (Figure 3) [98].

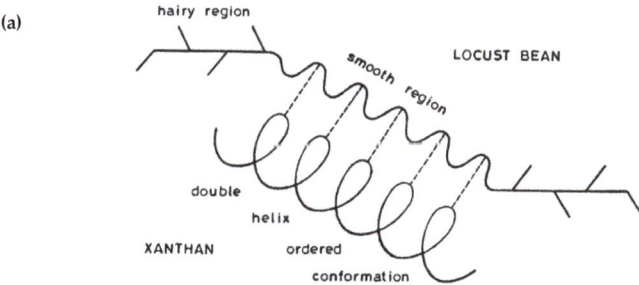

Figure 3. *Cont.*

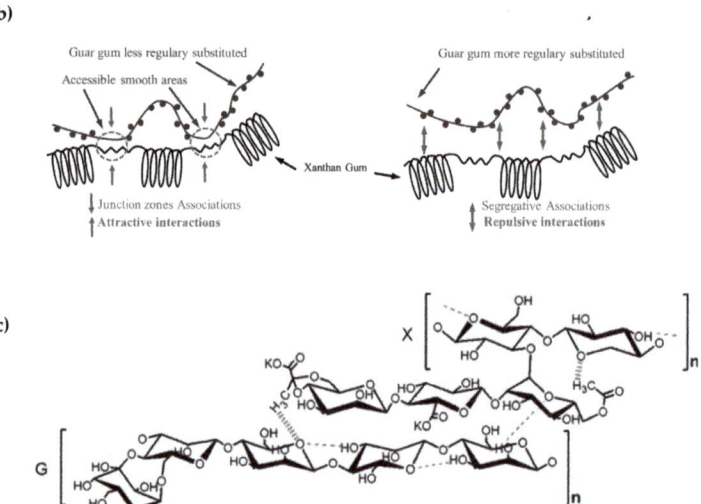

Figure 3. Different models of Xanthan Gum—Galactomannan synergies [98]: (**a**) Unilever model [99], (**b**) Norwich model [100], (**c**) Tako model [101] (G, galactomannan; X, xanthan gum).

This synergistic interaction is responsible for the modifications found in the rheological properties of the solutions, such as the viscosity, depending on the XG:GM ratio [102], pH, or the GM fine structure [103,104] and on the temperature [105]. Schreiber et al. showed, by atomic force microscopy measurements, the synergies between XG and different types of GM, and explained this phenomenon by the length and the flexibility of polysaccharide chains. A separation of phase between XG and GM occurs after 2 days when both are mixed at room temperature, and a change of the mechanical properties takes place after two weeks. The addition of salt reduces the synergism by protecting the anion charge of xanthan and by lowering the gelation strength [106]. These viscosity effects are useful in the food industry [107] for the improvement of sauces formulations, for example [108,109].

Kurt and al. studied the interactions between these two polysaccharides and glycerol in order to create a biodegradable edible film [110]. The optimization of the film formulation shows a nonlinear behavior for the mechanical properties, meaning that LBG:XG:glycerol mixtures possess some synergism. Films were successfully created and potentially used for this purpose. By mixing LBG, XG and potato starch, Yu et al. improved the final product made by 3D printing, as they observed differences by changing the proportions of LBG:XG. They reported that XG improves the printing performance and gel fineness but shows more printing deviations and low shape retention ability. LBG, on the other hand, produces better mechanical properties and printing accuracy but a lower fluidity and a bad quality of the final product [111].

The LBG:XG mixtures are also applicable for drug delivery, as Sharma et al. considered [112]. The modification of LBG:XG proportions in their experimental design allowed them to prepare microparticles with more favorable delivery kinetics for celecoxib. The encapsulation of tea polyphenols has been studied by Tian et al. [113]. The polymeric beads were made by using a w/o emulsion and the tea polyphenol release was tested in PBS solutions. A sustained release was obtained, and good stability of the LBG:XG matrix was assessed without the use of synthetic emulsifiers, which is a useful innovation in the field of new delivery materials. Bektas et al. showed also the feasibility of using these mixtures in tissue engineering [114]. In that investigation, they added mastic gum to prepare cryogels. The mechanical properties, the porosity and the cytocompatibility of the matrices formed, make it useful as a bioactive agent delivery system or as scaffolds for cartilages, for example.

As mentioned above, the use of locust bean gum is viable as a binder in green batteries. Yang et al. decided to study the effect of the LBG:XG mixture and reported its effectiveness and low-cost production [115].

Carrageenan and galactomannans show, in general, synergistic effects also. Turquois et al. showed the synergies between LBG and κ-carrageenan. They demonstrated the influence of the polysaccharide solution concentrations, the κ-carrageenan:LBG ratio and the influence of molecular weight [116]. Rheological tests of galactomannans/κ-carrageenan mixtures have also been performed by Pinheiro et al. (Figure 4) [117]. They demonstrated the effect of the M:G ratio and the polysaccharide microstructure on viscosity. Some previous works with ^{13}C NMR showed the interactions between κ-carrageenan and LBG, which are influenced by the distribution of the galactose branched units on the backbone [118]. As for the XG:GM mixtures, the smooth regions of LBG are responsible for the entanglement between both polysaccharides. For a low total concentration, the gelation is produced by a bi-continuous two-phase system [119], one being the LBG, the other constituted by droplets or a secondary phase of κ-carrageenan. Potassium chloride also influences the gelation process, producing it more easily [120]. Light scattering [121] and small angle X-ray scattering [122] put in evidence the entanglements of carrageenan with itself and explain the role played by locust bean gum on the interaction with the charged polymer. Additionally, the extent of the double helix conformation of carrageenan, the M:G ratio and the distribution of galactose units affect the synergism. This particular synergism is useful for medical purposes such as wound healing and tissue-repairing devices [123]. As studied by Mendes de Moraes et al., the mixture κ-carrageenan:LBG permits the transdermal delivery of hydrophilic compounds [124]. The release of arbutin by this hydrogel is better than that from a commercial cream, also permitting improved skin hydration and the reduction in the melanin index while being non-toxic. The preparation of 3D printable food is feasible by combining κ-carrageenan:LBG:XG, which has resulted as particularly useful for dysphagic patients [125]. Adding alginate and through a $w_1/o/w_2$ double emulsion, Wang et al. succeeded in creating sustained and controlled delivery devices [126]. The formulation κC:LBG:chitosan:PVA, obtained by mixing all the components after dissolving each one individually, was studied by Yong et al. to prepare intelligent packaging films [127]. The product was efficient for the immobilization of anthocyanins, allowing the film to be pH and ammonia sensitive.

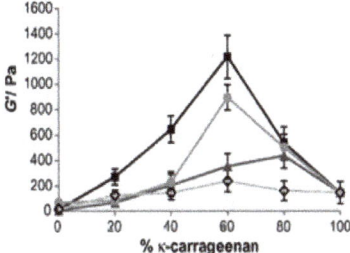

Figure 4. G' elastic component of galactomannan/κ-carrageenan mixed gels (guar gum (▲); locust bean gum (■); *Gleditsia triacanthos* galactomannan (◇) and *Sophora japonica* galactomannan (●)) [117].

Although polyvinyl alcohol (PVA) is a synthetic polymer, PVA:LBG is also a well-studied combination [128]. Double-layer films made with LBG:PVA and agar:PVA to be used as shrimp freshness indicators were studied by Yao et al. [129]. They integrated red pitaya betacyanins in the LBG layer to provide a "sensitive layer" and TiO$_2$ in the agar section as the "protective layer". The film was prepared by placing an LBG:PVA solution onto the agar:PVA layer. A good light and water vapor barrier capacity have been demonstrated, as the sensitive layer worked well when exposed to different atmospheres, and TiO$_2$ prevented the color changing, permitting a better color contrast when used with shrimp. The combination of LBG:carboxymethyl chitosan has also been studied by

Yu et al. in order to prepare films [130]. The incorporation of natural essential oil in the structure permits a better elongation at break, a higher water resistance, and oxygen barrier properties, consequently increasing the antioxidant and antibacterial activities and improving its hydrophobicity but decreasing its water vapor barrier capacity and tensile strength. Chitosan can also be incorporated into LBG as nanoparticles to produce another type of biobased film [131]. The final product possesses a great resistance but it diminished when a natural deep eutectic solvent plasticizer was added. The combination LBG:PVA has been used with extracts of *Loropetalum chinense* var. *rubrum* petals to form another type of smart packaging [132]. Yun et al. demonstrated the antioxidant and antimicrobial activity, pH and ammonia sensitivity and its capability as a good freshness indicator. By adding betacyanins into film production, smart packagings are also feasible [133].

Locust bean gum and alginate form an interpenetrated complex, useful for drug delivery purposes because of their swelling behavior and drug release control [134–137]. The matrix is usually prepared by the ionotropic-gelation technique, using the coacervation of alginate with a divalent cation as Ca^{2+}, which yields edible beads from natural raw materials. Aclofenac [135,137], capecitabine [134] and captopril [136] releases were analyzed in these types of systems. The drug delivery study showed a controlled release resulting from these matrices. This technique can also be employed for the encapsulation of tea polyphenols [138]. The preparation of edible packaging is feasible simply by mixing LBG with sodium alginate, integrating daphnetin [139] and using a CO_2 atmosphere [140]. A low bacteria development was observed and a tasteful product was obtained.

Another polysaccharide that can be found mixed with LBG is inulin. This carbohydrate is mainly used in drug delivery, but it is valid for food applications, as studied by Góral et al. [141]. They found that the concentration of polysaccharides will decrease the cryoscopic temperature, the melting time of coconut milk-based ice cream and its hardness. A higher inulin ratio over LBG produced a higher overrun and tasteful ice cream. In the food industry, it is possible to use carboxymethyl cellulose combined with LBG to stabilize unfizzy doogh [142]. Numerous patents between locust bean gum, varied polysaccharides and proteins have been published, mostly for developing sauces or other food recipes. Table 1 summarizes the examples of the mixtures between LBG and other polymers collected in this review.

Table 1. Examples of synergies between LBG and other biobased polymers, method of preparation of the mixtures and their uses.

Biobased Polymer Coupled	Preparation Method	Use	Reference
XG	Mixture	Food industry	[107–109]
XG, Glycerol	Mixture	Edible film	[110]
XG, potato starch	Mixture	3D printing	[111]
XG	Emulsion w/o	Drug delivery	[112]
XG	Emulsion w/o	Encapsulation	[113]
XG, mastic gum	Freeze dried from Mix	Tissue engineering	[114]
XG	Mixture	Binder in green battery	[115]
ι-, κ-Carrageenan, Gelatin	Mixture	Wound healing, Tissue repairing	[123]
κ-Carrageenan	Mixture	Transdermal delivery	[124]
κ-Carrageenan, XG	Mixture	Food 3D printing	[125]
κ-Carrageenan, alginate	Double emulsion $w_1/o/w_2$	Delivery device	[126]
κ-C: Chitosan: PVA	Mixture	Film packaging	[127]
PVA + agar: PVA	Mixture and double layer	Film packaging	[129]
Carboxymethyl Chitosan	Mixture	Film packaging	[130]
Chitosan	Mixture	Biobased films	[131]
PVA	Mixture	Smart packaging	[132,133]
Alginate	Mixture + ionic gelation	Drug delivery	[134–137]
Alginate	Mixture + ionic gelation	Encapsulation	[138]
Alginate	Mixture	Edible packaging	[139,140]
Carboxymethyl Cellulose	Mixture	Food	[142]

It is also feasible to prepare aerogels based on LBG and graphene oxide for water purification applications [143]. Their 3D structure has allowed the sorption of rhodamine-B and successfully remove this dye over indigo carmine thanks to the dye charge.

3. LBG Derivatives

3.1. Modifications of Functional Groups

As specified by Barreto Santos et al. and by Yadav and Maiti, galactomannan polysaccharides can easily be chemically derivatized by functions such as sulfation, carboxylation, or acetylation, for example [13,144]. This derivatization usually employs hazardous chemicals and solvents but produces non-toxic and biodegradable matrices.

Braz et al. decided to prepare different types of modified LBG (Figure 5) [145]. The sulfation of LBG can be performed with SO_3DMF as reported by Braz et al. [146]. The resulting modified polysaccharide can be mixed with chitosan to form solid and compact spherical beads as a promising antigen delivery material. The carboxylation of LBG is also feasible, by using TEMPO and NaBr after an organic reaction. The last described is the grafting of a quaternary ammonium salt thanks to GTMA and HCl. The different LBG materials formed were complexed with a reverse-charged polysaccharide: for carboxylic LBG and sulfated LBG, chitosan was used. Ammonium LBG was complexed with sulfated LBG. The main purpose is to use these complexes for drug delivery applications; their toxicological evaluation shows that the ammonium derivative presents severe cytotoxicity but it reverted when complexed with sulfated LBG.

The carboxymethylation of LBG is possible by using monochloroacetic acid to generate an efficient drug delivery matrix [147]. The synthesis permits a good degree of substitution but with a decrease in the viscosity and the molar mass. As reported by Katy et al., the CMLBG is "safe enough for internal use" and recommends the usage of CMLBG:PVA interpenetrated network microbeads crosslinked with glutaraldehyde for controlled oral drug delivery. A greener way to employ CMLBG is to combine it with alginate and use Al^{3+} as a crosslinker [148]. The resulting IPN features depend on the gelation time, the higher, the better the release behavior. Al^{3+} is also an ionic crosslinker for CMLBG and CMC by single water-in-water emulsion gelation processes with applicability for drug delivery [149]. Glipizide [148,150] and diclofenac [149] releases have been studied using these networks.

LBG can also be derivatized with inorganic components, such as palladium to transform it into a green catalyst [151]. The Pd insertion is made thanks to $Pd(OAc)_2$ reacting with LBG in water and through ultrasonic irradiation at 80 °C. The precipitate after cooling, recovered by adding ethanol, is then filtrated and isolated. Following this procedure, the Pd is inserted on the polysaccharide-reduced ends. Ben Romdhane et al. tried different reactions using Pd@LBG as the catalyst and succeeded in recuperating a good yield and regenerating the catalyst five times. Another example has been prepared by Tagad et al.: an LBG derivatized by gold nanoparticles [152]. The $HAuCl_4$ is introduced into a solution of LBG and autoclaved at 120 °C and 15 psi. 4-nitrophenol to 4-aminophenol reductions were efficiently catalyzed by Au@LBG. When doped with SnO_2, it shows a fast response and good ethanol-sensing behavior.

Singh et al. have successfully grafted polyacrylamide functions onto LBG by microwave-initiated graft copolymerization techniques [153]. They proved its non-morbidity and toxicity in different organs and a controlled release of budesonide in the colon. Jin et al. grafted methyl acrylate and acrylic acid from LBG by Fenton reactions [154]. The grafting affects the viscosity, contact angle, water solubility and mechanical properties. Adhesion to polyester fiber makes it useful for textile applications. Another polymer family grafted from crosslinked LBG by means of divinyl sulfate are polyethyleneimines [155]. Good blood compatibility, an easily modifiable polysaccharide and a good, controlled release behavior made it a promising drug carrier.

Finally, the network structure constituted by sodium acrylate:LBG:N,N'-methylenebisacrylamide has been prepared by irradiation in order to create a superabsorbent polymer [156].

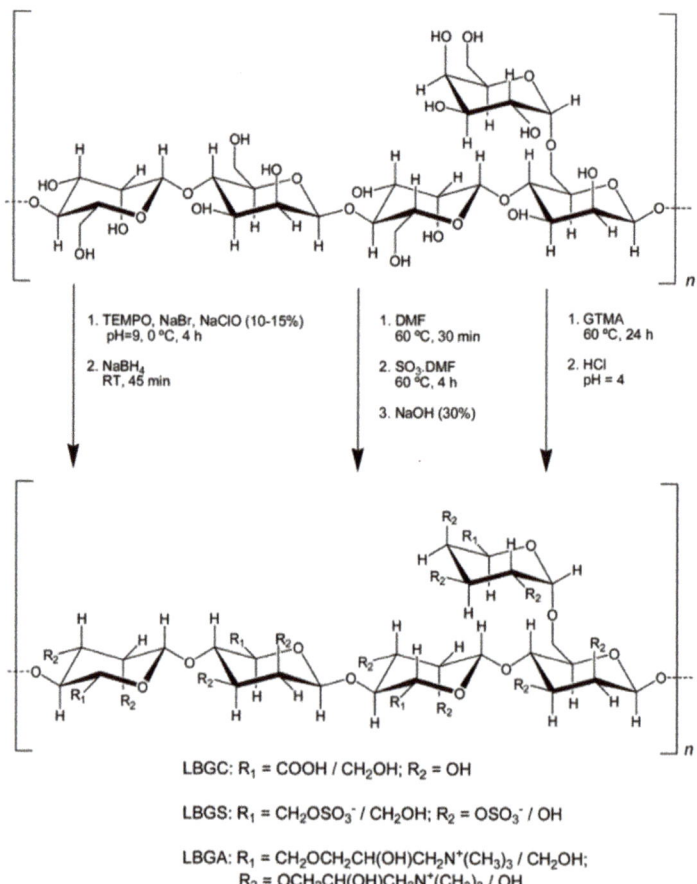

Figure 5. Chemical modification of locust bean gum: LBG carboxylate, LBG sulfate, LBG trimethylammonium [145].

3.2. Crosslinking Reactions

It has been reported that glutaraldehyde is a potential crosslinker for several polysaccharides with chitosan [157]. Jana et al. showed the possibility to form LBG:CS matrices for drug delivery. It suppressed the burst release allowing a sustained release [158]. In addition, glutaraldehyde can crosslink a single LBG to produce drug delivery matrices [159].

Citric acid, used as an LBG crosslinker through solventless reactions with basic catalysts, has been studied by Petitjean et al. [160]. Provided that a temperature above 170 °C and a sufficient reaction time (>20 min) are used, the crosslinking process was produced with a good yield, resulting in a remarkable swelling behavior and dye sorption capabilities. By functionalization of these LBG networks with β-cyclodextrin, specific interactions between some sorbates and LBG are enhanced [161]. Besides, by adding lignin to the initial mixtures, the amounts of polyphenolic compounds sorbed were also significantly increased [162]. Another crosslinking process has been reported by Hadinugroho et al. [163]. LBG was swollen before being UV-cured with citric acid and an acidic catalyst. Then, the resulting product was washed with acetone and dried at ambient temperature. They showed that, under acidic conditions, the protonation of the C6 hydroxyl group of mannose and galactose was easier, and so they concluded that the crosslinking was taking place at this carbon site [163,164]. The disintegration of the tablets produced by these materials has been also studied by this group [165].

4. Conclusions and Perspectives

Among the galactomannans, locust bean gum has a mannose-to-galactose ratio of about 4:1 and a minimally branched structure and it needs heat to fully hydrate. Although it does not gel on its own, LBG forms gels with other hydrocolloids. LBG is non-digestible and may be classified as a soluble fiber. An efficient stabilizer in the food industry, its water-binding and thickening properties promote also its uses to improve the gel properties of some hydrocolloids.

In addition to those well-known applications in the food sector, the feasibility of producing several types of 'green' matrices using LBG, either by itself or derivatized or combined, has also been explored recently. Those materials, produced either by physical entanglements or chemical crosslinking reactions, can be used in other fields, including packaging, biopharmaceutical devices, batteries, catalysts, etc.

As for the hydrocolloid market, LBG prices were similar in 2020 to those of alginates, pectin or agar (ca. 18 $/kg) [9]. Nevertheless, the excellent properties of LBG for some applications in the food industry and its relatively low production volume (well below that of guar gum, for instance) have produced a shortage in the market. Carob trees take over a decade to become productive, so an increasing demand cannot be met simply by planting more trees.

Other alternatives start being used as LBG replacements, and these gum hydrocolloids are often employed in combinations. The question is whether those replacements are satisfactory enough to meet the food and beverage industry needs, so a higher-priced LBG can be used for more "selective" purposes in other fields. In this review, two interesting characteristics of locust bean gum have been highlighted. On the one hand, its synergisms with other biobased polymers are remarkable, widening its range of potential applications. On the other hand, the possibilities of derivatizing its chains and/or creating crosslinking bridges are also of great interest in order to explore other possibilities.

Author Contributions: Conceptualization, M.P. and J.R.I.; methodology, M.P.; writing—original draft preparation, M.P.; writing—review and editing, J.R.I.; supervision, J.R.I. All authors have read and agreed to the published version of the manuscript.

Funding: This work was funded by PIUNA (University of Navarra, ref. 2018-15).

Institutional Review Board Statement: Not applicable.

Informed Consent Statement: Not applicable.

Acknowledgments: M.P. is grateful to the Asociación de Amigos de la Universidad de Navarra for his doctoral grant.

Conflicts of Interest: The authors declare no conflict of interest.

References

1. Gioxari, A.; Amerikanou, C.; Nestoridi, I.; Gourgari, E.; Pratsinis, H.; Kalogeropoulos, N.; Andrikopoulos, N.K.; Kaliora, A.C. Carob: A Sustainable Opportunity for Metabolic Health. *Foods* **2022**, *11*, 2154. [CrossRef]
2. Frühbauerová, M.; Červenka, L.; Hájek, T.; Pouzar, M.; Palarčík, J. Bioaccessibility of Phenolics from Carob (*Ceratonia siliqua* L.) Pod Powder Prepared by Cryogenic and Vibratory Grinding. *Food Chem.* **2022**, *377*, 131968. [CrossRef]
3. Gregoriou, G.; Neophytou, C.M.; Vasincu, A.; Gregoriou, Y.; Hadjipakkou, H.; Pinakoulaki, E.; Christodoulou, M.C.; Ioannou, G.D.; Stavrou, I.J.; Christou, A.; et al. Anti-Cancer Activity and Phenolic Content of Extracts Derived from Cypriot Carob (*Ceratonia siliqua* L.) Pods Using Different Solvents. *Molecules* **2021**, *26*, 5017. [CrossRef]
4. Christou, A.; Martinez-Piernas, A.B.; Stavrou, I.J.; Garcia-Reyes, J.F.; Kapnissi-Christodoulou, C.P. HPLC-ESI-HRMS and Chemometric Analysis of Carobs Polyphenols—Technological and Geographical Parameters Affecting Their Phenolic Composition. *J. Food Compos. Anal.* **2022**, *114*, 104744. [CrossRef]
5. Richane, A.; Ismail, H.B.; Darej, C.; Attia, K.; Moujahed, N. Potential of Tunisian Carob Pulp as Feed for Ruminants: Chemical Composition and In Vitro Assessment. *Trop. Anim. Health Prod.* **2022**, *54*, 58. [CrossRef]

6. Brassesco, M.E.; Brandão, T.R.S.; Silva, C.L.M.; Pintado, M. Carob Bean (*Ceratonia siliqua* L.): A New Perspective for Functional Food. *Trends Food Sci. Technol.* **2021**, *114*, 310–322. [CrossRef]
7. Zhu, B.J.; Zayed, M.Z.; Zhu, H.X.; Zhao, J.; Li, S.P. Functional Polysaccharides of Carob Fruit: A Review. *Chin. Med.* **2019**, *14*, 40. [CrossRef]
8. Deuel, H.; Neukom, H. Some Properties of Locust Bean Gum. In *Natural Plant Hydrocolloids*; American Chemical Society: Washington, DC, USA, 1954; pp. 51–61. [CrossRef]
9. Seisun, D.; Zalesny, N. Strides in Food Texture and Hydrocolloids. *Food Hydrocoll.* **2021**, *117*, 106575. [CrossRef]
10. Majee, S.B.; Avlani, D.; Biswas, G.R. Non-Starch Plant Polysaccharides: Physicochemical Modifications and Pharmaceutical Applications. *J. Appl. Pharm. Sci.* **2016**, *6*, 231–241. [CrossRef]
11. Ahmad, S.; Ahmad, M.; Manzoor, K.; Purwar, R.; Ikram, S. A Review on Latest Innovations in Natural Gums Based Hydrogels: Preparations & Applications. *Int. J. Biol. Macromol.* **2019**, *136*, 870–890. [CrossRef]
12. Gillet, S.; Blecker, C.; Paquot, M.; Richel, A. Relationship between Chemical Structure and Physical Properties in Carob Galactomannans. *Comptes Rendus Chim.* **2014**, *17*, 386–401. [CrossRef]
13. Yadav, H.; Maiti, S. Research Progress in Galactomannan-Based Nanomaterials: Synthesis and Application. *Int. J. Biol. Macromol.* **2020**, *163*, 2113–2126. [CrossRef]
14. Dea, I.C.M.; Morrison, A. Chemistry and Interactions of Seed Galactomannans. *Adv. Carbohydr. Chem. Biochem.* **1975**, *31*, 241–312. [CrossRef]
15. Blakeney, A.B.; Harris, P.J.; Henry, R.J.; Stone, B.A. A Simple and Rapid Preparation of Alditol Acetates for Monosaccharide Analysis. *Carbohydr. Res.* **1983**, *113*, 291–299. [CrossRef]
16. McCleary, B.V.; Clark, A.H.; Dea, I.C.M.; Rees, D.A. The Fine Structures of Carob and Guar Galactomannans. *Carbohydr. Res.* **1985**, *139*, 237–260. [CrossRef]
17. Daas, P.J.H.; Schols, H.A.; De Jongh, H.H.J. On the Galactosyl Distribution of Commercial Galactomannans. *Carbohydr. Res.* **2000**, *329*, 609–619. [CrossRef]
18. Wielinga, W.C.; Meyhall, A.G. Galactomannans. In *Handbook of Hydrocolloids*, 2nd ed.; Woodhead Publishing: Sawston, UK, 2009; pp. 228–251. [CrossRef]
19. Baker, C.W.; Whistler, R.L. Distribution of D-Galactosyl Groups in Guaran and Locust-Bean Gum. *Carbohydr. Res.* **1975**, *45*, 237–243. [CrossRef]
20. Painter, T.J.; González, J.J.; Hemmer, P.C. The Distribution of D-Galactosyl Groups in Guaran and Locust-Bean Gum: New Evidence from Periodate Oxidation. *Carbohydr. Res.* **1979**, *69*, 217–226. [CrossRef]
21. Gidley, M.J.; McArthur, A.J.; Underwood, D.R. 13C NMR Characterization of Molecular Structures in Powders, Hydrates and Gels of Galactomannans and Glucomannans. *Top. Catal.* **1991**, *5*, 129–140. [CrossRef]
22. Hoffman, J.; Lindberg, B.; Painter, T. The Distribution of the D-Galactose Residues in Guaran and Locust Bean Gum. *Acta Chem. Scand.* **1976**, *30b*, 365–376. [CrossRef]
23. Smith, F. The Constitution of Carob Gum. *J. Am. Chem. Soc.* **1948**, *70*, 3249–3253. [CrossRef]
24. Chudzikowski, R.J. Guar Gum and Its Applications. *J. Soc. Cosmet. Chem.* **1971**, *22*, 43–60.
25. Maier, H.; Anderson, M.; Karl, C.; Magnuson, K.; Whistler, R.L. Guar, Locust Bean, Tara and Fenugreek Gums. In *Industrial Gums*; Elsevier: Amsterdam, The Netherlands, 1993; pp. 181–226. ISBN 9780127462530.
26. Mirhosseini, H.; Amid, B.T. A Review Study on Chemical Composition and Molecular Structure of Newly Plant Gum Exudates and Seed Gums. *Food Res. Int.* **2012**, *46*, 387–398. [CrossRef]
27. Ray, P.M.; Shininger, T.L.; Ray, M.M. Isolation of β-Glucan Synthetase Particles From Plant Cells and Identification With Golgi Membranes. *Proc. Natl. Acad. Sci. USA* **1969**, *64*, 605–612. [CrossRef]
28. Sharma, P.; Sharma, S.; Ramakrishna, G.; Srivastava, H.; Gaikwad, K. A Comprehensive Review on Leguminous Galactomannans: Structural Analysis, Functional Properties, Biosynthesis Process and Industrial Applications. *Crit. Rev. Food Sci. Nutr.* **2020**, *62*, 443–465. [CrossRef]
29. Dhugga, K.S.; Barreiro, R.; Whitten, B.; Stecca, K.; Hazebroek, J.; Randhawa, G.S.; Dolan, M.; Kinney, A.J.; Tomes, D.; Nichols, S.; et al. Guar Seed β-Mannan Synthase Is a Member of the Cellulose Synthase Super Gene Family. *Science* **2004**, *303*, 363–366. [CrossRef]
30. Edwards, M.; Bulpin, P.V.; Dea, I.C.M.; Reid, J.S.G. Biosynthesis of Legume-Seed Galactomannans in Vitro. *Planta* **1989**, *178*, 41–51. [CrossRef]
31. Reid, J.S.G.; Edwards, M.; Dea, I.C.M. Biosynthesis of Galactomannan in the Endosperms of Developing Fenugreek (*Trigonella foenum-graecum* L.) and Guar (*Cyamopsis tetragonoloba* [L.] Taub.) Seeds. *Top. Catal.* **1987**, *1*, 381–385. [CrossRef]
32. Edwards, M.; Scott, C.; Gidley, M.; Reid, J.S.G. Control of Mannose/Galactose Ratio during Galactomannan Formation in Developing Legume Seeds. *Planta* **1992**, *187*, 67–74. [CrossRef]
33. Bouzouita, N.; Khaldi, A.; Zgoulli, S.; Chebil, L.; Chekki, R.; Chaabouni, M.M.; Thonart, P. The Analysis of Crude and Purified Locust Bean Gum: A Comparison of Samples from Different Carob Tree Populations in Tunisia. *Food Chem.* **2007**, *101*, 1508–1515. [CrossRef]
34. Haddarah, A.; Bassal, A.; Ismail, A.; Gaiani, C.; Ioannou, I.; Charbonnel, C.; Hamieh, T.; Ghoul, M. The Structural Characteristics and Rheological Properties of Lebanese Locust Bean Gum. *J. Food Eng.* **2014**, *120*, 204–214. [CrossRef]
35. Yousif, A.K.; Alghzawi, H.M. Processing and Characterization of Carob Powder. *Food Chem.* **2000**, *69*, 283–287. [CrossRef]

36. Dakia, P.A.; Blecker, C.; Robert, C.; Wathelet, B.; Paquot, M. Composition and Physicochemical Properties of Locust Bean Gum Extracted from Whole Seeds by Acid or Water Dehulling Pre-Treatment. *Food Hydrocoll.* **2008**, *22*, 807–818. [CrossRef]
37. Biner, B.; Gubbuk, H.; Karhan, M.; Aksu, M.; Pekmezci, M. Sugar Profiles of the Pods of Cultivated and Wild Types of Carob Bean (*Ceratonia siliqua* L.) in Turkey. *Food Chem.* **2007**, *100*, 1453–1455. [CrossRef]
38. McCleary, B.V.; Matheson, N.K. α-d-Galactosidase Activity and Galactomannan and Galactosylsucrose Oligosaccharide Depletion in Germinating Legume Seeds. *Phytochemistry* **1974**, *13*, 1747–1757. [CrossRef]
39. Anderson, E. Endosperm Mucilages of Legumes. *Ind. Eng. Chem.* **1949**, *41*, 2887–2890. [CrossRef]
40. Rafique, C.M.; Smith, F. The Constitution of Guar Gum. *J. Am. Chem. Soc.* **1950**, *72*, 4634–4637. [CrossRef]
41. McCleary, B.V.; Matheson, N.K.; Small, D.M. Galactomannans and a Galactoglucomannan in Legume Seed Endosperms: Structural Requirements for β-Mannanase Hydrolysis. *Phytochemistry* **1976**, *15*, 1111–1117. [CrossRef]
42. Kapoor, V.P. A Galactomannan from the Seeds of Delonix Regia. *Phytochemistry* **1972**, *11*, 1129–1132. [CrossRef]
43. Azero, E.G.; Andrade, C.T. Testing Procedures for Galactomannan Purification. *Polym. Test.* **2002**, *21*, 551–556. [CrossRef]
44. Grimaud, F.; Pizzut-Serin, S.; Tarquis, L.; Ladevèze, S.; Morel, S.; Putaux, J.L.; Potocki-Veronese, G. In Vitro Synthesis and Crystallization of β-1,4-Mannan. *Biomacromolecules* **2019**, *20*, 846–853. [CrossRef] [PubMed]
45. Petkowicz, C.L.O.; Milas, M.; Mazeau, K.; Bresolin, T.; Reicher, F.; Ganter, J.L.M.S.; Rinaudo, M. Conformation of Galactomannan: Experimental and Modelling Approaches. *Food Hydrocoll.* **1999**, *13*, 263–266. [CrossRef]
46. Wu, Y.; Li, W.; Cui, W.; Eskin, N.A.M.; Goff, H.D. A Molecular Modeling Approach to Understand Conformation-Functionality Relationships of Galactomannans with Different Mannose/Galactose Ratios. *Food Hydrocoll.* **2012**, *26*, 359–364. [CrossRef]
47. Zhang, X.; Dai, L.; Li, P.; Wang, T.; Qin, L.; Xiang, J.; Chang, H. Characterization of Hydrophobic Interaction of Galactomannan in Aqueous Solutions Using Fluorescence-Based Technique. *Carbohydr. Polym.* **2021**, *267*, 118183. [CrossRef]
48. Petkowicz, C.L.O.; Reicher, F.; Mazeau, K. Conformational Analysis of Galactomannans: From Oligomeric Segments to Polymeric Chains. *Carbohydr. Polym.* **1998**, *37*, 25–39. [CrossRef]
49. Mansel, B.W.; Ryan, T.M.; Chen, H.L.; Lundin, L.; Williams, M.A.K. Polysaccharide Conformations Measured by Solution State X-ray Scattering. *Chem. Phys. Lett.* **2020**, *739*, 136951. [CrossRef]
50. Feng, L.; Yin, J.; Nie, S.; Wan, Y.; Xie, M. Structure and Conformation Characterization of Galactomannan from Seeds of *Cassia obtusifolia*. *Food Hydrocoll.* **2018**, *76*, 67–77. [CrossRef]
51. Xu, W.; Liu, Y.; Zhang, F.; Lei, F.; Wang, K.; Jiang, J. Physicochemical Characterization of Gleditsia Triacanthos Galactomannan during Deposition and Maturation. *Int. J. Biol. Macromol.* **2020**, *144*, 821–828. [CrossRef]
52. Gaisford, S.E.; Harding, S.E.; Mitchell, J.R.; Bradley, T.D. A Comparison between the Hot and Cold Water Soluble Fractions of Two Locust Bean Gum Samples. *Carbohydr. Polym.* **1986**, *6*, 423–442. [CrossRef]
53. Kök, M.S.; Hill, S.E.; Mitchell, J.R. Viscosity of Galactomannans during High Temperature Processing: Influence of Degradation and Solubilisation. *Food Hydrocoll.* **1999**, *13*, 535–542. [CrossRef]
54. Pollard, M.; Kelly, R.; Illmann, S.; Fischer, P. Improved Rheological Properties of Biopolymers through Innovations in Raw Material Processing. In Proceedings of the 6th Annual European Rheology Conference (AERC 2010), Göteborg, Sweden, 7–9 April 2010.
55. Pollard, M.A.; Fischer, P. Partial Aqueous Solubility of Low-Galactose-Content Galactomannans—What Is the Quantitative Basis? *Curr. Opin. Colloid Interface Sci.* **2006**, *11*, 184–190. [CrossRef]
56. Brummer, Y.; Cui, W.; Wang, Q. Extraction, Purification and Physicochemical Characterization of Fenugreek Gum. *Food Hydrocoll.* **2003**, *17*, 229–236. [CrossRef]
57. Marguerite, R. Relation between the Molecular Structure of Some Polysaccharides and Original Properties in Sol and Gel States. *Food Hydrocoll.* **2001**, *15*, 433–440.
58. Dea, I.C.M.; Clark, A.H.; McCleary, B.V. Effect of the Molecular Fine Structure of Galactomannans on Their Interaction Properties—The Role of Unsubstituted Sides. *Top. Catal.* **1986**, *1*, 129–140. [CrossRef]
59. Richardson, P.H.; Willmer, J.; Foster, T.J. Dilute Solution Properties of Guar and Locust Bean Gum in Sucrose Solutions. *Food Hydrocoll.* **1998**, *12*, 339–348. [CrossRef]
60. Doublier, J.L.; Launay, B. Rheology of Galactomannan Solutions: Comparative Study of Guar Gum and Locust Bean Gum. *J. Texture Stud.* **1981**, *12*, 151–172. [CrossRef]
61. Morris, E.R.; Cutler, A.N.; Ross-Murphy, S.B.; Rees, D.A.; Price, J. Concentration and Shear Rate Dependence of Viscosity in Random Coil Polysaccharide Solutions. *Carbohydr. Polym.* **1981**, *1*, 5–21. [CrossRef]
62. Clark, A.H.; Ross-Murphy, S.B. Structural and Mechanical Properties of Biopolymer Gels. In *Biopolymers*; Springer: Berlin/Heidelberg, Germany, 1987; pp. 57–192. [CrossRef]
63. De Gennes, P.G. Brownian Motions of Flexible Polymer Chains. *Nature* **1979**, *282*, 367–370. [CrossRef]
64. Sittikijyothin, W.; Torres, D.; Gonçalves, M.P. Modelling the Rheological Behaviour of Galactomannan Aqueous Solutions. *Carbohydr. Polym.* **2005**, *59*, 339–350. [CrossRef]
65. Rizzo, V.; Tomaselli, F.; Gentile, A.; La Malfa, S.; Maccarone, E. Rheological Properties and Sugar Composition of Locust Bean Gum from Different Carob Varieties (*Ceratonia siliqua* L.). *J. Agric. Food Chem.* **2004**, *52*, 7925–7930. [CrossRef]
66. Izydorczyk, M.S.; Biliaderis, C.G. Gradient Ammonium Sulphate Fractionation of Galactomannans. *Food Hydrocoll.* **1996**, *10*, 295–300. [CrossRef]

67. Andrade, C.T.; Azero, E.G.; Luciano, L.; Gonçalves, M.P. Solution Properties of the Galactomannans Extracted from the Seeds of Caesalpinia Pulcherrima and Cassia Javanica: Comparison with Locust Bean Gum. *Int. J. Biol. Macromol.* **1999**, *26*, 181–185. [CrossRef]
68. Garcia-Ochoa, F.; Casas, J. Viscosity of Locust Bean Gum Solutions. *J. Sci. Food Agric.* **1992**, *59*, 97–100. [CrossRef]
69. Kapoor, V.P.; Milas, M.; Taravel, F.R.; Rinaudo, M. Rheological Properties of Seed Galactomannan from *Cassia nodosa* Buch.-Hem. *Carbohydr. Polym.* **1994**, *25*, 79–84. [CrossRef]
70. Pollard, M.A.; Kelly, R.; Wahl, C.; Fischer, P.; Windhab, E.; Eder, B.; Amadó, R. Investigation of Equilibrium Solubility of a Carob Galactomannan. *Food Hydrocoll.* **2007**, *21*, 683–692. [CrossRef]
71. Gidley, M.; Grant Reid, J. Galactomannans and Other Cell Wall Storage Polysaccharides in Seeds. In *Food Polysaccharides and Their Applications*; CRC Press: Boca Raton, FL, USA, 2006; pp. 181–215.
72. Alves, M.M.; Antonov, Y.A.; Gonçalves, M.P. The Effect of Structural Features of Gelatin on Its Thermodynamic Compatibility with Locust Bean Gum in Aqueous Media. *Food Hydrocoll.* **1999**, *13*, 157–166. [CrossRef]
73. Bahramparvar, M.; Tehrani, M.M. Application and Functions of Stabilizers in Ice Cream. *Food Rev. Int.* **2011**, *27*, 389–407. [CrossRef]
74. Blibech, M.; Maktouf, S.; Chaari, F.; Zouari, S.; Neifar, M.; Besbes, S.; Ellouze-Ghorbel, R. Functionality of Galactomannan Extracted from Tunisian Carob Seed in Bread Dough. *J. Food Sci. Technol.* **2015**, *52*, 423–429. [CrossRef]
75. Rojas-Argudo, C.; del Río, M.A.; Pérez-Gago, M.B. Development and Optimization of Locust Bean Gum (LBG)-Based Edible Coatings for Postharvest Storage of "Fortune" Mandarins. *Postharvest Biol. Technol.* **2009**, *52*, 227–234. [CrossRef]
76. Abulyazid, I.; Abd Elhalim, S.A.; Sharada, H.M.; Aboulthana, W.M.; Abd Elhalim, S.T.A. Hepatoprotective Effect of Carob Pods Extract (*Ceratonia siliqua* L.) against Cyclophosphamide Induced Alterations in Rats. *Int. J. Curr. Pharm. Rev. Res.* **2017**, *8*, 149–162. [CrossRef]
77. Chait, Y.A.; Gunenc, A.; Bendali, F.; Hosseinian, F. Simulated Gastrointestinal Digestion and In Vitro Colonic Fermentation of Carob Polyphenols: Bioaccessibility and Bioactivity. *LWT Food Sci. Technol.* **2020**, *117*, 108623. [CrossRef]
78. Macho-González, A.; Garcimartín, A.; Naes, F.; López-Oliva, M.E.; Amores-Arrojo, A.; González-Muñoz, M.J.; Bastida, S.; Benedí, J.; Sánchez-Muniz, F.J. Effects of Fiber Purified Extract of Carob Fruit on Fat Digestion and Postprandial Lipemia in Healthy Rats. *J. Agric. Food Chem.* **2018**, *66*, 6734–6741. [CrossRef] [PubMed]
79. Xie, J.; Wang, Z.; Cui, H.; Nie, H.; Zhang, T.; Gao, X.; Qiao, Y. Effects of Enzymatic Hydrolysate of Locust Bean Gum on Digestibility, Intestinal Morphology and Microflora of Broilers. *J. Anim. Physiol. Anim. Nutr.* **2020**, *104*, 230–236. [CrossRef]
80. Prajapati, V.D.; Jani, G.K.; Moradiya, N.G.; Randeria, N.P.; Nagar, B.J. Locust Bean Gum: A Versatile Biopolymer. *Carbohydr. Polym.* **2013**, *94*, 814–821. [CrossRef] [PubMed]
81. Verma, A.; Tiwari, A.; Panda, P.K.; Saraf, S.; Jain, A.; Jain, S.K. Locust Bean Gum in Drug Delivery Application. In *Natural Polysaccharides in Drug Delivery and Biomedical Applications*; Elsevier: Amsterdam, The Netherlands, 2019; pp. 203–222.
82. Sudhakar, Y.; Kuotsu, K.; Bandyopadhyay, A.K. Buccal Bioadhesive Drug Delivery—A Promising Option for Orally Less Efficient Drugs. *J. Control. Release* **2006**, *114*, 15–40. [CrossRef]
83. Sujja-Areevath, J.; Munday, D.L.; Cox, P.J.; Khan, K.A. Relationship between Swelling, Erosion and Drug Release in Hydrophillic Natural Gum Mini-Matrix Formulations. *Eur. J. Pharm. Sci.* **1998**, *6*, 207–217. [CrossRef]
84. Alonso, M.J.; Torres, D.; Cun, M. Preparation and in Vivo Evaluation of Mucoadhesive Microparticles Containing Amoxycillin ± Resin Complexes for Drug Delivery to the Gastric Mucosa. *Eur. J. Pharm. Biopharm.* **2001**, *51*, 199–205.
85. Hirsch, S.; Binder, V.; Schehlmann, V.; Kolter, K.; Bauer, K.H. Lauroyldextran and Crosslinked Galactomannan as Coating Materials for Site-Specific Drug Delivery to the Colon. *Eur. J. Pharm. Biopharm.* **1999**, *47*, 61–71. [CrossRef]
86. Suzuki, S.; Lim, J. Mixture in a Multiphase Emulsification Technique for Sustained Drug Release. *J. Microencapsul.* **1994**, *1*, 197–203. [CrossRef]
87. Marianecci, C.; Carafa, M.; di Marzio, L.; Rinaldi, F.; di Meo, C.; Alhaique, F.; Matricardi, P.; Coviello, T. A New Vesicle-Loaded Hydrogel System Suitable for Topical Applications: Preparation and Characterization. *J. Pharm. Pharm. Sci.* **2011**, *14*, 336–346. [CrossRef]
88. Henderson, T.M.A.; Ladewig, K.; Haylock, D.N.; McLean, K.M.; O'Connor, A.J. Cryogels for Biomedical Applications. *J. Mater. Chem. B* **2013**, *1*, 2682–2695. [CrossRef] [PubMed]
89. Liu, B.; Huang, Y.; Wang, J.; Li, Z.; Yang, G.; Jin, S.; Iranmanesh, E.; Hiralal, P.; Zhou, H. Highly Conductive Locust Bean Gum Bio-Electrolyte for Superior Long-Life Quasi-Solid-State Zinc-Ion Batteries. *RSC Adv.* **2021**, *11*, 24862–24871. [CrossRef] [PubMed]
90. Dionísio, M.; Grenha, A. Locust Bean Gum: Exploring Its Potential for Biopharmaceutical Applications. *J. Pharm. Bioallied Sci.* **2012**, *4*, 175–185. [CrossRef]
91. Rocks, J. Xanthan Gum. *Enzym. Food Technol.* **1971**, *25*, 476–483.
92. Schorsch, C.; Garnier, C.; Doublier, J.L. Microscopy of Xanthan/Galactomannan Mixtures. *Carbohydr. Polym.* **1995**, *28*, 319–323. [CrossRef]
93. Shatwell, K.P.; Sutherland, I.W.; Ross-Murphy, S.B. Influence of Acetyl and Pyruvate Substituents on the Solution Properties of Xanthan Polysaccharide. *Int. J. Biol. Macromol.* **1990**, *12*, 71–78. [CrossRef]
94. Dea, I.C.M.; Morris, E.R.; Rees, D.A.; Welsh, E.J.; Barnes, H.A.; Price, J. Associations of like and Unlike Polysaccharides: Mechanism and Specificity in Galactomannans, Interacting Bacterial Polysaccharides, and Related Systems. *Carbohydr. Res.* **1977**, *57*, 249–272. [CrossRef]

95. Cuvelier, G.; Tonon, C.; Launay, B. Xanthan—Carob Mixtures at Low Concentration: Viscosimetric Study. *Top. Catal.* **1987**, *1*, 583–585. [CrossRef]
96. Morris, E.R.; Rees, D.A.; Young, G.; Walkinshaw, M.D.; Darke, A. Order-Disorder Transition for a Bacterial Polysaccharide in Solution. A Role for Polysaccharide Conformation in Recognition between Xanthomonas Pathogen and Its Plant Host. *J. Mol. Biol.* **1977**, *110*, 1–16. [CrossRef]
97. Cairns, P.; Miles, M.J.; Morris, V.J. Intermolecular Binding of Xanthan Gum and Carob Gum. *Nature* **1986**, *322*, 89–90. [CrossRef]
98. Takemasa, M.; Nishinari, K. Solution Structure of Molecular Associations Investigated Using NMR for Polysaccharides: Xanthan/Galactomannan Mixtures. *J. Phys. Chem. B* **2016**, *120*, 3027–3037. [CrossRef] [PubMed]
99. García-Ochoa, F.; Santosa, V.E.; Casas, J.A.; Gómez, E. Xanthan gum: Production, recovery, and properties. *Biotechnol. Adv.* **2000**, *18*, 549–579. [CrossRef] [PubMed]
100. Grisel, M.; Aguni, Y.; Renou, F.; Malhiac, C. Impact of fine structure of galactomannans on their interactions with xanthan: Two co-existing mechanisms to explain the synergy. *Food Hydrocoll.* **2015**, *51*, 449–458. [CrossRef]
101. Tako, M.; Teruya, T.; Tamaki, Y.; Ohkawa, K. Co-gelation mechanism of xanthan and galactomannan. *Colloid Poly. Sci.* **2010**, *288*, 1161–1166. [CrossRef]
102. Jo, W.; Bak, J.H.; Yoo, B. Rheological Characterizations of Concentrated Binary Gum Mixtures with Xanthan Gum and Galactomannans. *Int. J. Biol. Macromol.* **2018**, *114*, 263–269. [CrossRef]
103. Hayta, M.; Dogan, M.; Aslan Türker, D. Rheology and Microstructure of Galactomannan–Xanthan Gum Systems at Different PH Values. *J. Food Process Eng.* **2020**, *43*, e13573. [CrossRef]
104. Schreiber, C.; Ghebremedhin, M.; Zielbauer, B.; Dietz, N.; Vilgis, T.A. Interaction of Xanthan Gums with Galacto- And Glucomannans. Part I: Molecular Interactions and Synergism in Cold Gelled Systems. *J. Phys. Mater.* **2020**, *3*, 034013. [CrossRef]
105. Craig, D.Q.M.; Kee, A.; Tamburic, S.; Barnes, D. An Investigation into the Temperature Dependence of the Rheological Synergy between Xanthan Gum and Locust Bean Gum Mixtures. *J. Biomater. Sci. Polym. Ed.* **1997**, *8*, 377–389. [CrossRef]
106. Ghebremedhin, M.; Schreiber, C.; Zielbauer, B.; Dietz, N.; Vilgis, T.A. Interaction of Xanthan Gums with Galacto- And Glucomannans. Part II: Heat Induced Synergistic Gelation Mechanism and Their Interaction with Salt. *J. Phys. Mater.* **2020**, *3*, 034014. [CrossRef]
107. Valencia, G.A.; Zare, E.N.; Makvandi, P.; Gutiérrez, T.J. Self-Assembled Carbohydrate Polymers for Food Applications: A Review. *Compr. Rev. Food Sci. Food Saf.* **2019**, *18*, 2009–2024. [CrossRef]
108. Dolz, M.; Hernández, M.J.; Delegido, J. Oscillatory Measurements for Salad Dressings Stabilized with Modified Starch, Xanthan Gum, and Locust Bean Gum. *J. Appl. Polym. Sci.* **2006**, *102*, 897–903. [CrossRef]
109. Arocas, A.; Sanz, T.; Fiszman, S.M. Food Hydrocolloids Improving Effect of Xanthan and Locust Bean Gums on the Freeze-Thaw Stability of White Sauces Made with Different Native Starches. *Food Hydrocoll.* **2009**, *23*, 2478–2484. [CrossRef]
110. Kurt, A.; Toker, O.S.; Tornuk, F. Effect of Xanthan and Locust Bean Gum Synergistic Interaction on Characteristics of Biodegradable Edible Film. *Int. J. Biol. Macromol.* **2017**, *102*, 1035–1044. [CrossRef]
111. Yu, H.; Chi, S.; Li, D.; Wang, L.; Wang, Y. Effect of Gums on the Multi-Scale Characteristics and 3D Printing Performance of Potato Starch Gel. *Innov. Food Sci. Emerg. Technol.* **2022**, *80*, 103102. [CrossRef]
112. Sharma, N.; Deshpande, R.D.; Sharma, D.; Sharma, R.K. Development of Locust Bean Gum and Xanthan Gum Based Biodegradable Microparticles of Celecoxib Using a Central Composite Design and Its Evaluation. *Ind. Crops Prod.* **2016**, *82*, 161–170. [CrossRef]
113. Tian, H.; Xiang, D.; Wang, B.; Zhang, W.; Li, C. Using Hydrogels in Dispersed Phase of Water-in-Oil Emulsion for Encapsulating Tea Polyphenols to Sustain Their Release. *Colloids Surfaces A Physicochem. Eng. Asp.* **2021**, *612*, 125999. [CrossRef]
114. Bektas, E.I.; Gurel Pekozer, G.; Kök, F.N.; Torun Kose, G. Evaluation of Natural Gum-Based Cryogels for Soft Tissue Engineering. *Carbohydr. Polym.* **2021**, *271*, 118407. [CrossRef]
115. Yang, S.; Zhou, C.; Wang, Q.; Chen, B.; Zhao, Y.; Guo, B.; Zhang, Z.; Gao, X.; Chowdhury, R.; Wang, H.; et al. Highly Aligned Ultra-Thick Gel-Based Cathodes Unlocking Ultra-High Energy Density Batteries. *Energy Environ. Mater.* **2022**, *5*, 1332–1339. [CrossRef]
116. Turquois, T.; Rochas, C.; Taravel, F.R. Rheological Studies of Synergistic Kappa Carrageenan-Carob Galactomannan Gels. *Carbohydr. Polym.* **1992**, *17*, 263–268. [CrossRef]
117. Pinheiro, A.C.; Bourbon, A.I.; Rocha, C.; Ribeiro, C.; Maia, J.M.; Gonalves, M.P.; Teixeira, J.A.; Vicente, A.A. Rheological Characterization of κ-Carrageenan/Galactomannan and Xanthan/Galactomannan Gels: Comparison of Galactomannans from Non-Traditional Sources with Conventional Galactomannans. *Carbohydr. Polym.* **2011**, *83*, 392–399. [CrossRef]
118. Rochas, C.; Taravel, F.R.; Turquois, T.N.m.r. Studies of Synergistic Kappa Carrageenan-Carob Galactomannan Gels. *Int. J. Biol. Macromol.* **1990**, *12*, 353–358. [CrossRef] [PubMed]
119. Fernandes, P.B.; Gonçalves, M.P.; Doublier, J.L. Influence of Locust Bean Gum on the Rheological Properties of Kappa-Carrageenan Systems in the Vicinity of the Gel Point. *Carbohydr. Polym.* **1993**, *22*, 99–106. [CrossRef]
120. Fernandes, P.B.; Gonçales, M.P.; Doublier, J.-L. Rheological Behaviour of Kappa-Carrageenan/Galactomannan Mixtures At a Very Low Level of Kappa-Carrageenan. *J. Texture Stud.* **1994**, *25*, 267–283. [CrossRef]
121. Viebke, C. A Light Scattering Study of Carrageenan/Galactomannan Interactions. *Carbohydr. Polym.* **1995**, *28*, 101–105. [CrossRef]
122. Turquois, T.; Rochas, C.; Taravel, F.-R.; Doublier, J.L.; Axelos, M.-A.-V. Small-angle X-ray Scattering of K-carrageenan Based Systems: Sols, Gels, and Blends with Carob Galactomannan. *Biopolymers* **1995**, *36*, 559–567. [CrossRef]

123. Pettinelli, N.; Rodríguez-Llamazares, S.; Bouza, R.; Barral, L.; Feijoo-Bandín, S.; Lago, F. Carrageenan-Based Physically Crosslinked Injectable Hydrogel for Wound Healing and Tissue Repairing Applications. *Int. J. Pharm.* **2020**, *589*, 119828. [CrossRef]
124. Mendes de Moraes, F.; Trauthman, S.C.; Zimmer, F.; Pacheco, P.P.; Pont Morisso, F.D.; Ziulkoski, A.L.; Kanis, L.A.; Modolon Zepon, K.M. A Polysaccharide-Based Hydrogel as a Green Platform for Enhancing Transdermal Delivery. *Sustain. Chem. Pharm.* **2022**, *25*, 100604. [CrossRef]
125. Pant, A.; Lee, A.Y.; Karyappa, R.; Lee, C.P.; An, J.; Hashimoto, M.; Tan, U.X.; Wong, G.; Chua, C.K.; Zhang, Y. 3D Food Printing of Fresh Vegetables Using Food Hydrocolloids for Dysphagic Patients. *Food Hydrocoll.* **2021**, *114*, 106546. [CrossRef]
126. Wang, W.; Sun, R.; Xia, Q. Influence of Gelation of Internal Aqueous Phase on In Vitro Controlled Release of W1/O/W2 Double Emulsions-Filled Alginate Hydrogel Beads. *J. Food Eng.* **2022**, *337*, 111246. [CrossRef]
127. Yong, H.; Liu, J.; Kan, J.; Liu, J. Active/Intelligent Packaging Films Developed by Immobilizing Anthocyanins from Purple Sweetpotato and Purple Cabbage in Locust Bean Gum, Chitosan and κ-Carrageenan-Based Matrices. *Int. J. Biol. Macromol.* **2022**, *211*, 238–248. [CrossRef]
128. Matar, G.H.; Andac, M.; Elmas, A. Locust Bean Gum-Polyvinyl Alcohol Hydrogels: Synthesis, Characterization, Swelling Behaviors, and Mathematical Models. *J. Appl. Polym. Sci.* **2022**, *139*, 51498. [CrossRef]
129. Yao, X.; Yun, D.; Xu, F.; Chen, D.; Liu, J. Development of Shrimp Freshness Indicating Films by Immobilizing Red Pitaya Betacyanins and Titanium Dioxide Nanoparticles in Polysaccharide-Based Double-Layer Matrix. *Food Packag. Shelf Life* **2022**, *33*, 100871. [CrossRef]
130. Yu, H.; Zhang, C.; Xie, Y.; Mei, J.; Xie, J. Effect of *Melissa officinalis* L. Essential Oil Nanoemulsions on Structure and Properties of Carboxymethyl Chitosan/Locust Bean Gum Composite Films. *Membranes* **2022**, *12*, 568. [CrossRef]
131. Grala, D.; Biernacki, K.; Freire, C.; Kuźniarska-Biernacka, I.; Souza, H.K.S.; Gonçalves, M.P. Effect of Natural Deep Eutectic Solvent and Chitosan Nanoparticles on Physicochemical Properties of Locust Bean Gum Films. *Food Hydrocoll.* **2022**, *126*, 107460. [CrossRef]
132. Yun, D.; He, Y.; Zhu, H.; Hui, Y.; Li, C.; Chen, D.; Liu, J. Smart Packaging Films Based on Locust Bean Gum, Polyvinyl Alcohol, the Crude Extract of *Loropetalum chinense* Var. *rubrum* Petals and Its Purified Fractions. *Int. J. Biol. Macromol.* **2022**, *205*, 141–153. [CrossRef] [PubMed]
133. Wu, Y.; Tang, P.; Quan, S.; Zhang, H.; Wang, K.; Liu, J. Preparation, Characterization and Application of Smart Packaging Films Based on Locust Bean Gum/Polyvinyl Alcohol Blend and Betacyanins from Cockscomb (*Celosia cristata* L.) Flower. *Int. J. Biol. Macromol.* **2021**, *191*, 679–688. [CrossRef]
134. Upadhyay, M.; Adena, S.K.R.; Vardhan, H.; Yadav, S.K.; Mishra, B. Development of Biopolymers Based Interpenetrating Polymeric Network of Capecitabine: A Drug Delivery Vehicle to Extend the Release of the Model Drug. *Int. J. Biol. Macromol.* **2018**, *115*, 907–919. [CrossRef] [PubMed]
135. Prajapati, V.D.; Jani, G.K.; Moradiya, N.G.; Randeria, N.P.; Maheriya, P.M.; Nagar, B.J. Locust Bean Gum in the Development of Sustained Release Mucoadhesive Macromolecules of Aceclofenac. *Carbohydr. Polym.* **2014**, *113*, 138–148. [CrossRef]
136. Pawar, H.A.; Lalitha, K.G.; Ruckmani, K. Alginate Beads of Captopril Using Galactomannan Containing Senna Tora Gum, Guar Gum and Locust Bean Gum. *Int. J. Biol. Macromol.* **2015**, *76*, 119–131. [CrossRef]
137. Jana, S.; Gandhi, A.; Sheet, S.; Sen, K.K. Metal Ion-Induced Alginate-Locust Bean Gum IPN Microspheres for Sustained Oral Delivery of Aceclofenac. *Int. J. Biol. Macromol.* **2015**, *72*, 47–53. [CrossRef]
138. Belščak-Cvitanović, A.; Jurić, S.; Đorđević, V.; Barišić, L.; Komes, D.; Ježek, D.; Bugarski, B.; Nedović, V. Chemometric Evaluation of Binary Mixtures of Alginate and Polysaccharide Biopolymers as Carriers for Microencapsulation of Green Tea Polyphenols. *Int. J. Food Prop.* **2017**, *20*, 1971–1986. [CrossRef]
139. Liu, W.; Mei, J.; Xie, J. Effect of Locust Bean Gum-Sodium Alginate Coatings Incorporated with Daphnetin Emulsions on the Quality of Scophthalmus Maximus at Refrigerated Condition. *Int. J. Biol. Macromol.* **2021**, *170*, 129–139. [CrossRef]
140. Cao, J.; Liu, W.; Mei, J.; Xie, J. Effect of Locust Bean Gum-Sodium Alginate Coatings Combined with High CO_2 Modified Atmosphere Packaging on the Quality of Turbot (*Scophthalmus maximus*) during Refrigerated Storage. *Polymers* **2021**, *13*, 4376. [CrossRef] [PubMed]
141. Góral, M.; Kozłowicz, K.; Pankiewicz, U.; Góral, D.; Kluza, F.; Wójtowicz, A. Impact of Stabilizers on the Freezing Process, and Physicochemical and Organoleptic Properties of Coconut Milk-Based Ice Cream. *LWT* **2018**, *92*, 516–522. [CrossRef]
142. Khanniri, E.; Yousefi, M.; Khorshidian, N.; Sohrabvandi, S.; Mortazavian, A.M. Development of an Efficient Stabiliser Mixture for Physical Stability of Nonfat Unfizzy Doogh. *Int. J. Dairy Technol.* **2019**, *72*, 8–14. [CrossRef]
143. Li, K.; Lei, Y.; Liao, J.; Zhang, Y. A Facile Synthesis of Graphene Oxide/Locust Bean Gum Hybrid Aerogel for Water Purification. *Carbohydr. Polym.* **2021**, *254*, 117318. [CrossRef]
144. Santos, M.B.; Garcia-Rojas, E.E. Recent Advances in the Encapsulation of Bioactive Ingredients Using Galactomannans-Based as Delivery Systems. *Food Hydrocoll.* **2021**, *118*, 106815. [CrossRef]
145. Braz, L.; Grenha, A.; Corvo, M.C.; Lourenço, J.P.; Ferreira, D.; Sarmento, B.; Rosa da Costa, A.M. Synthesis and Characterization of Locust Bean Gum Derivatives and Their Application in the Production of Nanoparticles. *Carbohydr. Polym.* **2018**, *181*, 974–985. [CrossRef]
146. Braz, L.; Grenha, A.; Ferreira, D.; Rosa da Costa, A.M.; Gamazo, C.; Sarmento, B. Chitosan/Sulfated Locust Bean Gum Nanoparticles: In Vitro and in Vivo Evaluation towards an Application in Oral Immunization. *Int. J. Biol. Macromol.* **2017**, *96*, 786–797. [CrossRef]

147. Kaity, S.; Ghosh, A. Carboxymethylation of Locust Bean Gum: Application in Interpenetrating Polymer Network Microspheres for Controlled Drug Delivery. *Ind. Eng. Chem. Res.* **2013**, *52*, 10033–10045. [CrossRef]
148. Dey, P.; Sa, B.; Maiti, S. Impact of Gelation Period on Modified Locust Bean-Alginate Interpenetrating Beads for Oral Glipizide Delivery. *Int. J. Biol. Macromol.* **2015**, *76*, 176–180. [CrossRef] [PubMed]
149. Bhattacharya, S.S.; Ghosh, A.K.; Banerjee, S.; Chattopadhyay, P.; Ghosh, A. Al^{3+} Ion Cross-Linked Interpenetrating Polymeric Network Microbeads from Tailored Natural Polysaccharides. *Int. J. Biol. Macromol.* **2012**, *51*, 1173–1184. [CrossRef] [PubMed]
150. Dey, P.; Maiti, S.; Sa, B. Gastrointestinal Delivery of Glipizide from Carboxymethyl Locust Bean Gum-Al^{3+}-Alginate Hydrogel Network: In Vitro and in Vivo Performance. *J. Appl. Polym. Sci.* **2013**, *128*, 2063–2072. [CrossRef]
151. Ben Romdhane, R.; Atoui, D.; Ketata, N.; Dali, S.; Moussaoui, Y.; Ben Salem, R. Pd Supported on Locust Bean Gum as Reusable Green Catalyst for Heck and Sonogashira Coupling Reactions and 4-nitroaniline Reduction under Ultrasound Irradiation. *Appl. Organomet. Chem.* **2022**, *36*, e6340. [CrossRef]
152. Tagad, C.K.; Rajdeo, K.S.; Kulkarni, A.; More, P.; Aiyer, R.C.; Sabharwal, S. Green Synthesis of Polysaccharide Stabilized Gold Nanoparticles: Chemo Catalytic and Room Temperature Operable Vapor Sensing Application. *RSC Adv.* **2014**, *4*, 24014–24019. [CrossRef]
153. Singh, I.; Rani, P.; Gazali, B.S.P.; Kaur, S. Microwave Assisted Synthesis of Acrylamide Grafted Locust Bean Gum for Colon Specific Drug Delivery. *Curr. Microw. Chem.* **2018**, *5*, 46–53. [CrossRef]
154. Jin, E.; Wang, S.; Song, C.; Li, M. Influences of Monomer Compatibility on Sizing Performance of Locust Bean Gum-g-P(MA-Co-AA). *J. Text. Inst.* **2022**, *113*, 1083–1092. [CrossRef]
155. Sagbas, S.; Sahiner, N. Modifiable Natural Gum Based Microgel Capsules as Sustainable Drug Delivery Systems. *Carbohydr. Polym.* **2018**, *200*, 128–136. [CrossRef]
156. Şen, M.; Hayrabolulu, H. Radiation Synthesis and Characterisation of the Network Structure of Natural/Synthetic Double-Network Superabsorbent Polymers. *Radiat. Phys. Chem.* **2012**, *81*, 1378–1382. [CrossRef]
157. Nayak, A.K.; Hasnain, M.S.; Aminabhavi, T.M. Drug Delivery Using Interpenetrating Polymeric Networks of Natural Polymers: A Recent Update. *J. Drug Deliv. Sci. Technol.* **2021**, *66*, 102915. [CrossRef]
158. Jana, S.; Sen, K.K. Chitosan—Locust Bean Gum Interpenetrating Polymeric Network Nanocomposites for Delivery of Aceclofenac. *Int. J. Biol. Macromol.* **2017**, *102*, 878–884. [CrossRef] [PubMed]
159. Coviello, T.; Alhaique, F.; Dorigo, A.; Matricardi, P.; Grassi, M. Two Galactomannans and Scleroglucan as Matrices for Drug Delivery: Preparation and Release Studies. *Eur. J. Pharm. Biopharm.* **2007**, *66*, 200–209. [CrossRef] [PubMed]
160. Petitjean, M.; Aussant, F.; Vergara, A.; Isasi, J.R. Solventless Crosslinking of Chitosan, Xanthan, and Locust Bean Gum Networks Functionalized with β-Cyclodextrin. *Gels* **2020**, *6*, 51. [CrossRef] [PubMed]
161. Petitjean, M.; Isasi, J.R. Chitosan, Xanthan and Locust Bean Gum Matrices Crosslinked with β-Cyclodextrin as Green Sorbents of Aromatic Compounds. *Int. J. Biol. Macromol.* **2021**, *180*, 570–577. [CrossRef]
162. Petitjean, M.; Lamberto, N.; Zornoza, A.; Isasi, J.R. Green Synthesis and Chemometric Characterization of Hydrophobic Xanthan Matrices: Interactions with Phenolic Compounds. *Carbohydr. Polym.* **2022**, *288*, 119387. [CrossRef]
163. Hadinugroho, W.; Martodihardjo, S.; Fudholi, A.; Riyanto, S. Study of a Catalyst of Citric Acid Crosslinking on Locust Bean Gum. *J. Chem. Technol. Metall.* **2017**, *52*, 1086–1091.
164. Hadinugroho, W.; Martodihardjo, S.; Fudholi, A.; Riyanto, S. Esterification of Citric Acid with Locust Bean Gum. *Heliyon* **2019**, *5*, e02337. [CrossRef]
165. Hadinugroho, W.; Martodihardjo, S.; Fudholi, A.; Riyanto, S. Preparation of Citric Acid-Locust Bean Gum (CA-LBG) for the Disintegrating Agent of Tablet Dosage Forms. *J. Pharm. Innov.* **2021**. [CrossRef]

Article

Extrusion Based 3D Printing of Sustainable Biocomposites from Biocarbon and Poly(trimethylene terephthalate)

Elizabeth Diederichs [1,2], Maisyn Picard [1,2], Boon Peng Chang [1], Manjusri Misra [1,2,*] and Amar Mohanty [1,2,*]

[1] Bioproducts Discovery and Development Centre, Department of Plant Agriculture, University of Guelph, Crop Science Building, 50 Stone Road East, Guelph, ON N1G 2W1, Canada; ediederi@uoguelph.ca (E.D.); maisyncpicard@gmail.com (M.P.); cbpchang@gmail.com (B.P.C.)

[2] School of Engineering, University of Guelph, Thornbrough Building, 50 Stone Road East, Guelph, ON N1G 2W1, Canada

* Correspondence: mmisra@uoguelph.ca (M.M.); mohanty@uoguelph.ca (A.M.)

Abstract: Three-dimensional (3D) printing manufactures intricate computer aided designs without time and resource spent for mold creation. The rapid growth of this industry has led to its extensive use in the automotive, biomedical, and electrical industries. In this work, biobased poly(trimethylene terephthalate) (PTT) blends were combined with pyrolyzed biomass to create sustainable and novel printing materials. The *Miscanthus* biocarbon (BC), generated from pyrolysis at 650 °C, was combined with an optimized PTT blend at 5 and 10 wt % to generate filaments for extrusion 3D printing. Samples were printed and analyzed according to their thermal, mechanical, and morphological properties. Although there were no significant differences seen in the mechanical properties between the two BC composites, the optimal quantity of BC was 5 wt % based upon dimensional stability, ease of printing, and surface finish. These printable materials show great promise for implementation into customizable, non-structural components in the electrical and automotive industries.

Keywords: biobased polymers; mechanical properties; thermal properties

1. Introduction

Three-dimensional (3D) printing, an additive manufacturing technique, is rapidly gaining popularity due to reduced material requirements and tooling time, as compared to alternative processing methods. Traditionally, computerized number control (CNC) machining has been used to make complete parts through a subtractive process. However, 3D printing offers the ability to fabricate complex geometries in an additive layer-by-layer fashion with limited to no post-print modifications [1]. Fused filament fabrication (FFF) is a relatively low-cost method of 3D printing that has extensive applications in industry, such as the biomedical [2] and aerospace [3,4] industries. FFF also functions well in rapid prototyping [5] and personal home-based printing [1]. FFF works off of the basic process of extruding a polymer-based filament through a heated nozzle to build parts layer-by-layer in the z-direction [6]. This process allows for high customization as the part shape is defined by a computer 3D model and is not limited by mold fabrication, as in the case of injection molding [7].

There are many commercial polymers commonly used for FFF. They can often be categorized as engineering thermoplastics, bioplastics, or commodity plastics [8–10], such as acrylonitrile butadiene styrene (ABS), polylactic acid (PLA), or high-density polyethylene (HDPE), respectively. Engineering thermoplastics show favor in many FFF applications due to superior mechanical performances and increased thermal stability as compared to commodity plastics [11]. Engineering thermoplastics can be further categorized as either biobased or petroleum based. Biobased polymers, or bioplastics, are materials that are biologically sourced, biodegradable, or a combination of both [12]. There has been a substantial drive in the industry to develop 3D printing bioplastics and engineering bioplastics.

The most commonly studied bioplastics for extrusion based 3D printing include PLA [13], polycaprolactone [14], or blends of bioplastics [15]. However, many of these bioplastics have comparatively low melting temperatures and limited mechanical performances, as compared to engineering thermoplastics [16]. To counter these limitations, engineering thermoplastics derived from organic sources are gaining popularity. One of the best examples is poly(trimethylene terephthalate) (PTT), which is partially biobased, due to one reactant being made from corn derivatives [6,17].

There is a global demand to generate products that conserve resources, recycle or refurbish products, or transform wastes into value-added products [18,19]. In the polymer industry, one of the methods to improve the sustainability of composite production is to valorize biomass waste as natural fillers. Wastes from agricultural residues [20,21], forestry residues [22], or food industries [17,23] can be diverted from landfills and instead pyrolyzed to become a carbonaceous natural filler. Pyrolysis is the thermochemical conversion of organic matter to biocarbon (BC), bio-oil, and syngas [17]. Characteristics of BC are dependent on the pyrolysis temperature, residence time, heating rate, and environmental conditions such as inert or oxygenated conditions [24]. The diversity of BC's surface characteristics, thermal, and mechanical properties have led to its use in polymer applications in the automotive [17] and electronic industries [25]. Biocarbon can function as a composite filler suspended in a continuous polymer phase. The implementation of BC offers benefits such as reduced initial materials cost, increased thermal stability over other natural fillers, limited to no odor during processing, and increased biocontent compared to inorganic fillers [17,26,27].

The use of BC in composites is more commonly studied in injection molding practices than in 3D printing processes. However, preliminary successes have suggested BC as a potential filler in additive manufacturing applications. Idrees et al. [28] performed work to incorporate BC with recycled poly(ethylene terephthalate) PET. Significant increases in tensile properties were observed with the addition of BC. Biocarbon was successful in creating printable parts and had effects on important parameters such as the glass transition temperature (T_g) and the coefficient of linear thermal expansion (CLTE) [28]. A reduction in CLTE was also seen with the addition of graphene into ABS, therefore increasing the thermal stability of the polymer, which is beneficial for FFF [29]. Ertane et al. [30] worked to implement BC into PLA to increase its potential applications. Interestingly, they found that FFF processing increased the interaction between the PLA matrix and the BC.

This work focuses on the implementation of BC to increase the biocontent within polymer filaments for FFF technologies. The use of a partially biobased engineering thermoplastic in addition to the natural filler was intentionally selected to offer a potential substitute for petroleum-based engineering thermoplastics with carbon black or other inorganic fillers. It is important to highlight that this work is an intermediary work. This work uses PTT derived blends containing a chain extender (CE) and impact modifier (IM) that were optimized for 3D printing dimensionally stable, complete, and warpage-free parts [6].

This work focuses on combining the aforementioned polymer blends with bioderived *Miscanthus* BC. The strategy of BC has been investigated and proven to improve the thermal stability and reduce CLTE of the polymer blends in injection molding type works. There are additional benefits to using BC, such as its use as a colorant and reduced costs from decreased quantities of expensive polymer required. Similar principles applied and the content of BC was optimized for printing complete and warpage-free parts. The goal of this work was to create novel printing materials that could be used in big-area-additive manufacturing. This PTT blend with the addition of BC has potential for use as 3D printing composite materials for non-structural automotive parts and electrical components.

2. Results and Discussion

2.1. Characterization of Biocarbon

The surface characteristics and general morphological shapes of the BC were determined via Raman spectroscopic analysis (Figure 1). The Raman analysis compared the graphitic content in the G band to the disordered content in the D band. The D and G peaks of BC reached maximal values at 1350 and 1589 cm^{-1}, respectively. These peaks corresponded with similar peaks reported in the literature for other BC samples [25]. A ratio of intensities from the D/G bands was 1.3. Graphene aerogels with a similar intensity ratio have also found success in 3D printing [31]. This ratio suggests that there is more disordered content than organized carbon. The difference in peaks is a result of in-plane vibrations associated with sp^2 carbon in graphite lattices and plane stretching [32].

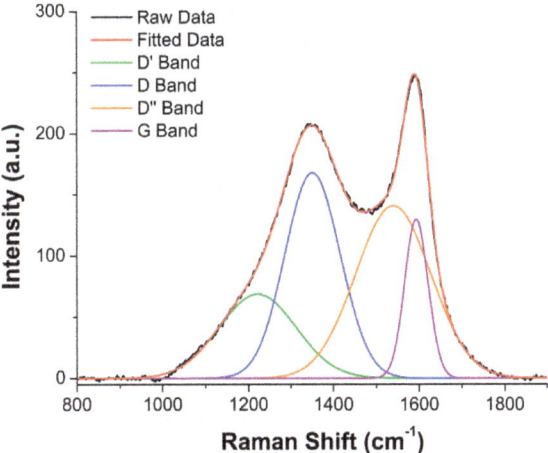

Figure 1. Raman spectra of *Miscanthus* BC pyrolyzed at 650 °C and after 24 h of ball milling.

Based on the SEM analysis of the same *Miscanthus* BC studied by our lab [20], it was determined that the ball milling caused the BC to have a more uniform shape and smaller size distribution [20]. This *Miscanthus* BC with 24 h of ball milling was studied previously and found to have an average particle size of 0.9 µm [20]. Since the size of the particle determines the precision of the print [33], it was important to have small particles. The smaller sized filler was also used to reduce potential clogging at the nozzle. The next steps for this research would be to determine the correlation between graphitic structure and intrinsic properties connecting Raman spectroscopy and surface morphology through X-ray diffraction (XRD) analysis.

Miscanthus BC at 650 °C was chosen since lignocellulosic biomass derived BC offers a higher modulus than other sources of BC due to a high stiffness of samples. The increased stiffness is further associated with increased polymer performance [21]. Since the BC was pyrolyzed at a lower temperature, the BC was able to maintain more surface functionalities than samples pyrolyzed at higher temperatures [17]. These surface functionalities on the BC interact with the functional polymers in the blend [34].

2.2. 3D Printed Composites

The 3D printed composites are referenced as blend content/BC content. The blend is comprised of 90 wt % PTT, 10 wt % IM and 0.5 phr of CE. Further details can be found in Table 2 below as well as the materials section.

The challenges and analyses found of 3D printing with the BC composites are outlined in the sections below. A preliminary investigation on the mechanical properties were performed and reported with injection molded composites with 0, 2.5, 5, 7.5, and 10 wt % BC (refer Table S1, in the Supplementary Materials). Minimal differences were seen in flexural and tensile properties of the BC composites, therefore 5 and 10 wt % BC compositions were chosen for further testing with 3D printing.

2.2.1. Challenges of Biocarbon in Printing

Printing composites with BC can be challenging; for example, various works have reported the degradation of the printer nozzle due to the abrasiveness of BC [30]. Jamming at high BC loadings and poor surface finishes have also been reported [30]. The challenges with interlay adhesion and due to the addition of a filler were also examined and are discussed in Section 2.2.6 below. Figure 2A depicts the extrudate from the FFF printed nozzle comparing the polymer blend to that of the composites. To further clarify, a schematic was drawn to highlight that filaments containing BC tended to be less consistent and had a rougher appearance. The slight swelling seen after the material exits the nozzle could have been responsible for the rougher, poorer quality surface finishes on the higher BC content FFF parts, as seen in Figure 2B. The 100/0 and 95/5 samples displayed smooth, cleanly defined rasters; however, the 90/10 printed samples were very rough and contained clumps of polymer and BC on the surface. This is represented in lower left section of the image below (Figure 2B) for the polymer blend and composites containing 5 and 10 wt % of BC. To examine the aforementioned parts on a microscopic level, SEM surface images were captured at a 300 times magnification (Figure 2B). Voids, identified by the blue circle on the Figure 2C, were most prominent on the surface of the 90/10 samples. Ertane et al. [30] also observed an increase in surface voids with increased BC content. The changes in surface finish of the BC composite materials likely relate to the rougher surface finish of the *Miscanthus* fiber. This is more prominent at the 10 wt % fiber since more fibers are present overall. The challenges and impacts of the natural filler surface and its correlation to print quality and mechanical performance are summarized by Duigou et al. [35] from combining literary works. Although challenges exist with the inclusion of BC into FFF feedstocks, low additions of BC did not decrease the printability of the filament. From its effects on printability and surface finish, 5 wt % biocarbon content is optimal as comparable to 10 wt %.

2.2.2. Thermogravimetric Analysis (TGA)

TGA analysis is important for looking at the degradation characteristics of a material and the thermal stability over a range of temperatures. The maximal degradation temperature of PTT was about 350 °C [36,37]. There was no significant shifting in the maximal degradation temperature peak, but there was an improvement in the thermal stability as can be seen clearly in the TGA curves. The increase in thermal stability was noted from the curve relating to the change in weight (%) as there were less losses over the higher temperatures (Figure 3). The derivative curves highlight the presence of BC through the elevated section at 500 °C. This again is associated with reduction in degradation at this temperature. When BC is pyrolyzed at temperatures greater than that of the maximal polymer degradation temperature, BC offers increased thermal stability to the composites [38]. The improvement in thermal stability was further confirmed in the CLTE analysis. This is critical for higher temperature applications such as internal automotive components that operate at higher temperatures.

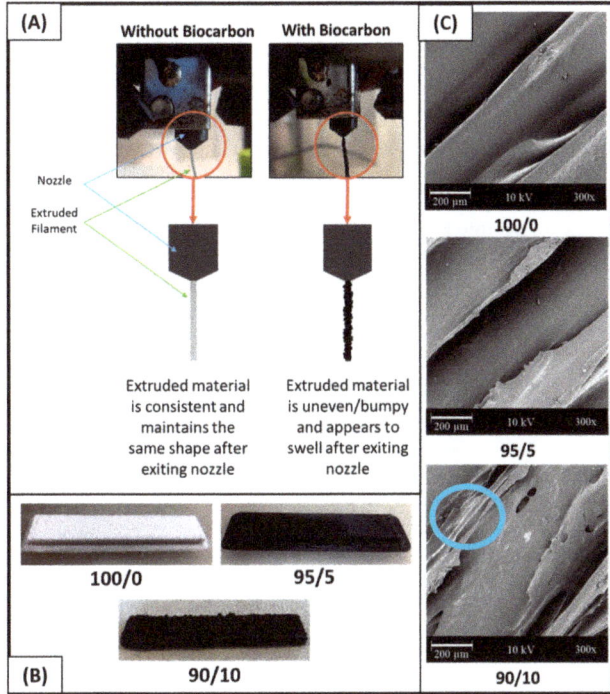

Figure 2. PTT/BC biocomposites in FFF: (**A**) the upper left section provides a comparison of an extrudate with and without biocarbon to depict challenges with consistency, (**B**) the lower left section compares FFF samples with different BC content for surface roughness, and (**C**) the right side of the image examines SEM images of the FFF surface of FFF printed samples. See Table 2 for nomenclature.

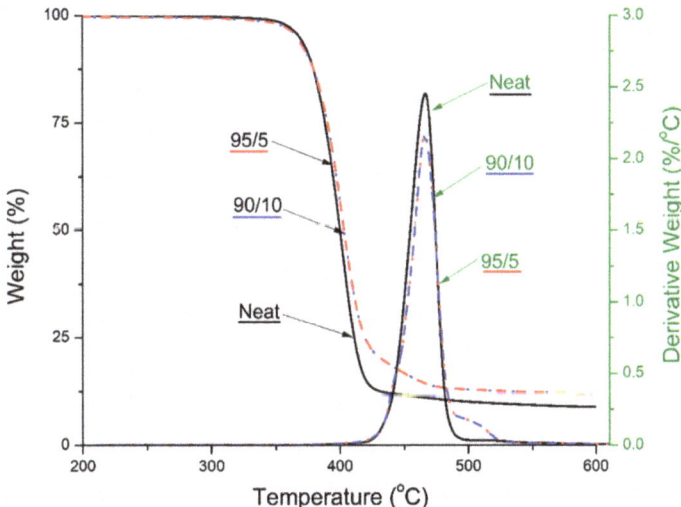

Figure 3. Thermogravimetric analysis (TGA) curves (weight and derivative weight) of the PTT blends and its biocomposites. See Table 2 for nomenclature.

2.2.3. Thermomechanical Analysis (TMA)

CLTE and the T_g are important properties to investigate for success in FFF. CLTE is a measure of the expansion of the material at elevated temperatures, which can be used to approximate the swelling or shrinking that could occur during 3D printing processes. In a study by Fitzharris et al. [39], CLTE was found to have a substantial impact on the success of printing due to its contributions to warpage and reduced thermal stability. A direct correlation was observed between CLTE and warpage, and many studies have shown that fillers tend to decrease CLTE [21,27,39]. Therefore, it is important to note that the addition of BC into PTT decreased the CLTE, as seen in Figure 4A. The lowest CLTE was seen in the 90/10 composite, at a value of 13% less than the 100/0 value. CLTE decreases with filler addition due to their effect of hindering polymer chain movement [40]. T_g is also important to FFF as it dictates what build platform (or bed) temperature range will be successful. A bed temperature slightly above T_g has been found to be optimal in a study by Spoerk et al. [41]. Insignificant changes in T_g were seen due to the influence of BC, as shown in Figure 4B. The polymer blend and composites had a T_g over the range of 59–61 °C. This is increased from neat PTT's T_g of 45–55 °C [42], which indicated the additives and BC had an effect on the neat PTT's thermal properties. Printing of these composites was done with bed temperatures ranging from 55 °C in the 100/0 to 80 °C in the 90/10. Bed temperatures were kept similar to the T_g, however increasing BC content in the samples resulted in a necessary increase in bed temperature to ensure good bed adhesion.

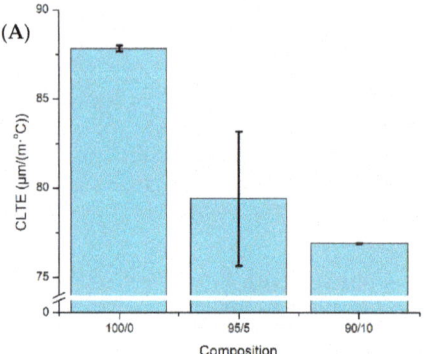 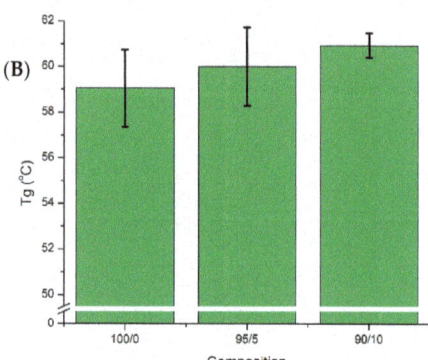

Figure 4. Thermomechanical analysis of the blend and composites: (**A**) coefficient of linear thermal expansion (CLTE) and (**B**) glass transition temperature (T_g) where injection molded samples were used. See Table 2 for nomenclature.

2.2.4. Rheology

The three rheological properties of the PTT blend and its biocomposites were determined from rheological analysis including complex viscosity (Figure 5A), storage modulus (Figure 5B), and loss modulus (Figure 5C). The complex viscosity of the PTT blend decreased as the angular frequency increased, which showing a typical shear thinning behavior. The complex viscosity was found to decrease after incorporation of BC and further decreased as the BC content increased from 5 to 10 wt %. Similar results were found with engineering thermoplastic polyetheretherketone (PEEK) and carbon fibers [43]. This observation suggests that the composites are more sensitive to high shearing action [6]. Nguyen et al. [44] describe the ideal zone for 3D printing composites; based on their research, the challenges with printing the 90/10 composites were a result of too low of a complex viscosity [44]. The shear thinning behavior of the material can be used to determine the required pressure and force to extrude the molten polymer through the FFF nozzle [45].

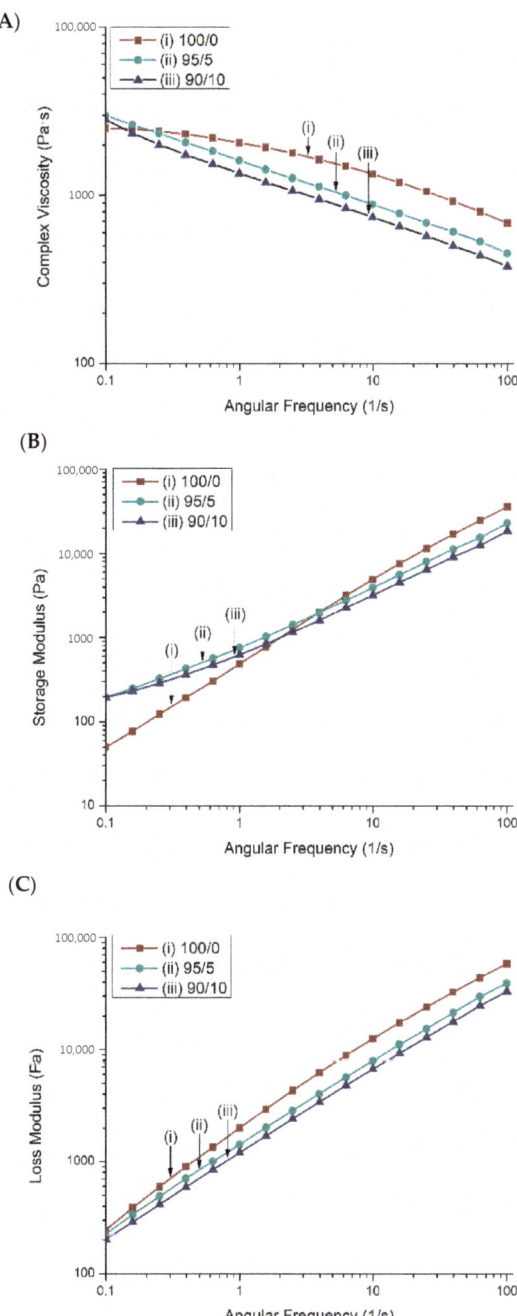

Figure 5. Rheological analysis of composites: (**A**) complex viscosity (η), (**B**) storage modulus (G′), and (**C**) loss modulus (G″) for injection molded samples of the PTT blend and *Miscanthus* BC composites. See Table 2 for nomenclature.

The storage modulus defines the potential energy stored in the materials and thus describes its elastic response under deformation [46]. The PTT blend and its biocomposites were considered to be frequency dependent since the storage modulus and loss modulus value increases with increasing frequency. This has also been found in other composites [46].

2.2.5. FFF Mechanical Properties

The mechanical performances of the FFF printed samples are presented in Table 1. As compared to the blend (100/0), the 95/5 and 90/10 composites both experienced a loss in mechanical performance. There is not a significant difference in the mechanical performance between the 5 wt % and 10 wt % BC. The reduction in mechanical performance was attributed to poor layer adhesion and agglomeration of particles, as noted in SEM images below (Figure 5). The printing parameters were optimized through systematic trials to produce parts that were free of warpage and delamination. Poor layer adhesion was exacerbated by limitations with the FFF printer used, such as the lack of an environmental temperature control chamber. Furthermore, it was likely that the agglomeration of particles at the nozzle of FFF printers, also discussed by Zhang et al. [47], resulted in poor flow of the materials. Further works with large scale additive manufacturing technologies with heated chambers will be investigated as it is anticipated that there would be improvements in the mechanical performance of such parts. For the current work, it is recommended that materials be used for non-structural components in FFF as there are some limitations with the mechanical performances. An example of non-structural parts would be interior car parts that are non-load bearing.

Table 1. Mechanical properties of FFF samples.

Sample Composition	Tensile Strength (MPa)	Tensile Modulus (GPa)	Elongation at Break (%)	Impact Strength (J/m)
100/0	35.3 ± 2.41	1.77 ± 0.10	6.34 ± 0.79	61.26 ± 11.87
95/5	26.4 ± 3.70	1.31 ± 0.22	4.01 ± 0.78	34.05 ± 4.71
90/10	28.3 ± 1.09	1.49 ± 0.09	3.54 ± 0.53	32.25 ± 3.85

2.2.6. Scanning Electron Microscopy

Micrographs, taken via SEM, show that all the FFF samples experienced poor layer adhesion, noted by the voids and gaps in the samples (Figure 6). There were substantially more noticeable voids in the 10 wt % BC samples, as noted by the blue circles on Figure 5C. Similar voids have been found when PLA and BC composites were 3D printed [30]. The pullout voids were likely one aspect that led to the decreases in mechanical performance. In work with other 3D printed composites, researchers have suggested that the extrudate is only heated to a semi-molten state rather than completely molten state, which has reduced the fusion of layers before cooling [48]. It is likely these same phenomena were seen in this work due to the presence of BC and with limitations of the nozzle temperature.

Table 2. Nomenclature and composition of composites.

Composition of Samples	Blend * Content (wt %)	Biocarbon Content (wt %)
100/0	100	-
95/5	95	5
90/10	90	10

* Blend: 90 wt % PTT, 10 wt % IM, and 0.5 phr CE.

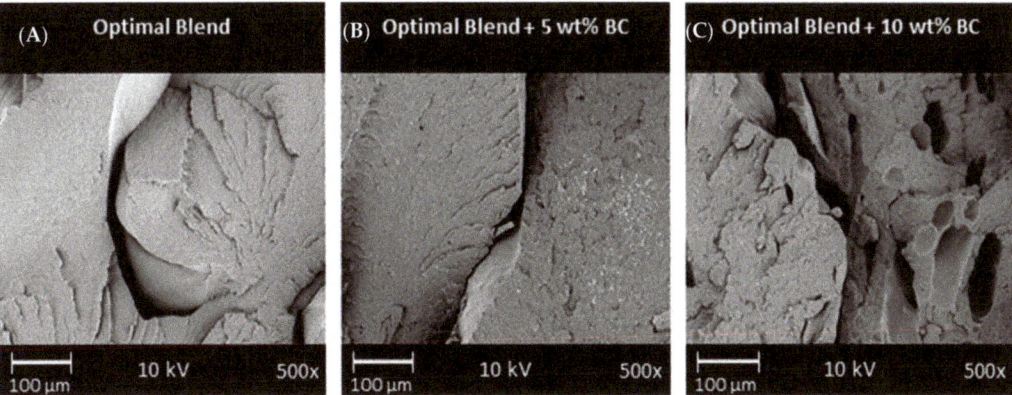

Figure 6. SEM images of the impact fracture surface for FFF samples for the (**A**) optimal polymer blend, (**B**) optimal blend with 5 wt % biocarbon, and (**C**) optimal blend with 10 wt % biocarbon. See Table 2 for nomenclature for optimal blend.

One of the goals of this project was to determine which composite offered superior performance. Superior performance was determined via comparisons of the surface finish, mechanical performance, CLTE data, and the SEM morphological analysis of the layers for both the 5 and 10 wt %. The ideal content of BC would be suggested at 5 wt % since this content maintains a smooth surface finish and offers comparable mechanical performance and CLTE to 10 wt % BC. Printing samples with 5 wt % BC are superior at maintaining dimensional accuracy. With higher BC content than 10 wt %, samples had extremely poor printability with this FFF printer configuration.

3. Materials and Methods

3.1. Materials

Sorona® PTT is a biobased engineering thermoplastic and was sourced from DuPont (Wilmington, DE, USA). It has a 37% biobased content that comes from corn-derived 1,3-propanediol [31]. The chain extender (CE) used was poly(styrene-acrylic-co-glycidyl methacrylate), known as Joncryl 4368, and was purchased from BASF (Ludwigshafen, Germany). An impact modifier (IM), poly(ethylene-n-butylene-acrylate-co-glycidyl methacrylate) trademarked Elvaloy PTW, was also incorporated. PTT, the CE and the IM composed the matrix of the composites. The biocarbon was provided by Competitive Green Technologies (Leamington, ON, Canada). The BC, made from *Miscanthus*, was pyrolyzed at 650 °C and then ball milled for 24 h. This BC was then dried in an oven at 100 °C until the moisture was less than 0.2% before compounding. A Sartorius infrared moisture analyzer (Gottingen, Germany) was used to confirm the moisture content.

3.2. Characterization of Biocarbon

Raman Spectroscopic Analysis

To determine the relative concentrations of disordered carbon versus ordered carbon, Raman spectroscopic analysis was performed. A Thermo Fisher Scientific DXR™ 2 Raman microscope (Waltham, MA, USA) was used at a magnification of 10 times. A 532 nm laser at 5 mW power was operated through a 50 µm pinhole to collect data over the range of 50–3000 cm^{-1}. For peak deconvolution, four peaks were used in combination with the Gaussian–Lorentzian area mode as this is common practice in the literature [49].

3.3. Processing Methods

3.3.1. Reactive Extrusion

Pellets of PTT were dried at 80 °C for 24 h (moisture content <0.2%) and were mixed by hand with the CE, IM, and dried BC. This was then fed into a Liestritz-Micro-27 co-rotating twin screw extruder (Nuremberg, Germany) and pellets and filaments were collected. Materials were processed at 240 °C with a 100 rpm screw speed and a 7 kg/h feed rate. The processing conditions were established from previous work [6].

3.3.2. Injection Molding

Pellets with a moisture content of <0.2% were fed into an Arburg AllRounder 77 Ton (Loßburg, Germany) coinjection molding machine. Processing settings were a 240 °C barrel temperature, 80 °C mold temperature, and a 20 s hold time. Samples were created for thermomechanical and rheological analyses.

3.3.3. 3D Printing

Samples were 3D printed using FFF technology on a Lulzbot Taz 6 (Fargo, ND, USA) printer system. The 3D models were generated using SolidWorks (Dassault Systems, Vélizy-Villacoublay, France) and the printer was run using the Cura Lulzbot Edition (2.6.69) software (Ultimaker, Geldermalsen, The Netherlands). A 0.5 mm diameter brass nozzle and a borosilicate glass/polythyleneimine bed were used. The implemented printing parameters were 65 °C bed temperature, 280 °C nozzle temperature, 0.3 mm layer height, and 35 mm/s print speed.

3.4. Nomenclature and Composition

The matrix used for the composites in this study was a blend of PTT, CE, and IM. The blend composition used was the optimal composition found in work done on printing PTT by Diederichs et al. [6]. This optimal blend was 90 wt %, 10 wt %, and 0.5 phr of the PTT, IM, and CE, respectfully. The blend was then combined with *Miscanthus* BC at varying weight percentages. The nomenclature and compositions of the composites can be found in Table 2.

The amount of BC incorporated into the blends was based on previous research. In another work with an engineering thermoplastic polyester, PET, authors had success in printing samples with BC loadings from 0.5 to 5% [28]. Incremental increases in BC were performed to determine a maximal print content at 10 wt % for the printer configuration used in this work. Further details are described below.

3.5. Characterization

3.5.1. Mechanical Testing

Flexural and tensile properties were tested on an Instron Universal Testing Machine model 3382 (Norwood, MA, USA). The crosshead speed for the tensile test was 5 mm/min, while the crosshead speed for the flexural test was 14 mm/min that is analogous to ASTM D368 (type IV) and ASTM D790, respectively. Izod notched impact testing was performed on a Zwick/Roell HIT25P impact tester (Ulm, Germany) in accordance to ASTM D256.

3.5.2. Differential Scanning Calorimetry

A heat–cool–heat analysis was performed on a TA Instruments DSC Q200 (New Castle, DE, USA) with a nitrogen flow rate of 50 mL/min from 0 to 250 °C at a rate of 10 °C/min. From this, the heat of fusion, melting temperature, and degree of crystallinity were analyzed. Equation (1) was used to calculate the degree of crystallinity 1:

$$Degree\ of\ Crystalinity = \frac{\Delta H_m}{W_{f,\ PTT} \times \Delta H_m^o} \quad (1)$$

where ΔH_m is the heat of fusion of the sample, W_f is the weight fraction of PTT in the composite, and ΔH_m^o is the heat of fusion for 100% crystalline PTT, which is 30 kJ/mol [50]. The 0.9, 0.81, and 0.855 were the W_f values for the neat polymer blend, 10 wt % BC, and 5 wt % BC, respectfully. The DSC section is included in the Supplementary Information.

3.5.3. Thermogravimetric Analysis

A TGA Q500 (TA Instruments, New Castle, DE, USA) was used to analyze the degradation temperature of the composites. The samples were heated to 500 °C at a rate of 10 °C/min in a nitrogen environment.

3.5.4. Thermomechanical Analysis

A TMA Q400 (TA Instruments, New Castle, DE, USA) was used to analyze the glass transition temperature (T_g) and coefficient of linear thermal expansion (CLTE) of the injection molded samples in the flow direction. The thermal history was removed from the samples by heating to 70 °C and cooled prior to recording data. The samples were then heated from −60 to 170 °C at a rate of 5 °C/min. A force of 0.05 N was applied to the sample. The CLTE was recorded before T_g in the range of −30–40 °C. The formula for calculating the CLTE can be found in Equation (2) below:

$$CLTE = \frac{\frac{\Delta L}{L_o}}{\Delta T} \qquad (2)$$

where ΔL is the change of sample length, L_o is the original length of the sample, and ΔT is the change in temperature.

A DMA Q800 (TA Instruments, New Castle, DE, USA) was used to analyze the heat deflection temperatures (HDTs) of the composites. The samples were heated from 0 to 180 °C and the reported temperature is when the samples reached a deflection of 250 µm, in accordance with ASTM D648.

3.5.5. Rheology

A Modular Compact Rheometer 302 manufactured by Anton Paar (Graz, Austria) was used to analyze the viscoelastic properties of the composites. Complex viscosity (η), storage modulus (G′), and loss modulus (G″) were tested at 250 °C using a frequency sweep test from 0.1 to 100 rad/s. Rheological properties determine how the materials behaves under a given force. It has been suggested in the literature that rheological studies of 3D printing materials can lead to successful printing and identification of printable materials [51].

3.5.6. SEM of Composites

A Phenom-world ProX SEM (Eindhoven, The Netherlands) was used to image the fractured Izod impact surfaces and the FFF surface finish of the samples. This was also used to analyze the surface of the BC. The SEM images were taken at 10 kV and variable magnifications. Prior to imaging, samples were gold-coated for 10 s with a Cressington Sputter Coater 108 (Watford, England) auto vacuum sputter chamber.

4. Conclusions

To increase the use of FFF in industries such as the automotive and electrical, there is demand for thermally stable, sustainable printing materials. Such materials could be used in customizable jig and fixtures for product assembly and non-structural components or electrical housings. This work is a short communication on the adaption of PTT blends from previous work and their combination of *Miscanthus* BC for extrusion-based 3D printing. The PTT blend when combined with 5 wt % BC was tougher than that of 10 wt % BC and was able to maintain better dimensional accuracy during the print, therefore leaving a more visually appealing print. Further benefits of this new combination of materials were the reduction in CLTE and increased glass transition temperature over that of the PTT blend. This is more favorable for higher temperature applications, especially for use in big area additive manufacturing equipment where the thermal exposure to materials is greater. The combination of materials and printing parameters resulted in successful production of complete and warpage free samples in small scale extrusion 3D printing. The samples could be further studied for their potential in commercialized filament. The goal of increasing biocontent in engineering thermoplastics was achieved. In due time, materials like these PTT biocomposites will lead the way in sustainable product development in industries like the automotive, aerospace, and electronics industries.

Supplementary Materials: The following are available online, Table S1: Mechanical properties of injection moulded composites.

Author Contributions: Conceptualization, M.M. and A.M.; Methodology, M.M. and A.M.; Software, E.D., M.P. and B.P.C.; Validation, B.P.C.; Formal Analysis, E.D., B.P.C. and M.P.; Investigation, E.D. and M.P.; Resources, M.M. and A.M.; Data Curation, M.P. and E.D.; Writing—Original Draft Preparation, E.D. and M.P.; Writing—Review and Editing, B.P.C., M.M. and A.M.; Visualization, E.D. and M.P.; Supervision, B.P.C., M.M. and A.M.; Project Administration, M.M.; Funding Acquisition, M.M and A.M. All authors have read and agreed to the published version of the manuscript.

Funding: This research was funded by: (i) Ontario Ministry of Economic Development, Job Creation and Trade ORF-RE09-078 (Project Nos. 053970 and 054345); (ii) the Ontario Ministry of Agriculture, Food and Rural Affairs (OMAFRA), University of Guelph, Bioeconomy Industrial Uses Research Program Theme Project Nos. 030252 and 030485; and (iii) the Natural Sciences and Engineering Research Council of Canada (NSERC), Canada Research Chair (CRC) program Project No. 460788.

Institutional Review Board Statement: Not applicable.

Informed Consent Statement: Not applicable.

Data Availability Statement: The data presented in this study are available in Supplementary Material.

Conflicts of Interest: The authors declare no conflict of interest.

Acronyms

ABS	Acrylonitrile butadiene styrene
BC	Biocarbon
CE	Chain extender
CLTE	Coefficient of linear thermal expansion
CNC	Computer numbered control
DSC	Differential scanning calorimetry
DMA	Dynamic mechanical analysis
FFF	Fused filament fabrication
HDT	Heat deflection temperature
IM	Impact modifier
PEEK	Polyetheretherketone
PET	Poly(ethylene terephthalate)
PLA	Poly(lactic acid)
PTT	Poly(trimethylene terephthalate)
SEM	Scanning electron microscopy
TGA	Thermogravimetric analysis
XRD	X-ray diffraction
3D	Three-dimensional

References

1. Brans, K. 3D Printing, a Maturing Technology. *IFAC Proc. Vol.* **2013**, *46*, 468–472. [CrossRef]
2. Hedayati, S.K.; Behravesh, A.H.; Hasannia, S.; Bagheri Saed, A.; Akhoundi, B. 3D printed PCL scaffold reinforced with continuous biodegradable fiber yarn: A study on mechanical and cell viability properties. *Polym. Test.* **2020**, *83*, 106347. [CrossRef]
3. Wong, J.Y.; Pfahnl, A.C. 3D printing of surgical instruments for long-duration space missions. *Aviat. Sp. Environ. Med.* **2014**, *85*, 758–763. [CrossRef] [PubMed]
4. Mazzei Capote, G.A.; Rudolph, N.M.; Osswald, P.V.; Osswald, T.A. Failure surface development for ABS fused filament fabrication parts. *Addit. Manuf.* **2019**, *28*, 169–175. [CrossRef]

5. Dwivedi, G.; Srivastava, S.K.; Srivastava, R.K. Analysis of barriers to implement additive manufacturing technology in the Indian automotive sector. *Int. J. Phys. Distrib. Logist. Manag.* **2017**, *47*, 972–991. [CrossRef]
6. Diederichs, E.V.; Picard, M.C.; Chang, B.P.; Misra, M.; Mielewski, D.F.; Mohanty, A.K. Strategy to improve printability of renewable resource based engineering plastic tailored for FDM applications. *ACS Omega* **2019**, 1–11. [CrossRef] [PubMed]
7. Bogue, R. 3D printing: The dawn of a new era in manufacturing? *Assem. Autom.* **2013**, *33*, 307–311. [CrossRef]
8. Stansbury, J.W.; Idacavage, M.J. 3D printing with polymers: Challenges among expanding options and opportunities. *Dent. Mater.* **2016**, *32*, 54–64. [CrossRef] [PubMed]
9. Benwood, C.; Anstey, A.; Andrzejewski, J.; Misra, M.; Mohanty, A.K. Improving the Impact Strength and Heat Resistance of 3D Printed Models: Structure, Property, and Processing Correlations during Fused Deposition Modeling (FDM) of Poly(Lactic Acid). *ACS Omega* **2018**, *3*, 4400–4411. [CrossRef]
10. Melnikova, R.; Ehrmann, A.; Finsterbusch, K. 3D printing of textile-based structures by Fused Deposition Modelling (FDM) with different polymer materials. *IOP Conf. Ser. Mater. Sci. Eng.* **2014**, *62*. [CrossRef]
11. Thomas, S.; Visakh, P.M. *Handbook of Engineering and Specialty Thermoplastics*; Thomas, S.P.M.V., Ed.; John Wiley & Sons, Ltd.: New York, NY, USA; Scrivener Publishing: Beverly, MA, USA, 2012; Volume 4, ISBN 9780470639252.
12. Nagarajan, V.; Mohanty, A.K.; Misra, M. Perspective on Polylactic Acid (PLA) based Sustainable Materials for Durable Applications: Focus on Toughness and Heat Resistance. *ACS Sustain. Chem. Eng.* **2016**, *4*, 2899–2916. [CrossRef]
13. Torres, J.; Cole, M.; Owji, A.; DeMastry, Z.; Gordon, A.P. An approach for mechanical property optimization of fused deposition modeling with polylactic acid via design of experiments. *Rapid Prototyp. J.* **2016**, *22*, 387–404. [CrossRef]
14. Muwaffak, Z.; Goyanes, A.; Clark, V.; Basit, A.W.; Hilton, S.T.; Gaisford, S. Patient-specific 3D scanned and 3D printed antimicrobial polycaprolactone wound dressings. *Int. J. Pharm.* **2017**, *527*, 161–170. [CrossRef]
15. Qahtani, M.; Wu, F.; Misra, M.; Gregori, S.; Mielewski, D.F.; Mohanty, A.K. Experimental Design of Sustainable 3D-Printed Poly(Lactic Acid)/Biobased Poly(Butylene Succinate) Blends via Fused Deposition Modeling. *ACS Sustain. Chem. Eng.* **2019**, *7*, 14460–14470. [CrossRef]
16. Picard, M.; Mohanty, A.K.; Misra, M. Recent advances in additive manufacturing of engineering thermoplastics: Challenges and opportunities Maisyn. *RSC Adv.* **2020**, *10*, 36058–36089. [CrossRef]
17. Picard, M.; Thukur, S.; Misra, M.; Mielewski, D.F.; Mohanty, A.K. Biocarbon from peanut hulls and their green composites with biobased poly(trimethylene terephthalate) (PTT). *Sci. Rep.* **2020**, *10*, 3310. [CrossRef]
18. Ellen MacArthur Foundation New Plastics Economy Global Commitment—June 2019 Report. 2019, pp. 28–39. Available online: https://www.newplasticseconomy.org/assets/doc/GC-Report-June19.pdf (accessed on 8 July 2021).
19. Dertinger, S.C.; Gallup, N.; Tanikella, N.G.; Grasso, M.; Vahid, S.; Foot, P.J.S.; Pearce, J.M. Technical pathways for distributed recycling of polymer composites for distributed manufacturing: Windshield wiper blades. *Resour. Conserv. Recycl.* **2020**, *157*, 104810. [CrossRef]
20. Wang, T.; Rodriguez-Uribe, A.; Misra, M.; Mohanty, A.K. Sustainable Carbonaceous Biofiller from Miscanthus: Size Reduction, Characterization, and Potential Bio-composites Applications. *BioResources* **2018**, *13*, 3720–3739. [CrossRef]
21. Li, Z.; Reimer, C.; Wang, T.; Mohanty, A.K.; Misra, M. Thermal and Mechanical Properties of the Biocomposites of Miscanthus Biocarbon and. *Polymers* **2020**, *12*, 1–13.
22. Demir, M.; Kahveci, Z.; Aksoy, B.; Palapati, N.K.R.; Subramanian, A.; Cullinan, H.T.; El-Kaderi, H.M.; Harris, C.T.; Gupta, R.B. Graphitic Biocarbon from Metal-Catalyzed Hydrothermal Carbonization of Lignin. *Ind. Eng. Chem. Res.* **2015**, *54*, 10731–10739. [CrossRef]
23. Pala, M.; Kantarli, I.C.; Buyukisik, H.B.; Yanik, J. Hydrothermal carbonization and torrefaction of grape pomace: A comparative evaluation. *Bioresour. Technol.* **2014**, *161*, 255–262. [CrossRef] [PubMed]
24. Arnold, S.; Rodriguez-Uribe, A.; Misra, M.; Mohanty, A.K. Slow pyrolysis of bio-oil and studies on chemical and physical properties of the resulting new bio-carbon. *J. Clean. Prod.* **2016**, *172*, 2748–2758. [CrossRef]
25. You, X.; Misra, M.; Gregori, S.; Mohanty, A.K. Preparation of an Electric Double Layer Capacitor (EDLC) Using Miscanthus-Derived Biocarbon. *ACS Sustain. Chem. Eng.* **2018**, *6*, 318–324. [CrossRef]
26. Li, Z.; Reimer, C.; Picard, M.C.; Misra, M.; Mohanty, A.K. Characterization of Chicken Feather Biocarbon for Use in Sustainable Biocomposites. *Front. Mater.* **2020**, *7*, 3. [CrossRef]
27. Watt, E.; Abdelwahab, M.A.; Snowdon, M.R.; Mohanty, A.K.; Khalil, H.; Misra, M. Hybrid biocomposites from polypropylene, sustainable biocarbon and graphene nanoplatelets. *Sci. Rep.* **2020**, *10*, 1–13. [CrossRef] [PubMed]
28. Idrees, M.; Jeelani, S.; Rangari, V. Three-Dimensional-Printed Sustainable Biochar-Recycled PET Composites. *ACS Sustain. Chem. Eng.* **2018**, *6*, 13940–13948. [CrossRef]
29. Dul, S.; Fambri, L.; Pegoretti, A. Fused deposition modelling with ABS-graphene nanocomposites. *Compos. Part A Appl. Sci. Manuf.* **2016**, *85*, 181–191. [CrossRef]
30. Ertane, E.G.; Dorner-Reisel, A.; Baran, O.; Welzel, T.; Matner, V.; Svoboda, S. Processing and Wear Behaviour of 3D Printed PLA Reinforced with Biogenic Carbon. *Adv. Tribol.* **2018**, *2018*, 1–11. [CrossRef]
31. Zhang, Q.; Zhang, F.; Medarametla, S.P.; Li, H.; Zhou, C.; Lin, D. 3D Printing of Graphene Aerogels. *Small* **2016**, *12*, 1702–1708. [CrossRef]
32. Yu, M.; Saunders, T.; Su, T.; Gucci, F.; Reece, M. Effect of Heat Treatment on the Properties of Wood-Derived Biocarbon Structures. *Materials* **2018**, *11*, 1588. [CrossRef]

33. Acquah, S.F.A.; Leonhardt, B.E.; Nowotarski, M.S.; Magi, J.M.; Chambliss, K.A.; Venzel, T.E.S.; Delekar, S.D.; Al-Hariri, L.A. Carbon Nanotubes and Graphene as Additives in 3D Printing. In *Carbon Nanotubes—Current Progress of their Polymer Composites*; Berber, M.R., Hafez, I.H., Eds.; IntechOpen: Rijeka, Croatia, 2016; pp. 227–250.
34. Behazin, E.; Misra, M.; Mohanty, A.K. Compatibilization of toughened polypropylene/biocarbon biocomposites: A full factorial design optimization of mechanical properties. *Polym. Test.* **2017**, *61*, 364–372. [CrossRef]
35. Le Duigou, A.; Correa, D.; Ueda, M.; Matsuzaki, R.; Castro, M. A review of 3D and 4D printing of natural fibre biocomposites. *Mater. Des.* **2020**, *194*, 108911. [CrossRef]
36. Liu, Z.; Chen, K.; Yan, D. Nanocomposites of poly(trimethylene terephthalate) with various organoclays: Morphology, mechanical and thermal properties. *Polym. Test.* **2004**, *23*, 323–331. [CrossRef]
37. Reddy, J.P.; Misra, M.; Mohanty, A. Renewable resources-based PTT [poly(trimethylene terephthalate)]/switchgrass fiber composites: The effect of compatibilization. *Pure Appl. Chem.* **2012**, *85*, 521–532. [CrossRef]
38. Umerah, C.O.; Kodali, D.; Head, S.; Jeelani, S.; Rangari, V.K. Synthesis of carbon from waste coconutshell and their application as filler in bioplast polymer filaments for 3D printing. *Compos. Part B Eng.* **2020**, *202*, 108428. [CrossRef]
39. Fitzharris, E.R.; Watanabe, N.; Rosen, D.W.; Shofner, M.L. Effects of material properties on warpage in fused deposition modeling parts. *Int. J. Adv. Manuf. Technol.* **2018**, *95*, 2059–2070. [CrossRef]
40. Codou, A.; Misra, M.; Mohanty, A.K. Sustainable biocarbon reinforced nylon 6/polypropylene compatibilized blends: Effect of particle size and morphology on performance of the biocomposites. *Compos. Part A Appl. Sci. Manuf.* **2018**, *112*, 1–10. [CrossRef]
41. Spoerk, M.; Gonzalez-Gutierrez, J.; Sapkota, J.; Schuschnigg, S.; Holzer, C. Effect of the printing bed temperature on the adhesion of parts produced by fused filament fabrication. *Plast. Rubber Compos.* **2018**, *47*, 17–24. [CrossRef]
42. Kurian, J.V. Sorona® Polymer: Present Status and Future Perspectives. In *Natural fibers, Biopolymers, and Biocomposites*; Mohanty, A.K., Misra, M., Drzal, L.T., Eds.; Taylor & Francis: Boca Raton, FL, USA, 2005; pp. 503–530. ISBN 9780203508206.
43. Yan, M.; Tian, X.; Peng, G.; Li, D.; Zhang, X. High temperature rheological behavior and sintering kinetics of CF/PEEK composites during selective laser sintering. *Compos. Sci. Technol.* **2018**, *165*, 140–147. [CrossRef]
44. Nguyen, N.A.; Barnes, S.H.; Bowland, C.C.; Meek, K.M.; Littrell, K.C.; Keum, J.K.; Naskar, A.K. A path for lignin valorization via additive manufacturing of high-performance sustainable composites with enhanced 3D printability. *Sci. Adv.* **2018**, 1–16. [CrossRef]
45. Wang, S.; Capoen, L.; D'hooge, D.R.; Cardon, L. Can the melt flow index be used to predict the success of fused deposition modelling of commercial poly(lactic acid) filaments into 3D printed materials? *Plast. Rubber Compos.* **2018**, *47*, 9–16. [CrossRef]
46. Cunha, M.P.; Grisa, A.M.C.; Klein, J.; Poletto, M.; Brandalise, R.N. Preparation and Characterization of Hollow Glass Microspheres-Reinforced Poly (acrylonitrile-co-butadiene-co-styrene) Composites. *Mater. Res.* **2018**, *21*. [CrossRef]
47. Zhang, X.; Chen, L.; Kowalski, C.; Mulholland, T.; Osswald, T.A. Nozzle flow behavior of aluminum/polycarbonate composites in the material extrusion printing process. *J. Appl. Polym. Sci.* **2019**, *136*, 1–9. [CrossRef]
48. Zhang, J.; Yang, B.; Fu, F.; You, F.; Dong, X.; Dai, M. Resistivity and its anisotropy characterization of 3D-printed acrylonitrile butadiene styrene copolymer (ABS)/carbon black (CB) composites. *Appl. Sci.* **2017**, *7*, 20. [CrossRef]
49. Snowdon, M.R.; Mohanty, A.K.; Misra, M. A study of carbonized lignin as an alternative to carbon black. *ACS Sustain. Chem. Eng.* **2014**, *2*, 1257–1263. [CrossRef]
50. Pyda, M.; Boller, A.; Grebowicz, J.; Chuah, H.; Lebedev, B.V.; Wunderlich, B. Heat capacity of poly(trimethylene terephthalate). *J. Polym. Sci. Part B Polym. Phys.* **1998**, *36*, 2499–2511. [CrossRef]
51. Torres, M.D. Role of the Rheology in the New Emerging Technologies as 3D Printing. *Rheol. Open Access* **2017**, *1*, 3–4.

Article

Radical Formation in Sugar-Derived Acetals under Solvent-Free Conditions

Aleksandra A. Wróblewska [1], H. Y. Vincent Ching [2], Jurrie Noordijk [1], Stefaan M. A. De Wildeman [1] and Katrien V. Bernaerts [1,*]

[1] Faculty of Science and Engineering, Biobased Materials, Maastricht University, P.O. Box 616, 6200 MD Maastricht, The Netherlands; olaelwroblo@gmail.com (A.A.W.); jurrie.noordijk@maastrichtuniversity.nl (J.N.); sdw@b4plastics.com (S.M.A.D.W.)
[2] Department of Chemistry, University of Antwerp, Universiteitsplein 1, B-2610 Wilrijk, Belgium; HongYueVincent.Ching@uantwerpen.be
* Correspondence: katrien.bernaerts@maastrichtuniversity.nl

Abstract: The degradation of acetal derivatives of the diethylester of galactarate (GalX) was investigated by electron paramagnetic resonance (EPR) spectroscopy in the context of solvent-free, high-temperature reactions like polycondensations. It was demonstrated that less substituted cyclic acetals are prone to undergo radical degradation at higher temperatures as a result of hydrogen abstraction. The EPR observations were supported by the synthesis of GalX based polyamides via ester-amide exchange-type polycondensations in solvent-free conditions at high temperatures in the presence and in the absence of radical inhibitors. The radical degradation can be offset by the addition of a radical inhibitor. The radical is probably formed on the methylene unit between the oxygen atoms and subsequently undergoes a rearrangement.

Keywords: polycondensation; polymerization; EPR spectroscopy; GalX; polyamide; radical decomposition; sugar derived monomers

Citation: Wróblewska, A.A.; Ching, H.Y.V.; Noordijk, J.; De Wildeman, S.M.A.; Bernaerts, K.V. Radical Formation in Sugar-Derived Acetals under Solvent-Free Conditions. *Molecules* **2021**, *26*, 5897. https://doi.org/10.3390/molecules26195897

Academic Editor: Sylvain Caillol

Received: 31 August 2021
Accepted: 25 September 2021
Published: 29 September 2021

Publisher's Note: MDPI stays neutral with regard to jurisdictional claims in published maps and institutional affiliations.

Copyright: © 2021 by the authors. Licensee MDPI, Basel, Switzerland. This article is an open access article distributed under the terms and conditions of the Creative Commons Attribution (CC BY) license (https://creativecommons.org/licenses/by/4.0/).

1. Introduction

Acetal moieties constitute a recurring motif in chemical synthesis whether in the context of organic synthesis as labile protective groups [1]. or, more recently, in polymer synthesis [2,3]. A wide variety of acetal protective groups have been used to protect carbohydrate-based polyols prior to their polymerization, resulting in functional biobased polymers with tunable properties [4]. For example, the incorporation of biacetilized mannarates, glucarates or galactarates into polymers elevates their glass transition region, suppresses crystallization of polymeric domains and lowers melting temperatures [5,6]. These properties can be translated to material properties and result in transparent polymers, which sustain their shapes at higher temperatures and simultaneously are easier to process due to their lower melting points. Furthermore, the protective groups can be selectively removed [7] leading to OH⁻ functionalized polymers, which are widely used in coatings, dynamic networks or high added-value materials for biomedical applications, etc.

The incorporation of acetal motifs into polyesters is well-documented; however, until recently, attempts to incorporate them into polyamides have been rather scarce. Polyamides (PAs) are a wide-spread type of polymers with renowned performance and chemical resistance. Typical polycondensation reactions to obtain PAs utilize the diacid (Figure 1 with R=H) form of the molecule.

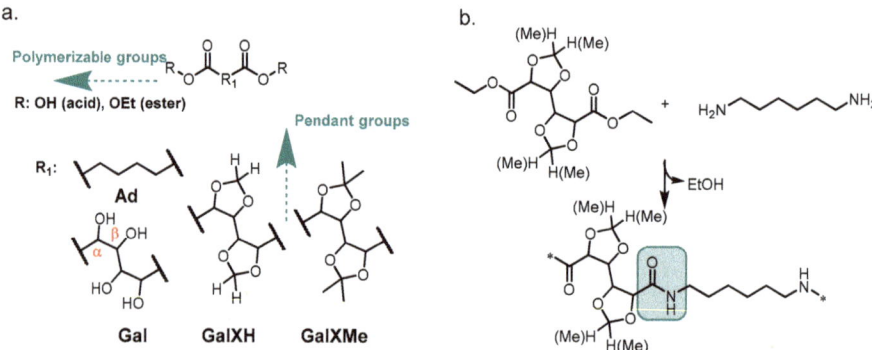

Figure 1. (a) The general structure of the investigated acetals (GalXH and GalXMe) and adipate motif (Ad); (b) the polymer obtained by reacting diethyl esters of GalXH and GalXMe with 1,6–hexamethylenediamine.

Upon reaction with diamines, polyamides are formed via intermediate salt formation. During the reaction, high temperatures (above 200 °C) are applied, releasing water and driving the equilibrium to the right. All these conditions (acid, water and heat) are incompatible with acetal chemistry. More precisely, substituted cyclic acetals (Figure 1a) can easily undergo acidic hydrolysis because stabilized tertiary carbocations are formed as intermediates [8]. This aspect has always been troubling for polyamide synthesis from GalX and consequently the majority of reported syntheses have been conducted in solution [9–11] through active ester methods utilizing, e.g., toxic pentachlorophenyl esters [3,12] and significant amounts of solvents, which are incompatible with the sustainable character of the monomers and could never find wide-spread application. In contrast, solvent-free polymer synthesis methods have received growing attention because of increasing environmental awareness and a greater focus on green chemistry principles [13], as well as economic aspects, e.g., the elimination of solvents and product isolation steps. Recent developments showed that in-melt polycondensation of cyclic acetal-bearing diacid monomers is possible if certain conditions are met [6]. Less substituted dioxolane-bearing molecules such as GalXH have been favored, due to their higher stabilities in acidic conditions. On the other hand, more substituted acetal rings (e.g., i-propylidene acetals like GalXMe) have been considered less stable and therefore neglected in polymer solvent-free methods using diacid monomers. However, we recently observed reversed stabilities under non-acidic conditions [5,13].

A recently published series of articles exhaustively describes the synthesis of polyamides from the diethyl esters (instead of the diacids) of sugar-derived cyclic acetal monomers (Figure 1a R=Et) in the melt [5,14]. Polymers with broader dispersities (branching) or even crosslinked networks were obtained with the less-substituted acetal monomer (GalXH) while better defined polymers were obtained with the more-substituted monomer (GalXMe). A similar polymerization method, but in solution, has been reported by Ogata et al. for diethyl adipate (Figure 1a Ad) and non-protected diethyl galactarate (Figure 1a Gal) [9–11,15]. They observed that polyhydroxy diethyl esters are activated towards diamines by the presence of oxygen atoms in the α and/or β position (see Figure 1a). The published findings were never translated to solvent-free methods and were never widely explored for the whole class of similar sugar-derived molecules. In addition, a similar dispersity trend has also been reported for polyester synthesis in the melt, which is less thermally demanding, using GalXH and GalXMe monomers, where broadening of the dispersity with GalXH was also observed and was attributed to transesterification [16]. This similarity suggests that the observed trend could be valid for all solvent-free polycondensations involving GalXH.

The aim of the present study is to carefully investigate the stability of acetal-containing monomers, mainly focusing on reaction conditions during melt polycondensation. The thermal stability of the GalX monomers has been elucidated using electron paramagnetic resonance (EPR) and supported by polymerization experiments in the melt (Figure 1b), which has shed new light on the underlying mechanism governing the observed behavior, thus aiding the development of solvent-free experiments, and later, material design.

2. Results and Discussion

2.1. Radical Degradation Study of the Acetal Fragments

Since the lower stability of less-substituted acetal rings in neutral reaction conditions cannot be explained by an ionic mechanism via intermediate carbocation formation [8], it was hypothesized that it proceeds via a radical pathway. To verify the proposed radical branching/cross-linking mechanism, EPR investigations were conducted. No paramagnetic species were detected when neat GalXH or GalXMe samples were heated and EPR spectra were subsequently recorded at room temperature or at 100 K. This meant either radical intermediates were not present, or they were highly reactive and could not be observed within the timeframe of the EPR measurements. Consequently, N-tert-butyl-α-phenyl nitrone (PBN) was added to the heating experiments as a spin trap. Spin traps react with radical intermediates, forming longer-living radical adducts that can be detected by EPR [17]. The EPR spectra of the adducts are characterized by the isotropic g-value (g_{iso}), isotropic hyperfine couplings of the nitroxide nitrogen ($A_{N\,iso}$), the α-proton ($A_{H\alpha\,iso}$), and in some cases also other protons, which are dependent on the spin-trapped radical. The mixtures were only heated to 140 °C because the concerted thermal decomposition of PBN at this temperature is still relatively slow [18]. A clear room temperature EPR signal was observed for the heated GalXH/PBN mixture that was subsequently dissolved in toluene, while, under the same conditions, no signals were observed for the heated GalXMe/PBN mixture or PBN alone (Figure 2). The observed signal is a triplet of doublets with g_{iso} = 2.0058, $A_{N\,iso}$ = 1.46 mT, and $A_{H\alpha\,iso}$ = 0.24 mT which are consistent with a carbon-centered radical trapped by PBN [17].

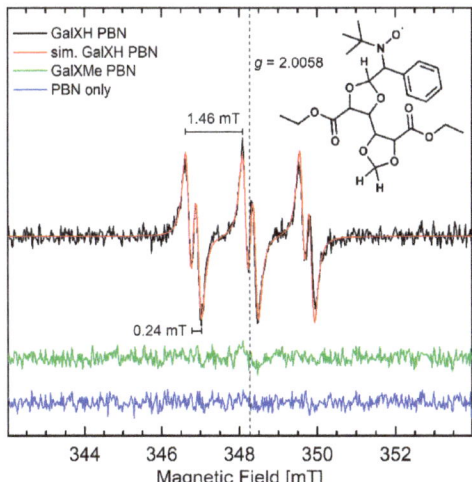

Figure 2. cw X-band (~9.7 GHz) solution EPR spectra of GalXH and GalXMe samples that had been heated in the presence of PBN and subsequently dissolved in toluene measured at room temperature using 10 mW microwave power, 0.1 mT modulation amplitude and 100 kHz modulation frequency. Simulation of the GalXH spectrum is shown in red, as well as the chemical structure of a possible spin-trapped adduct.

Since the only difference between GalXH and GalXMe are the methyl groups at the 2-position of the 1,3-dioxlane moiety, it is likely that the unpaired electron is centered at this position in the GalXH radical. We hypothesize that a radical is generated by hydrogen atom abstraction. The corresponding PBN-trapped dioxolanyl radical adduct is depicted in Figure 2. Furthermore, it should be noted that cyclic acetal radicals can undergo rearrangements, such as β-scission [19] giving aldehyde moieties. In our previous research on GalXH [8] aldehyde moieties were observed, which further supports the assignment of the dioxolanyl radical.

From the radical degradation studies, it appears that the *i*-propylidene acetals are resistant to thermal degradation via the radical mechanism, since they do not possess hydrogen atoms in the 2-position of the 1,3-dioxolane moiety for abstraction. Consequently, GalXMe can be freely used in the solvent-free high-temperature polymerization conditions, at least if acids are not present in a significant amount since they degrade *i*-propylidene acetals via a cationic mechanism [8]. The utilization of methylene acetals should be avoided. It is worth mentioning that multiple studies in the past used this type of acetal for polymerization, but the reports only include solution polymerization [3,20]. This choice could be motivated by the observed degradation and could be avoided if another acetal was chosen.

2.2. Polymerization

The relevance of the findings to the field were verified by conducting the melt polymerization of the diethyl esters of the two presented acetals (GalXH and GalXMe) with 1,6-hexamethylene diamine (HMDA), resulting in polyamides (Figure 1b). Such reactions require temperatures above 200 °C, mechanical stirring, and the removal of by-products (alcohol). During the reactions two parameters were investigated: the concentration of a radical inhibitor (Irganox 1330—a sterically hindered phenolic antioxidant) and the structure of the acetal. If radical degradation mechanisms were interfering with the polymerization, narrower dispersities of the final polymers would be expected in the presence of the inhibitor than in the absence of the inhibitor. Therefore, the polymerization reactions were evaluated based on the dispersity of the product. The theoretical value of the dispersity for polycondensates is 2 [21], however, in practice the value might vary due to limitations of the theory [22]. as well as side reactions [8]. Typically, if cyclization of polymeric chains is observed, or in the case of limited conversions (< 100%), the dispersity drops below 2. High dispersity typically means that side reactions occur that lead to increased functionality and consequently to branching or cross-linking at high conversions.

The reactions were performed using GalXH (**PA1** and **PA2** in Table 1) or GalXMe (**PA3** and **PA4** in Table 1), and diethyl esters with (**PA2** and **PA4** in Table 1) or without (**PA1** and **PA3** in Table 1) the radical inhibitor. The molecular weight distribution of GalXH polymers strongly depended on the presence of the inhibitor and, even in that case, an increase with respect to the generally accepted value was observed. The molecular weight of GalXMe polymers was not affected by the addition of the inhibitor, where a dispersity value close to 2 was observed in both entries.

Table 1. The list of prepared polymers and their properties.

Diethyl Acetal Type	Diamine Type	Inhibitor [b] (wt%)	Polymer Code	Polymer				
				GPC [c]		NMR [d]		
				M_n (kg·mol^{-1})	Đ	$M_{n,NMR}$ (g·mol^{-1})	DP (−)	p_{GalX} (%)
GalXH	HMDA [a]	0	PA1	23.7	4.53	8300	26	96.3
GalXH	HMDA [a]	5	PA2	16.3	2.93	6900	22	95.6
GalXMe	HMDA [a]	0	PA3	15.0	1.83	8200	22	95.7
GalXMe	HMDA [a]	5	PA4	14.0	1.88	7000	19	95.0

[a] 1,6-hexamethylenediamine, [b] wt% relative to both monomers, [c] GPC with RI detection in 1,1,1,3,3,3-hexafluoro-2-propanol/0.019%. NaTFA calibrated with poly (methyl methacrylate) standards. [d] The molecular mass $M_{n,NMR}$, degree of polymerization (DP) and extent of the polymerization (p) were calculated based on the NMR resonances of reacted GalX 1,2,4,5 at 3.27–4.39 ppm and end group resonances of 1,2,3,4 (marked with * in Figure 3) at 4.66–4.65 ppm.

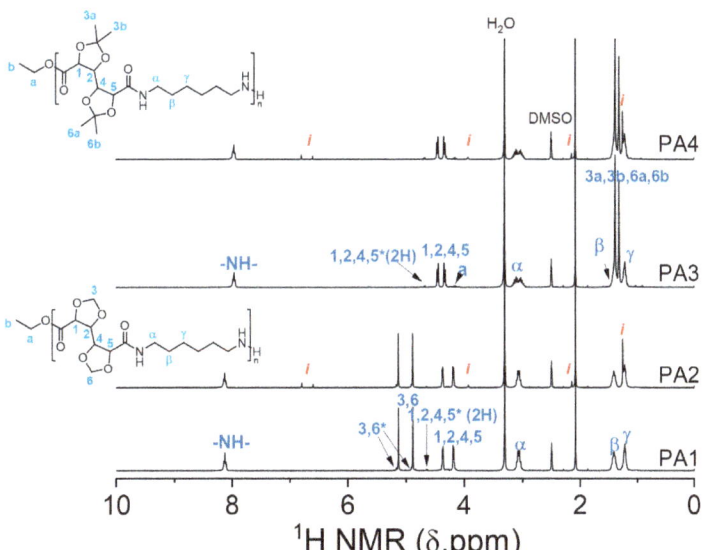

Figure 3. ^{1}H NMR spectra of **PA1**, **PA2**, **PA3** and **PA4**. The resonances of end groups are marked with a star "*" and resonances of Irganox 1330 with a letter "*i*".

NMR analysis revealed that, structurally, the polymers with and without the inhibitor are similar (Figure 3.). Furthermore, numerical analysis of the products according to the Carothers theory provided data about the molecular weight, degree of polymerization, and the extent of polymerization. The molecular weight of the polymers, though lower than obtained by GPC, follows the same trend. **PA1** has a higher degree of polymerization than the other polymers. The extent of polymerization is 95.0–96.3% with the highest value achieved for **PA1**. This experiment showed that in the case of GalXMe the addition of an inhibitor is not necessary because it only introduces impurities without any significant improvement in dispersity. For GalXH, on the other hand, the addition of the inhibitor is necessary since it improves the polymer dispersity by the reduction in radical side reactions.

The polyamides **PA1–4** were further analyzed via DSC and TGA. The DSC curves confirm that in the case of GalXMe (**PA3,4** in Figure 4a) the addition of the inhibitor does not cause any distinguishable changes in the polymer. The T_g's of **PA3** and **PA4** are 79.3 °C and 79.2 °C, respectively, and the curves do not show any additional thermal events which might point towards side reactions with an inhibitor. In contrast the GalXH polyamides (**PA1,2** in Figure 4a) show the opposite. The thermal profile of the polyamide obtained with the addition of the inhibitor (**PA2**) shows multiple thermal events which are hypothesized to originate from the reactions between the inhibitor and acetal fragmentation. Upon the addition of the inhibitor the T_g drops slightly from 71.3 °C to 70.1 °C which is in line with the fact that **PA1** has higher molar mass in comparison to **PA2**.

The TGA analysis (Figure 4b) does not show any influence of the inhibitor on the degradation profiles of the polymers. They all start to degrade above 260 °C in an inert atmosphere and show 5% weight loss above 300 °C. All curves show that even with a rigorous drying step there is still insignificant weight loss observed at 100 °C, which is in line with our previous findings regarding water sorption of GalX polyamides [14].

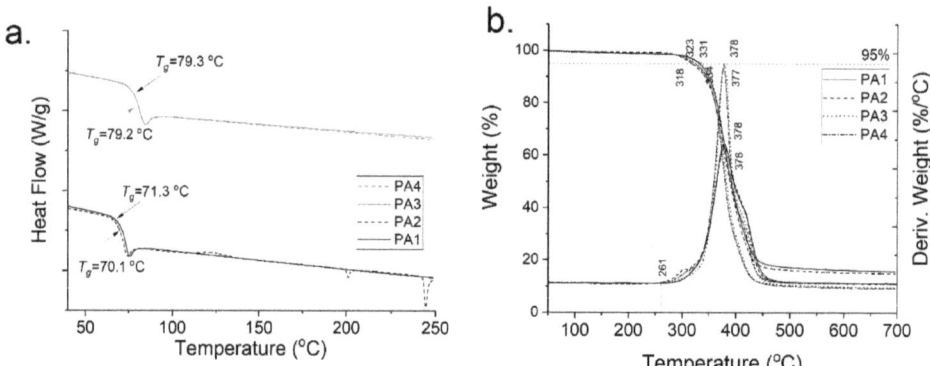

Figure 4. Thermal profiles of **PA1**, **PA2**, **PA3** and **PA4** (**a**) DSC curves and (**b**) TGA profiles.

3. Experimental Section

3.1. Materials and Methods

3.1.1. Materials

Diethyl 2,3:4,5-di-*O*-isopropylidene-galactarate > 99% (GalXMe) and diethyl 2,3:4,5-di-*O*-methylene-galactarate > 99% (GalXH) were supplied by Royal Cosun (Roosendaal, the Netherlands). 1,6-hexamethylene diamine (HMDA) 98%, Irganox 1330 and NaTFA were purchased from Sigma-Aldrich (Zwijndrecht, the Netherlands)and used as supplied. 1,1,1,3,3,3-hexafluoro-2-propanol (HFIP) and DMSO-d_6 were purchased from Acros Organics (The Hague, the Netherlands)and used as supplied. N-tert-butyl-α-phenyl nitrone (PBN) was purchased from TCI Europe N. V. (Zwijndrecht, Belgium).

3.1.2. Methods

Molecular weight of the polyamides was determined via gel permeation chromatography (GPC) supplied by Polymer Standards Service GmbH (PSS, Mainz, Germany). The polymers were dissolved in 1,1,1,3,3,3-hexafluoroisopropanol (HFIP) with 0.019% NaTFA salt. The sample for GPC measurement was prepared by dissolving 5.0 mg of the polymer in 1.5 mL of the solvent. The solutions were filtered over a 0.2 µm PTFE syringe filter before injection. The GPC apparatus was calibrated with poly(methyl methacrylate) standards. Two PFG combination medium microcolumns with 7 µm particle size (4.6 × 250 mm, separation range 100–1.000.000 Da) and a precolumn PFG combination medium with 7 µm particle size (4.6 × 30 mm) with refractive index detector (RI) were used in order to determine molecular weight and dispersities.

^{1}H NMR spectra were recorded in DMSO-d_6 on a Bruker Avance III HD Nanobay 300 MHz NMR spectrometer (Bruker Biospin GmbH, Rheinstetten, Germany).

Differential scanning calorimetry (DSC). DSC was performed on a TA Instruments Discovery DSC 250 (TA Instruments, New Castle, United States) equipped with a refrigerated cooling system (RCS). The samples were measured in Tzero pans with perforated Tzero hermetic lids to allow a nitrogen atmosphere around the sample. DSC thermograms were recorded with a heating rate of 10 °C min^{-1}. Only experimental data obtained from the second heating step are reported.

Thermal stabilities of the prepared polyamides were determined using thermogravimetric analysis (TGA) (TA Instruments Q500, TA Instruments, New Castle, DE, United States). Approximately 10 mg of the material was heated at 10 °C/min from 25 °C to 700 °C in a nitrogen atmosphere.

Electron paramagnetic resonance (EPR).

EPR sample preparation.

Neat GalXH or GalXMe (~50 mg) were loaded inside 4 mm quartz EPR tubes. The samples were heated at 200 °C under a nitrogen atmosphere. After 15 min the samples were rapidly cooled in liquid nitrogen and were transferred directly to the spectrometer. Mixtures of the spin-trap N-tert-Butyl-α-phenyl nitrone (30 mg, 0.17 mmol) with and without GalXH (45 mg, 0.15 mmol) or GalXMe (52 mg, 0.15 mmol) were heated at 140 °C under a nitrogen atmosphere. After 15 min the mixtures were cooled to room temperature and then each was dissolved in 300 µL of toluene. Aliquots of each solution were transferred to 2 mm capillaries for EPR measurements.

EPR

Room temperature EPR measurements were carried out using a Bruker E580 Elexys spectrometer (Bruker Biospin GmbH, Rheinstetten, Germany). The EPR spectra were recorded at X-band microwave frequency (~9.7 GHz) in continuous-wave (CW) mode with microwave power of 10 mW, 0.1 mT modulation amplitude and 100 kHz modulation frequency. Cryogenic temperature EPR measurements were carried out using a Bruker ESP300E spectrometer (Bruker Biospin GmbH, Rheinstetten, Germany) at X-band microwave frequency (~9.45 GHz). The EPR spectra were simulated with Matlab2018b (Mathworks, Natick, MA, United States) using the EasySpin-6.0 module, [23].

3.1.3. Synthesis of Polymers

A typical procedure adapted from Wroblewska et al. [5] is given below.

Synthesis of poly(hexamethylene-2,3:4,5-di-O-iso-propylidene-galactaramid) with 5 wt% inhibitor.

To a 100 mL three-necked round bottom flask (equipped with a vacuum-tight mechanical stirrer, a Vigreux column and a distillation condenser) diethyl 2,3:4,5-di-O-isopropylidene-galactarate (6.9276 g, 20 mmol), 1,6-hexamethylene diamine (2.3716 g, 20.4 mmol, slight excess taking into account the 98% purity of the starting product), and Irganox 1330 as an inhibitor (0.4626 g, 5 wt%) were added and slowly heated under nitrogen to 200–220 °C. After all ethanol had been fully distilled off (+/− 1.5 h), vacuum was applied for 3 h. The crude product was obtained as a yellowish material.

^{1}H NMR (300 MHz, DMSO-d_6) δ (ppm): 6.68 (2H, -NH-, s), 4.55 (2H, -CH-O-, m), 4.36 (2H, -CH-O-, m), 4.28 (4H, -CH_2-NH-, t), 1.58 (4H, -CH_2-CH_2, HMDA, m), 1.40 (6H, CH_3, GalX, s), 1.30 (6H, -CH_3, GalX, s). 1.26 (4H, -CH_2-CH_2-, HMDA, m).

4. Conclusions

Sugar-based cyclic acetal monomers are interesting monomers for solvent-free melt polycondensation reactions. They are abundant and easily obtained from waste streams, however the structure of the monomer should be carefully matched to the polymerization conditions. It was confirmed by EPR measurements that methylene acetals have the tendency to form radicals via thermally induced abstraction of hydrogen atoms and therefore participate in undesired side reactions. The degradation can be potentially offset using radical inhibitors, which allows control over polymerization of methylene acetals, but is redundant during the polymerization of i-propylidene acetals since they are not subject to radical degradation. Although these findings are based only on polyamide synthesis, which requires more demanding temperatures, they are applicable to other (melt) polycondensation reactions in general.

Author Contributions: Conceptualization, K.V.B. and A.A.W.; methodology, H.Y.V.C., K.V.B., J.N., A.A.W.; writing—original draft preparation, A.A.W. and H.Y.V.C., writing—review and editing, K.V.B., H.Y.V.C., A.A.W.; supervision, S.M.A.D.W., K.V.B.; funding acquisition, S.M.A.D.W. All authors have read and agreed to the published version of the manuscript.

Funding: This research was funded by the "Samenwerkingsverband Noord Nederland", Project T3006/Beets to Biopolymers. This research was funded by a H2020-MSCA-IF grant from the European Union, grant number 792946, iSPY.

Institutional Review Board Statement: Not applicable.

Informed Consent Statement: Not applicable.

Data Availability Statement: The data presented in this study are available on request from the corresponding author.

Acknowledgments: The authors are very grateful to Sabine Van Doorslaer of the University of Antwerp for her insightful advice and support.

Conflicts of Interest: There are no conflict of interest to declare.

Sample Availability: Samples of the compounds **PA1–PA4** are available from the authors.

References

1. Greene, T.W.; Wuts, P.G.M. *Greene's Protective Groups in Organic Synthesis*; John Wiley & Sons, Inc.: Hoboken, NJ, USA, 2007.
2. Capps, D.B. Linear polycyclospiroacetals and method for preparing them. U.S. Patent Application No. 2889290, 1 October 1956.
3. Munoz-Guerra, S. Carbohydrate-based polyamides and polyesters: An overview illustrated with two selected examples. *High Perform. Polym.* **2012**, *24*, 9–23. [CrossRef]
4. Wroblewska, A.; Wildeman, S.M.A.; Bernaerts, K.V. the consequences of the incorporation of (aliphatic/cycloaliphatic) sugar-based non-drop-in monomers into polyamides: In-depth study on representative examples. In *Recent Advances in Polyamides Research*; Correia, N.G., Ed.; NOVA: New York, NY, USA, 2019.
5. Wróblewska, A.A.; Bernaerts, K.V.; De Wildeman, S.M.A. Rigid, bio-based polyamides from galactaric acid derivatives with elevated glass transition temperatures and their characterization. *Polymer* **2017**, *124*, 252–262. [CrossRef]
6. Wroblewska, A.; Stevens, S.; Garsten, W.; Wildeman, S.M.A.; Bernaerts, K.V. A solvent-free method for the copolymerization of labile sugar-derived building blocks into polyamides. *ACS Sustain. Chem. Eng.* **2018**, *6*, 13504–13517. [CrossRef] [PubMed]
7. Picchioni, F.; Gavrila, I. Synthesis and Use of Carbohydrate-Based Linear Polyesters. WO Patent 2018/186744 Al, 11 October 2017.
8. Wróblewska, A.A.; De Wildeman, S.M.A.; Bernaerts, K.V. In-depth study of the synthesis of polyamides in the melt using biacetal derivatives of galactaric acid. *Polym. Degrad. Stab.* **2018**, *151*, 114–125. [CrossRef]
9. Ogata, N.; Sanui, K.; Hosoda, K.; Nakamura, H. Active polycondensation of diethyl 2,3,4,5-tetra hydroxy adipate with diamines. *J. Polym. Sci. Polym. Chem. Ed.* **1976**, *14*, 783–792. [CrossRef]
10. Ogata, N.; Sanui, K.; Hosoda, K.; Nakamura, H. Copolycondensation of hydroxyl diesters and active diesters with hexamethylene-diamine. *J. Polym. Sci. Polym. Chem. Ed.* **1977**, *15*, 1523–1526. [CrossRef]
11. Ogata, N.; Sanui, K.; Ohtake, T.; Nakamura, H. Solution polycondensation of diesters and diamines. *Polym. J.* **1979**, *11*, 827–833. [CrossRef]
12. Rodriquez-Galan, A.; Bou, J.J.; Munoz-Guerra, S. Stereoregular polyamides derived from methylene-l-tartaric acid and aliphatic diamines. *J. Polym. Sci. A Polym. Chem.* **1992**, *30*, 713–721. [CrossRef]
13. Anastas, P.T.; Warner, J.C. *Green Chemistry: Theory and Practice*; Oxford University Press: New York, NY, USA, 1998.
14. Wróblewska, A.A.; Noordijk, J.; Das, N.; Gerards, C.; De Wildeman, S.M.A.; Bernaerts, K.V. Structure—Property relations in new cyclic galactaric acid derived monomers and polymers therefrom: Possibilities and challenges. *Macromol. Rapid Commun.* **2018**, *34*, 1800077. [CrossRef] [PubMed]
15. Ogata, N.; Sanui, K.; Nakamura, H.; Kishi, H. Polycondensation of diethyl mucate with hexamethylenediamine in presence of poly(vinyl pyridine). *J. Polym. Sci. Polym. Chem. Ed.* **1980**, *18*, 933–938. [CrossRef]
16. Gavrila, I.; Raffa, P.; Picchioni, F. acetalised galactarate polyesters: Interplay between chemical structure and polymerisation kinetics. *Polymers* **2018**, *10*, 248. [CrossRef] [PubMed]
17. Buettner, G.R. Spin Trapping: ESR parameters of spin adducts. *Free. Radic. Biol. Med.* **1987**, *3*, 259–303. [CrossRef]
18. Goodrow, M.H.; Villarreal, J.A.; Grubbs, E.J. Kinetic study of the thermal decomposition of (Z)-N-tert-butyl-.alpha.-phenylnitrone. *J. Org. Chem.* **1974**, *39*, 3447–3449. [CrossRef]
19. Fielding, A.J.; Franchi, P.; Roberts, B.P.; Smits, T.M. EPR and computational studies of the formation and β-scission of cyclic and acyclic dialkoxyalkyl radicals. *J. Chem. Soc. Perkin Trans. 2* **2002**, 155–163. [CrossRef]
20. Rosu, C.; Negulescu, I.I.; Cueto, R.; Laine, R.; Daly, W.H. Synthesis and characterization of complex mixtures consisting of cyclic and linear polyamides from ethylbis-ketal galactarates. *J. Macromol. Sci. Part A Pure Appl. Chem.* **2013**, *50*, 940–952. [CrossRef]
21. Rogers, M.E.; Long, T.E. *Synthetic Methods in Step-Growth Polymers*; John Wiley & Sons, Inc.: Hoboken, NJ, USA, 2003.
22. Kéki, S.; Zsuga, M.; Kuki, Á. theoretical size distribution in linear step-growth polymerization for a small number of reacting species. *J. Phys. Chem. B* **2013**, *117*, 4151–4155. [CrossRef] [PubMed]
23. Stoll, S.; Schweiger, A. EasySpin, a comprehensive software package for spectral simulation and analysis in EPR. *J. Magn. Reson.* **2006**, *178*, 42–55. [CrossRef] [PubMed]

Article

Conditions to Control Furan Ring Opening during Furfuryl Alcohol Polymerization

Lucie Quinquet [†], Pierre Delliere [†] and Nathanael Guigo *

Institut de Chimie de Nice, Université Côte d'Azur, CNRS, UMR 7272, 06108 Nice, France; lucie.quinquet@univ-cotedazur.fr (L.Q.); pierre.delliere@univ-cotedazur.fr (P.D.)
* Correspondence: nathanael.guigo@univ-cotedazur.fr
† These authors contributed equally to this work.

Abstract: The chemistry of biomass-derived furans is particularly sensitive to ring openings. These side reactions occur during furfuryl alcohol polymerization. In this work, the furan ring-opening was controlled by changing polymerization conditions, such as varying the type of acidic initiator or the water content. The degree of open structures (DOS) was determined by quantifying the formed carbonyl species by means of quantitative ^{19}F NMR and potentiometric titration. The progress of polymerization and ring opening were monitored by DSC and FT-IR spectroscopy. The presence of additional water is more determining on ring opening than the nature of the acidic initiator. Qualitative structural assessment by means of ^{13}C NMR and FT-IR shows that, depending on the employed conditions, poly(furfuryl alcohol) samples can be classified in two groups. Indeed, either more ester or more ketone side groups are formed as a result of side ring opening reactions. The absence of additional water during FA polymerization preferentially leads to opened structures in the PFA bearing more ester moieties.

Keywords: biobased poly(furfuryl alcohol); ring-opening; degree of open structures

---

Citation: Quinquet, L.; Delliere, P.; Guigo, N. Conditions to Control Furan Ring Opening during Furfuryl Alcohol Polymerization. *Molecules* 2022, 27, 3212. https://doi.org/10.3390/molecules27103212

Academic Editor: Sylvain Caillol

Received: 19 April 2022
Accepted: 13 May 2022
Published: 17 May 2022

Publisher's Note: MDPI stays neutral with regard to jurisdictional claims in published maps and institutional affiliations.

Copyright: © 2022 by the authors. Licensee MDPI, Basel, Switzerland. This article is an open access article distributed under the terms and conditions of the Creative Commons Attribution (CC BY) license (https://creativecommons.org/licenses/by/4.0/).

1. Introduction

In recent years, research on biobased alternatives to fossil-based polymers has been developed as part of a general concern for sustainability [1]. Some, such as poly(lactic acid), can be obtained from starch [2]. However, arable lands are necessary in order to produce said starch, thus competing with food crops. Extensive studies have been conducted to substitute petroleum-based chemicals while avoiding such competition. Lignocellulosic biomass is an interesting resource since it can be obtained through agricultural wastes. It is constituted of lignin, cellulose, and hemi-cellulose [3] that can be processed into phenols, hexoses, and pentoses before being further converted into platform molecules [4]. Lignin may yield phenolic precursors that can partly substitute petroleum-based phenol for the preparation of phenol-formaldehyde resins [5,6]. The hexoses can be converted into hydroxymethylfurfural, a platform chemical that can be oxidized into 2,5-furandicarboxylic acid. This latter can be polymerized into poly(ethylene 2,5-furandicarboxylate), a serious candidate as a replacement for poly(ethylene terephthalate) [7]. Finally, pentoses can be dehydrated into furfural, another key platform molecule [8,9]. Furfural is most widely used for the production of furfuryl alcohol (FA) by catalytic hydrogenation [4]. FA can also be directly produced from xylose using one-pot systems [10]. The former can be polymerized into poly(furfuryl alcohol) (PFA) through acid catalyzed polycondensation (Scheme 1). This complex polymerization has been extensively studied [11–19]. From these studies, two main steps were identified in the polymerization. The first one consists in oligomer formation by FA polycondensation and the second is a branching via Diels–Alder cycloaddition, leading to a high crosslinking density. The tightness of the networks induces the strong brittle behavior of the material. This can limit its uses as a thermoset resin for structural applications when resilience and/or toughness are demanded. As a consequence, PFA is rarely used on its own and is mostly used in composites with reinforcement, such as flax fibers [20] or carbon fibers [21]. Currently, PFA is mostly used to bind sand particles together into foundry molds [22].

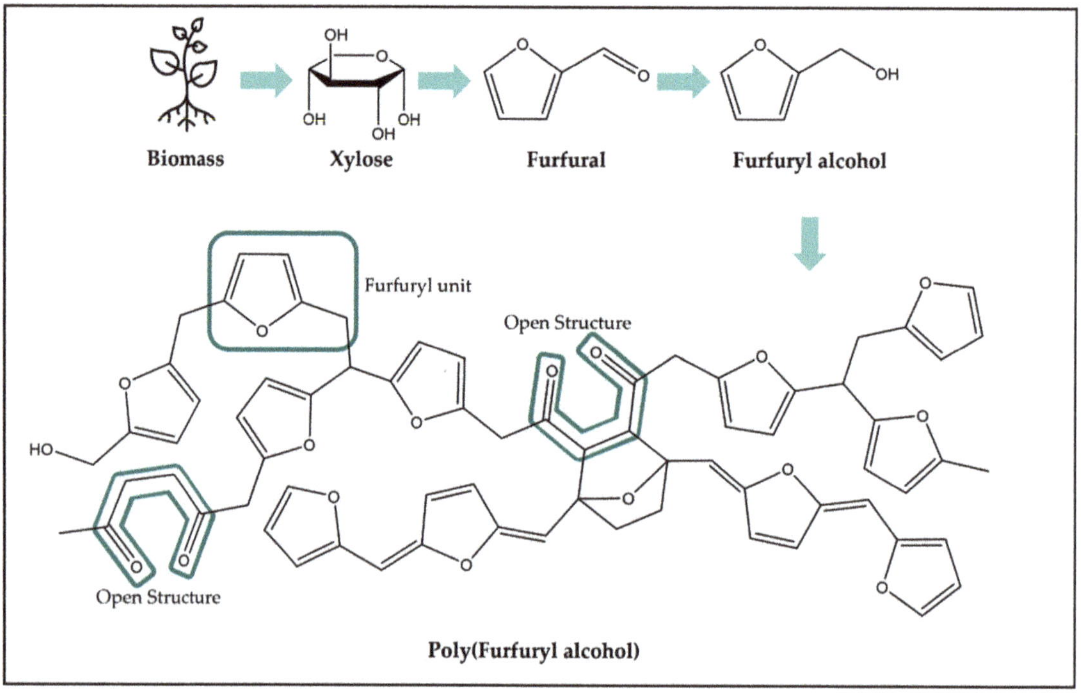

Scheme 1. Chain of biobased PFA's creation and highlight of the open structures.

The ring opening reaction of furfuryl alcohol yields levulinic acid (LA) or its related esters. The conversion of FA through this pathway has recently been a subject of interest since LA is an interesting platform molecule [23–26]. However, few studies have focused on the ring-opening reactions within PFA. Falco et al. [27] studied the effect of the furan ring opening on the structure and properties of PFA. Indeed, the furan ring can cleave into carbonyl containing moieties within the polymer [18,19,28]. They showed that the polymerization of FA with protic polar solvents (water, isopropanol) leads to more open structures, which have been associated to a lower cross-link density and lower glass transition temperature than the neat system.

Since carbonyl containing species such as ketones are formed during the furan ring opening, it is interesting to quantitively follow the amount of carbonyls in the reacting system. Several techniques have been developed to quantify the ketones in a product, such as the Faix method [29] modified by Black [30] via potentiometric titration and quantitative ^{19}F NMR [31]. These methods have been recently adapted to carbonyls formed within PFA in a study by Delliere et al. They investigated how the degree of opened structures progresses with the course of polymerization of FA [32] and proposed a new quantification metric for these opened structures, namely the 'degree of open structures (DOS)'. It is defined as the moles of opened structures from furan opening per moles of furan in the sample. It is assumed that each open structure corresponds to two ketones which were quantified using carbonyl quantification as formerly explained. Following the example depicted in Scheme 1, there is theoretically 14 furfuryl units but two of them have been opened into carbonyl containing species and two merged through a Diels Alder reaction. The resulting DOS of 0.14 was calculated as follows:

$$\text{DOS} = \frac{2 \text{ open structures}}{14 \text{ furfuryl units}} = 0.14 \tag{1}$$

The DOS will thus range from 0 to 1 assuming that one furfuryl unit will give a maximum of two ketonic species.

To our knowledge, no studies were devoted to the effects of the polymerization reaction conditions on the ring opening reaction within PFA.

Accordingly, in this work, the conditions of the furan ring opening during polymerization were studied.

First, the role of the acidic initiator on the furan ring opening was studied by comparing Brönsted initiators, such as citric acid, oxalic acid, and acetic acid, with Lewis initiators, such as boron trifluoride, iodine, alumina, or montmorillonite K10 (MMT-K10). Most have already been used in the past [33–35]. A Brönsted superacid, such as trifluoromethanesulfonic acid, as well as initiators combining the two acidities such as cation exchanged montmorillonite with alkyl-ammonium, were also employed. Indeed, cation exchanged montmorillonite (Org-MMT) displays a dual acidity which, as highlighted by Zavaglia et al. [34], leads to an acceleration of the FA polymerization rate. Therefore, the dual acidity of Org-MMT might arguably affect the ring opening occurring during polymerization. The effect of additional water on the ring opening was investigated as well.

In a final part, the nature of the carbonyl groups was investigated using PFAs from the aforementioned experiments as well as PFAs from syntheses using isopropanol as a solvent. They were then compared by displaying the $\text{DOS}_{\text{Titri}}$ as a function of the C=O area of the corresponding FTIR spectra.

2. Results and Discussion

The main purpose of this work was to study the furan ring opening, focusing on certain aspects, such as the effects of the acidic initiator, the presence or absence of additional water, and its ratio on the ring opening.

2.1. Screening of the Different Initiators

As stated previously, different initiators were investigated. For the Brönsted type, citric acid and oxalic acid were chosen as they have already been used for preparing PFA, as well as acetic acid for its industrial accessibility [35].

A Brönsted superacid, namely trifluoromethanesulfonic acid (TfOH), was also investigated.

As for the Lewis type, boron trifluoride in methanol solution was chosen for being in liquid state. Alumina, iodine, and MMT-K10 have been used in the past [11,33,36].

Montmorillonite (MMT) belongs to the family of smectites clays which are considered as selective, safe, efficient, and eco-friendly initiators. It is a layered aluminosilicate in which the silica and alumina form a sheet like structure. The hydrophilic nature of the clays is explained by the presence of alkali charge compensating counterions within the interlayer spaces. These counterions can be exchanged with organic or metallic cations. MMT exhibits both Brönsted and Lewis acid sites. The Brönsted acidity is mainly due to dissociation of the intercalated water molecules coordinated to cations. The Lewis acid sites are located on alumina sheets. Accordingly, MMT is of great interest in the case of FA polymerization. Indeed, its dual acidic character could catalyze both condensation and addition during the FA polymerization [34].

Therefore, Sodium MMT (MMT-K10) and an organically modified MMT obtained by exchanging the counterions with octadecyl ammonium cations (Org-MMT) were used in this study to separately study the effect of the Lewis acid site with MMT-K10 and the effect of the dual acidity with Org-MMT.

The results of the investigations on the role of the initiator are shown in Figure 1 for the titration results and Figure S1 for the NMR results. All results are indexed in the Supplementary Materials Table S1. The DOS varies significantly but there is no clear demarcation between the different types of acidity.

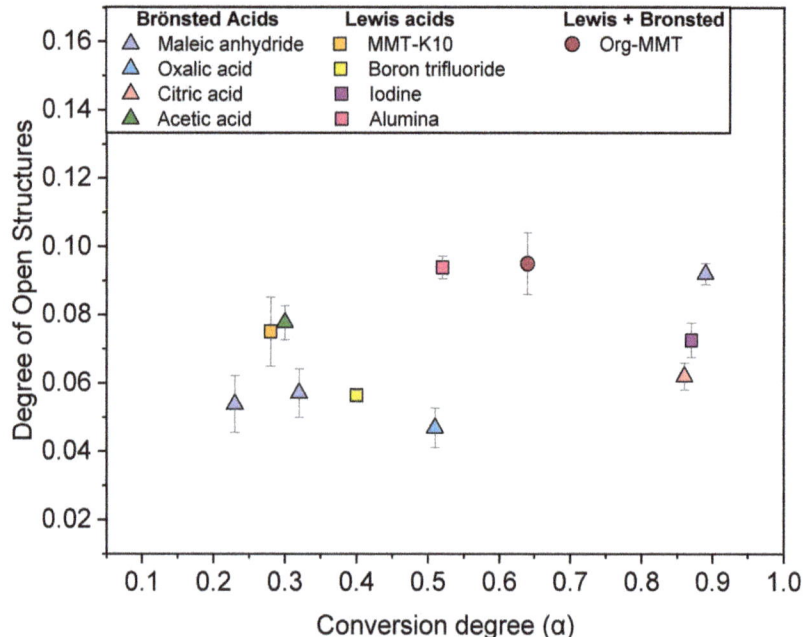

Figure 1. Comparison of the degree of open structures obtained from titration method synthesized with different initiators in function of the conversion degree.

The results plotted in Figure 1 suggest that the nature of acidic initiator (i.e., Lewis vs. Brönsted) does not particularly affect the DOS. Indeed, at a similar reaction progress, Lewis and Brönsted acids do not notably stand out from one another, and neither do the combined acidities.

The results from the ^{19}F NMR method are available in Figure S1. A point worth noting is the undeniable difference between the titration and NMR results. They follow the same trend. However, the DOS$_{Titri}$ is always superior to the DOS$_{NMR}$. As explained in a previous work [32], this gap may be explained by the difference in size, and therefore steric hindrance, of the derivatization agents used in both cases to react with the ketones. The potentiometric titration uses hydroxylamine, unlike the NMR titration which uses much larger hydrazines, such as 4-(trifluoromethyl)phenylhydrazine. The differences in the DOS of these two techniques demonstrate the interest in using both of them, giving in the end an overall range of the DOS.

These investigations showed that the DOS progressively increases with the conversion degree. However, the type of acidity does not particularly influence the DOS.

Another aspect worthy of examination was the role of added water on the DOS.

2.2. The Role of Additional Water

In order to determine whether the presence of additional water and its ratio affect the DOS, two approaches were implemented.

2.2.1. DOS Comparison with and without Additional Water

First, the effect of additional water was studied by conducting several syntheses in which an array of initiators was tested with and without additional water in a same amount in the system, i.e., the initial water amount, which will come in addition to the water generated during the polycondensation. The FA/water ratio was fixed to 50/50 w/w. The results of the first approach are indexed in Table S2. For each initiator, the water analogue has a higher DOS.

The results plotted in Figure 2 suggest that the presence of water does influence the DOS by increasing it. The ^{19}F NMR method in Figure S2 presents the same trend. Indeed, at a similar conversion degree, the DOS is higher when the reaction was performed in the presence of additional water.

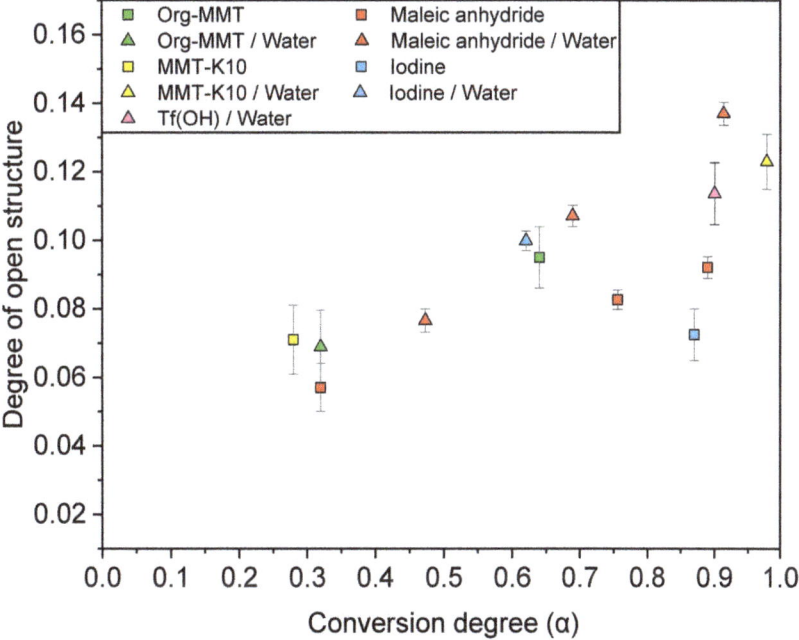

Figure 2. Comparison of the degree of open structures obtained by titration method and synthesized with different initiators with and without additional water (50/50 w/w) as function of the conversion degree.

As stated, the results of this investigation show that the added water plays a role in the furan ring opening in agreement with both Falco et al. and Delliere et al. [27,32]. In addition to this seminal investigation, it is clearly highlighted here that the increase in DOS is not related to the acidic initiator. In other words, all the different FA/initiator systems show a DOS increase with additional water.

2.2.2. The Role of the Additional Water Ratio

Secondly, several polymerizations were conducted by selecting only one initiator, namely maleic anhydride (MA), and changing the FA/additional water ratio, ranging from neat (100/0) to 40/60.

The results of the FA/additional water ratio investigations are indicated in Table S3. Figure 3 exhibits the DOS$_{Titri}$ as a function of the conversion degree for PFAs synthetized with more or less additional water. The ^{19}F NMR results are depicted in Figure S3.

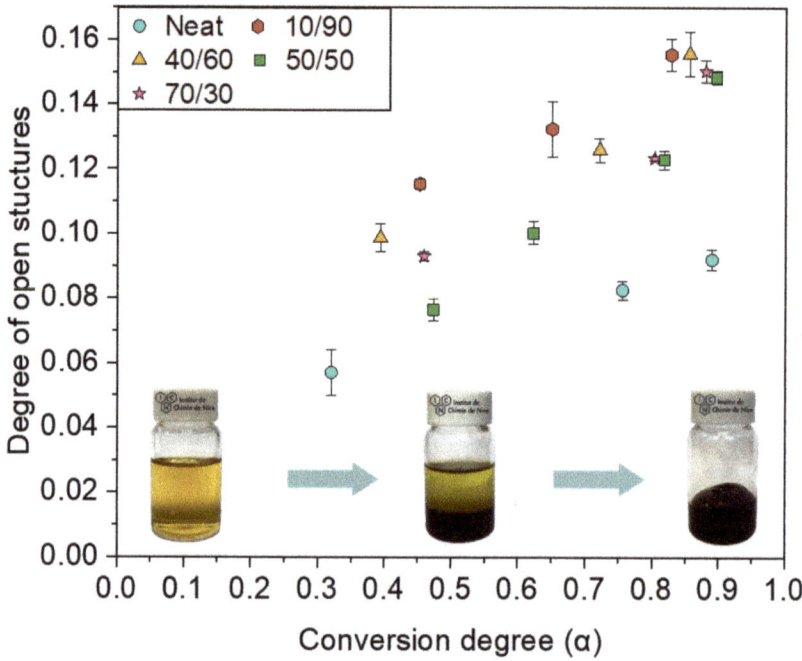

Figure 3. Comparison of the degree of open structures obtained with titration method and synthesized with different FA/Additional water ratios within function of the conversion degree.

Figure 3 and Figure S3 highlight that the DOS is continuously increasing with the conversion degree, independently of the FA/water ratio. The neat PFA shows the lowest DOS overall from the beginning to the near end of the polymerization, i.e., $\alpha \approx 0.9$. Regarding the water ratios, systems with less water (40/60 and 10/90) have slightly higher DOS than the 50/50 and 70/30 ratios at $\alpha < 0.80$. Nonetheless, at higher conversion degrees, i.e., $\alpha \approx 0.9$, the gap shrinks around a DOS of 0.15. This is most likely due to the insolubility of PFA in water. Indeed, FA and water are miscible, whereas above $\alpha \approx 0.20$ a demixion occurs, as depicted in Figure 3. Thus, small, open, and water-soluble oligomers might not be taken into account in the DOS at low conversion degrees.

Then, when the polymerization progresses, they would be integrated in the PFA phase, leading to a final DOS of 0.15.

Consequently, these investigations allowed to determine that adding water plays a role in the opening of the furan ring, but the ratio of added water does not impact the DOS that is reached at the end of polymerization.

2.2.3. Carbonyl Groups in PFA

A variety of PFAs were synthetized using the aforementioned catalysts with and without additional water. Secondly, syntheses using isopropanol (50/50 to FA) were performed. The DOS_{Titri} was measured as well as the FTIR C=O area, which is obtained by normalizing the FTIR spectra followed by an integration between 1840 and 1650 cm^{-1}. In a study of C=O content in lignins, Faix et al. used this concept of FTIR C=O area to quickly determine the C=O content of lignins [29]. They obtained excellent linear correlation between the C=O content and the FTIR C=O area ($R^2 > 0.98$). The goal of this study was to check if determining the DOS of PFAs can be somehow correlated with the FTIR C=O area.

The FTIR spectra of neat PFA and 50/50 water PFA catalyzed by 2%w MA at conversion degrees $\alpha \approx 0.5$ and $\alpha \approx 0.8$ are shown in Figure 4A. Both samples exhibit a C=O band at 1710 cm^{-1}. This band has been attributed to ketonic moieties. However, the neat PFA exhibits a shoulder around 1780 cm^{-1}, which has been attributed to ester moieties [19,27]. Figure 4B presents the ^{13}C NMR of the same samples. The peaks corresponding to ketones appear between 195 and 210 ppm and, as shown in Figure 4B, the sample PFA 50/50 exhibits more peaks in this region compared to the neat PFA. On the other hand, the resonance attributed to esters are located between 160 and 178 ppm. The neat PFA shows more peaks attributed to esters compared to PFA 50/50. In PFA 50/50, where α ≈ 0.8, the 174 ppm peak is attributed to the ester band from the levulinic acid. Indeed, low levels of acid can be detected in such PFAs [32].

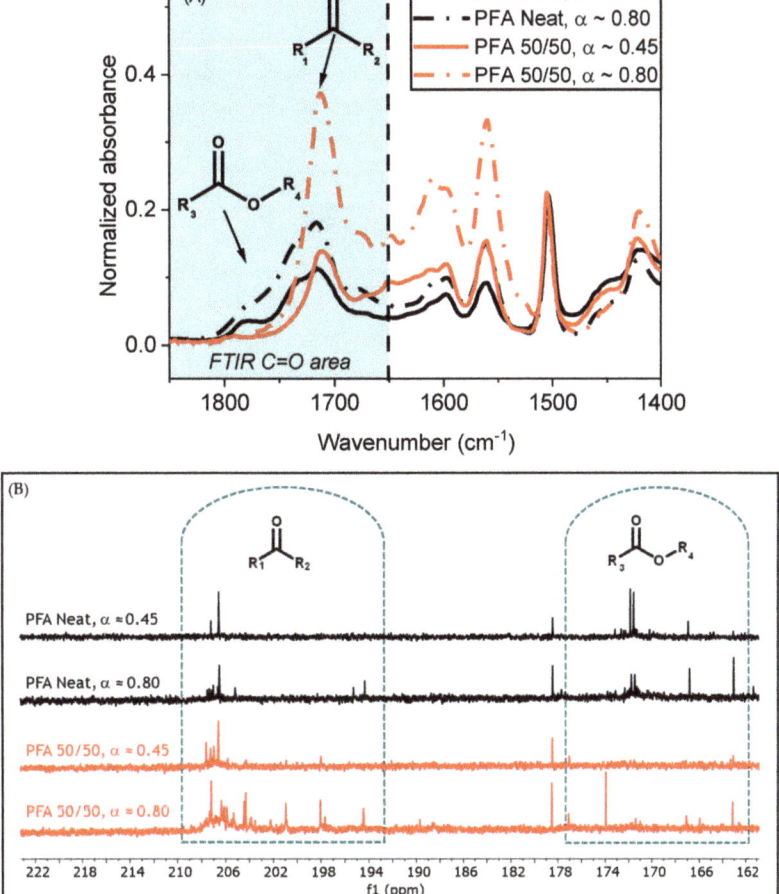

Figure 4. (**A**) Normalized FTIR spectra of PFA 50/50 water and neat PFA neat at α ≈ 0.5 and α ≈ 0.8. (**B**) ^{13}C NMR spectra of PFA 50/50 water and neat PFA at α ≈ 0.5 and α ≈ 0.8 with an accumulation of 6000 scans.

Figure 5 shows the DOS$_{Titri}$ as function of the FTIR C=O area for all the investigated PFAs. A separation between PFAs prepared with additional water, the ones prepared in neat conditions, and the ones prepared with additional IPA is portrayed in Figure 5. It is worth mentioning that both ketones and aldehydes functions are quantified by the oximation method, but this is not the case for the ester function. However, it is rather difficult to discriminate the ketones/aldehydes from the ester functions when integrating the more or less broad carbonyl peak in FTIR. Accordingly, if only ketones or aldehydes are present in the systems, the C=O content should increase linearly with the FTIR C=O area. However, if other functional groups such as esters are taken into account in the C=O area, the DOS vs. the C=O area will not be linear.

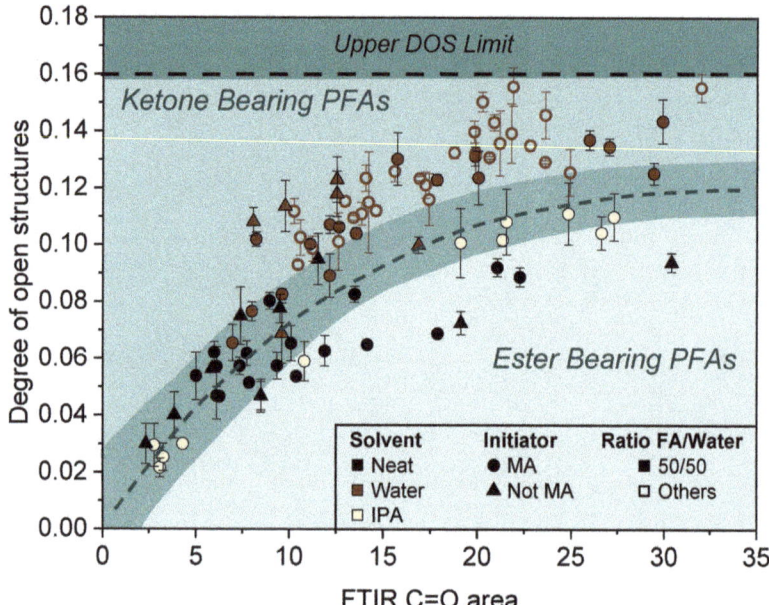

Figure 5. Degree of open structures obtained with titration method against FTIR C=O area for PFA synthetized in neat, aqueous and 50/50 IPA systems. PFA have been synthetized with aforementioned catalysts.

Moreover, in their work, Wewerka et al. used alumina as a catalyst and identified lactones in the system [11]. In Figure 5, the data point of alumina catalyzed PFA (0.094; 30.4) is indeed in the ester bearing group.

Regarding the PFAs prepared with 50/50 IPA, they also belong to the ester bearing groups in Figure 5 as observed by Falco et al. [27]. Yet, they have a higher DOS than the neat PFA. This suggests an opening of furan rings through the formation of compounds, such as levulinates or lactones, thus increasing both the DOS and FTIR C=O area.

Finally, the ^{13}C NMR spectra of PFA 50/50 in Figure 4B exhibits a significant number of carbonyl peaks from 194 to 210 ppm. Most of them are centered around 208 ppm, which matches with 1,4-diketones. Similar peaks can be found in the neat PFA. However, at $\alpha \approx 0.80$, new peaks rise in the 194–205 ppm area, in PFA 50/50 only. This confirms the existence of types of carbonyl containing species other than 1,4-diketones in this system.

These results suggest that additional water either prevents the formation of esters during FA polymerization or hydrolyses them or both.

3. Materials and Methods

3.1. Materials

Furfuryl alcohol (FA) (96%), 4-(trifluoromethyl)phenylhydrazine (96%), Hydroxylamine hydrochloride (99%), Maleic anhydride (99%), Citric acid (99.5%), Oxalic acid (99%), Acetic acid (99.5%), Isopropanol (98%)1,4-Bis(trifluoromethyl)benzene (98%), Triethanolamine (99%), trifluoromethanesulfonic acid (98%), Boron trifluoride-methanol solution (14%), and MMT K10 (Montmorillonite K10) were supplied by Sigma-Aldrich.

Organically modified montmorillonite (Org-MMT), obtained by the exchange of MMT counterions with octadecyl ammonium cations, was supplied from Nordmann, Rassmann GmbH (Nanomer I30E). DMSO-d6 (99.80%) was supplied Euristop. Chromium(III) acetylacetonate (97%) was supplied by Alfa aesar. Iodine resublimed was supplied by Carlo Erba. Aluminoxid 90 neutral was supplied by Carl Roth. Ethanol (96 v%) was supplied by VWR. Water was distilled using SI Analytics distillation unit purchased from Thermo Fischer Scientific, France (conductivity ~1 μS/cm). All chemicals were used as received.

3.2. Methods

3.2.1. Poly(furfuryl alcohol)'s Synthesis

Each synthesis was carried out in the following way: For syntheses without additional water, 25 g of FA were first added in a 100 mL round bottom flask. The appropriate amount of catalyst (Table 1) was then added under vigorous stirring. After complete solubilization of the catalyst, the mixture was then heated at 80 °C under reflux using a refrigerating tube and under vigorous stirring for an hour, samples were taken regularly and inspected via IR spectrometry.

Table 1. FA's polymerization conditions.

Initiator	Quantity of Initiator (Based on FA)	FA/Additional Water Ratio
Citric acid	2 mol%	100/0
Boron trifluoride in MeOH (14%)	0.07 mol%	100/0
Oxalic acid	2 mol%	100/0
Acetic acid	2 mol%	100/0
MMT K10	2 wt%	100/0; 50/50
Org-MMT	2 wt%	100/0; 50/50
Alumina	10 mol%	100/0
Iodine	1 mol%	100/0; 50/50
Trifluoroacetic acid	0.1 mol%	50/50
Maleic anhydride	2 mol%	100/0; 70/30; 50/50; 40/60; 10/90

After one hour, the refrigerating tube was taken off and the mixture was raised to 110 °C in order to remove the remaining water. The reaction was stopped when the mixture became a dark brown resinous rubber, still elastic, not hard. For syntheses with additional water, the appropriate amount of water (Table 1) was first added in a round bottom flask. The process is then identical to the one previously described. Depending on the initiator, the time of reaction varied. However, it never exceeded eight hours.

3.2.2. Liquid State Nuclear Magnetic Resonance (NMR)

NMR spectra were recorded on a Bruker AVANCE III (400 MHz for ^{19}F, 125.77 MHz for ^{13}C) using d$_6$-DMSO. ^{19}F NMR spectra were carried out using a standard fluorine 90° pulse program, a relaxation delay of 3 s, and 256 scans acquired between −35 and −85 ppm. The processing of spectra was conducted with MestReNova software, and baseline (Breinstein polynomial order 3) and phase correction were applied. Finally, the spectra were referenced to internal standard at −62.15 ppm. The ^{13}C NMR spectra were recorded with a 2 s relaxation delay and 6000 scans between −27.4 and 266.3 ppm. A high number of scans was used to ensure good signal/noise ratio in the carbonyl area.

3.2.3. Differential Scanning Calorimetry (DSC)

The DSC measurements were performed on a DSC1 from Mettler Toledo equipped with a FRS5 sensor (with 56 thermocouples Au-Au/Pd). The temperature and enthalpy calibrations were performed using indium and zinc standards and the data were analyzed with STARe software. The samples weighted from 5 mg to 10 mg. Volatilization of the polycondensation byproduct occurs during the curing of the different mixtures, which is why the samples were placed into 30 µL high-pressure gold-plated crucibles. According to previous work, the conversion degree (α) was calculated using the following Equation (2) [34,37,38]:

$$\alpha = \frac{Q_{Max} - Q_{residual}}{Q_{Max}} \qquad (2)$$

Here, Q_{Max} is the total heat released during the polymerization. $Q_{residual}$ corresponds to the heat generated when completing the polymerization of oligomers.

3.2.4. Fourier-Transform Infrared Spectroscopy (FTIR)

The FTIR (ATR) measurements were recorded on a Thermo scientific Nicolet iS20 spectrometer from 400 to 4000 cm^{-1} with 64 scans and a resolution of 4.0 cm^{-1}. A spectrum of air was used as background. The acquired spectra were analyzed with the OMNIC 9 software. First, a baseline was applied to the spectra. Then, they were normalized by using the C-O band at 1000 cm^{-1} as a reference band (absorbance = 1). The normalization is required to avoid variations caused by the refraction indexes. Indeed, it is the most intense band within both PFA and furfuryl alcohol [16,39]. The FTIR C=O area was set to (1650–1840 cm^{-1}), according to literature [40].

3.2.5. Carbonyl Quantification Methods and DOS Determination

In order to determine the DOS of the PFA resins, two carbonyl quantification methods adapted to furanic macromolecular systems were used as developed by Delliere et al. [32].

The first method, based on the work of Faix et al., uses the oximation and was conducted as follows [29,30]: About 100 mg of PFA were accurately weighted in a vial, followed by the addition of 2 mL of DMSO. The vials were then heated for 5 min at 80 °C in order to quickly dissolve the PFA. This step was followed by the addition of 2 mL of a 0.44 M hydroxylamine hydrochloride-EtOH solution (80 v%) and 2 mL of a 0.52 M triethanolamine-EtOH solution (96 v%). Moreover, blanks were performed, and the samples were triplicated. The last step consisted in heating the samples for 24 h at 80 °C in order to reach total oximation. The samples were then allowed to cool-down before titration with a 0.1 N HCl solution.

Consequently, the carbonyl content was calculated using the following equation:

$$CO\ content\ (mmol/g) = \frac{(V_0 - V)}{m_{PFA}} * [HCl] \qquad (3)$$

where V_0 is the endpoint of the blank, V is the endpoint of the sample, and m_{PFA} stands for the mass of PFA in the sample.

The other method employed for carbonyl quantification was NMR based on the work of Constant et al. [31]. Quantifications were conducted as follows:

About 100 mg of PFA, 140 mg of 4-(trifluoromethyl)phenylhydrazine (CF$_3$FNHNH$_2$) and 20 mg of 1,4-bis(trifluoromethyl)benzene were accurately weighted in a 1.5 mL vial. Thereafter, 400 µL of d_6-DMSO were added and the vial was shaken for 10 min before 5 s of centrifugation. Then, the content of the vials was transferred into NMR tubes and underwent heating for 24 h at 40 °C. Finally, 100 µL of a relaxing agent solution were added and homogenized just before analysis (28 mg/mL of chromium(III) acetylacetonate in d_6-DMSO). The carbonyl content can be calculated through the following equation using the integrated areas from the acquired spectra:

$$CO\ content\ (mmol/g) = \frac{\left[A_c - \left(\frac{A_{H°}*m_H}{m_{H°}}\right)\right] * m_{IS}}{A_{IS} * 0.5 * m_{PFA} * M_{IS}} * 10^3 \qquad (4)$$

where A_c stands for the hydrazones' areas. m_{IS}, M_{IS} and A_{IS} correspond respectively to the mass, molecular mass, and the integrated area of the standard. Since an impurity appears to be present in the commercial CF$_3$FNHNH$_2$ its amount was determined in every batch. This results in a correction factor where $A_{H°}$ is the integrated area of the impurity in the reference and $m_{H°}$ is the mass of hydrazine within it. m_H is the mass of hydrazine within the sample. Finally, the area of the standard was multiplied by 0.5 as it bears two CF_3 groups.

At last, as described in a previous work, PFA resins can be characterized by their degree of open structures [32]. Indeed, as described in earlier publications, carbonyls in PFAs emerge from the opening of a furan ring into 1,4-dicarbonyls [18,19]. Thus, the DOS can be defined as follows:

$$\text{Degree of open structures (DOS)} = \frac{1}{2} \cdot \frac{N_{C=O}}{N_{furanic}} \quad (5)$$

where $N_{C=O}$ is the moles of carbonyls and $N_{furanic}$ the moles of furan units, i.e., furan-CH_2–. This can be converted into Equation (6) for the carbonyl content to appear in the equation.

$$\text{DOS} = \frac{1}{2} \cdot 1000 \cdot CO\ content \cdot M_{(furfuryl\ unit)} \quad (6)$$

The DOS obtained from the NMR method will be called DOS_{NMR} while the one resulting from oximation will be DOS_{Titri}.

4. Conclusions & Prospects

In the present study, the conditions impacting the furan ring opening occurring during FA polymerization were investigated. Accordingly, the degree of open structures (DOS) was used as an indicator. Different acidic catalysts, spanning from Brönsted to Lewis acids, were employed to initiate FA polymerization. However, the results point towards the fact that no specific type of acidity particularly promotes or reduces the DOS.

Then, the influence of the additional water content was examined and different initial FA/water ratios ranging from 100/0 to 10/90 were employed. When FA polymerization is not too advanced (i.e., α < 0.8), the DOS is quite influenced by the water ratio. Logically, a higher water content will lead to a higher DOS. However, for α > 0.8, the DOS values are merging for all the FA/water ratio to an ultimate value of about 0.15. Only the neat FA (i.e., polymerized without any additional water) has a much lower DOS (i.e., 0.09).

Finally, insights on the nature of the carbonyls contained in the open structure of PFAs were given. The existence of esters in these open structured resins was highlighted, especially when FA polymerization was conducted without additional water. On the other hand, FA polymerization conducted with additional water rather highlighted the formation of ketonic groups in their opened structures and much fewer esters were identified.

To conclude, even a relatively low water content in FA (i.e., 30%) will enable to promote furan ring opening independently of the initiator employed. The absence of additional water on the other hand limits, but does not prevent, the formation of open structures that contain esters functions.

This investigation sheds new light on how side-reactions during FA polymerization can influence the final polymeric structure. The fact that FA polymerization is conducted under acidic catalysis in the presence of water (coming either from the condensation or added from the formulation) irrevocably leads to open structures containing ester and ketonic species. It might be interesting to investigate which already cross-linked PFA resins would be sensitive to ring opening occurring after complete polymerization.

Supplementary Materials: The following are available online at https://www.mdpi.com/article/10.3390/molecules27103212/s1, Figure S1: Comparison of the degree of open structures obtained from ^{19}F NMR method synthesized with different initiators in function of the conversion degree. Table S1: Index of the results of the different initiators used without water. Figure S2: Comparison of the degree of open structures obtained ^{19}F NMR method and synthesized with different initiators with and without additional water (50% w/w) in function of the conversion degree. Table S2: Index of the results of the different initiators used without water. Figure S3: Comparison of the degree of open structures obtained with ^{19}F NMR method and synthesized with different FA/Additional water ratios with in function of the conversion degree. Table S3: Index of the results of the different FA/additional water ratios used with maleic anhydride.

Author Contributions: Conceptualization, N.G.; methodology, P.D.; validation, N.G. and P.D.; formal analysis, L.Q. and P.D.; investigation, L.Q. and P.D.; data curation, L.Q. and P.D.; writing—original draft preparation, L.Q. and P.D.; writing—review and editing, N.G.; supervision, N.G.; project administration, N.G.; funding acquisition, N.G. All authors have read and agreed to the published version of the manuscript.

Funding: This research was funded by Agence Nationale de la Recherche (ANR), project FUTURES.

Institutional Review Board Statement: Not applicable.

Informed Consent Statement: Not applicable.

Data Availability Statement: Data is contained within the article or Supplementary Material.

Acknowledgments: The authors would like to acknowledge the COST Action FUR4Sustain- European network of FURan based chemicals and materials FOR a Sustainable development, CA18220, supported by COST (European Cooperation in Science and Technology, www.cost.eu, accessed on 12 April 2022).

Conflicts of Interest: The authors declare no conflict of interest. The funders had no role in the design of the study; in the collection, analyses, or interpretation of data; in the writing of the manuscript, or in the decision to publish the results.

Sample Availability: Samples of the compounds are not available from the authors.

References

1. Gandini, A.; Lacerda, T.M.; Carvalho, A.J.F.; Trovatti, E. Progress of Polymers from Renewable Resources: Furans, Vegetable Oils, and Polysaccharides. *Chem. Rev.* **2016**, *116*, 1637–1669. [CrossRef]
2. Castro-Aguirre, E.; Iñiguez-Franco, F.; Samsudin, H.; Fang, X.; Auras, R. Poly(lactic acid)—Mass production, processing, industrial applications, and end of life. *Adv. Drug Deliv. Rev.* **2016**, *107*, 333–366. [CrossRef] [PubMed]
3. Pettersen, R.C. The Chemical Composition of Wood. *Adv. Chem.* **1984**, *207*, 57–126. [CrossRef]
4. Corma Canos, A.; Iborra, S.; Velty, A. Chemical routes for the transformation of biomass into chemicals. *Chem. Rev.* **2007**, *107*, 2411–2502. [CrossRef] [PubMed]
5. Windeisen, E.; Wegener, G. Lignin as Building Unit for Polymers. *Polym. Sci. A Compr. Ref.* **2012**, *10*, 255–265. [CrossRef]
6. Foyer, G.; Chanfi, B.H.; Boutevin, B.; Caillol, S.; David, G. New method for the synthesis of formaldehyde-free phenolic resins from lignin-based aldehyde precursors. *Eur. Polym. J.* **2016**, *74*, 296–309. [CrossRef]
7. Sousa, A.F.; Silvestre, A.J.D. Plastics from renewable sources as green and sustainable alternatives. *Curr. Opin. Green Sustain. Chem.* **2022**, *33*, 100557. [CrossRef]
8. Ricciardi, L.; Verboom, W.; Lange, J.-P.; Huskens, J. Production of furans from C5 and C6 sugars in presence of polar organic solvents. *Sustain. Energy Fuels* **2021**, *6*, 11–28. [CrossRef]
9. Mariscal, R.; Maireles-Torres, P.; Ojeda, M.; Sádaba, I.; López Granados, M. Furfural: A renewable and versatile platform molecule for the synthesis of chemicals and fuels. *Energy Environ. Sci.* **2016**, *9*, 1144–1189. [CrossRef]
10. Sixta, H. Towards the Green Synthesis of Furfuryl Alcohol in A One-Pot System from Xylose: A Review. *Catalysts* **2020**, *10*, 1101. [CrossRef]
11. Wewerka, E.M. Study of the γ-alumina polymerization of furfuryl alcohol. *J. Polym. Sci. Part A-1 Polym. Chem.* **1971**, *9*, 2703–2715. [CrossRef]
12. Choura, M.; Belgacem, N.M.; Gandini, A. Acid-catalyzed polycondensation of furfuryl alcohol: Mechanisms of chromophore formation and cross-linking. *Macromolecules* **1996**, *29*, 3839–3850. [CrossRef]
13. Buchwalter, S.L. Polymerization of Furfuryl Acetate in Acetonitrole. *J. Polym. Sci.* **1985**, *23*, 2897–2911. [CrossRef]
14. Kim, T.; Assary, R.S.; Pauls, R.E.; Marshall, C.L.; Curtiss, L.A.; Stair, P.C. Thermodynamics and reaction pathways of furfuryl alcohol oligomer formation. *Catal. Commun.* **2014**, *46*, 66–70. [CrossRef]
15. Montero, A.L.; Montero, L.A.; Martínez, R.; Spange, S. Ab initio modelling of crosslinking in polymers. A case of chains with furan rings. *J. Mol. Struct. THEOCHEM* **2006**, *770*, 99–106. [CrossRef]
16. Barsberg, S.; Thygesen, L.G. Poly(furfuryl alcohol) formation in neat furfuryl alcohol and in cymene studied by ATR-IR spectroscopy and density functional theory (B3LYP) prediction of vibrational bands. *Vib. Spectrosc.* **2009**, *49*, 52–63. [CrossRef]
17. Bertarione, S.; Bonino, F.; Cesano, F.; Damin, A.; Scarano, D.; Zecchina, A. Furfuryl alcohol polymerization in H-Y confined spaces: Reaction mechanism and structure of carbocationic intermediates. *J. Phys. Chem. B* **2008**, *112*, 2580–2589. [CrossRef]
18. Conley, R.T.; Metil, I. An investigation of the structure of furfuryl alcohol polycondensates with infrared spectroscopy. *J. Appl. Polym. Sci.* **1963**, *7*, 37–52. [CrossRef]

19. Tondi, G.; Cefarin, N.; Sepperer, T.; D'Amico, F.; Berger, R.J.F.; Musso, M.; Birarda, G.; Reyer, A.; Schnabel, T.; Vaccari, L. Understanding the polymerization of polyfurfuryl alcohol: Ring opening and diels-alder reactions. *Polymers* **2019**, *11*, 2126. [CrossRef]
20. Toriz, G.; Arvidsson, R.; Westin, M.; Gatenholm, P. Novel cellulose ester-poly(furfuryl alcohol)-flax fiber biocomposites. *J. Appl. Polym. Sci.* **2003**, *88*, 337–345. [CrossRef]
21. Mak, K.; Fam, A. Fatigue Performance of Furfuryl Alcohol Resin Fiber-Reinforced Polymer for Structural Rehabilitation. *J. Compos. Constr.* **2020**, *24*, 04020012. [CrossRef]
22. Xi, X.; Wu, Z.; Pizzi, A.; Gerardin, C.; Lei, H.; Du, G. Furfuryl alcohol-aldehyde plywood adhesive resins. *J. Adhes.* **2020**, *96*, 814–838. [CrossRef]
23. Yan, K.; Wu, G.; La, T.; Jarvis, C. Production, properties and catalytic hydrogenation of furfural to fuel additives and value-added chemicals. *Renew. Sustain. Energy Rev.* **2014**, *38*, 663–676. [CrossRef]
24. Lange, J.P.; van de Graaf, W.D.; Haan, R.J. Conversion of furfuryl alcohol into ethyl levulinate using solid acid catalysts. *Chem. Sustain. Chem.* **2009**, *2*, 437–441. [CrossRef]
25. Chappaz, A.; Lai, J.; De Oliveira Vigier, K.; Morvan, D.; Wischert, R.; Corbet, M.; Doumert, B.; Trivelli, X.; Liebens, A.; Jérôme, F. Selective Conversion of Concentrated Feeds of Furfuryl Alcohol to Alkyl Levulinates Catalyzed by Metal Triflates. *ACS Sustain. Chem. Eng.* **2018**, *6*, 4405–4411. [CrossRef]
26. González Maldonado, G.M.; Assary, R.S.; Dumesic, J.; Curtiss, L.A. Experimental and theoretical studies of the acid-catalyzed conversion of furfuryl alcohol to levulinic acid in aqueous solution. *Energy Environ. Sci.* **2012**, *5*, 6981–6989. [CrossRef]
27. Falco, G.; Guigo, N.; Vincent, L.; Sbirrazzuoli, N. Opening Furan for Tailoring Properties of Bio-based Poly(Furfuryl Alcohol) Thermoset. *Chem. Sustain. Chem.* **2018**, *11*, 1805–1812. [CrossRef]
28. D'Amico, F.; Musso, M.E.; Berger, R.J.F.; Cefarin, N.; Birarda, G.; Tondi, G.; Bertoldo Menezes, D.; Reyer, A.; Scarabattoli, L.; Sepperer, T.; et al. Chemical constitution of polyfurfuryl alcohol investigated by FTIR and Resonant Raman spectroscopy. *Spectrochim. Acta—Part A Mol. Biomol. Spectrosc.* **2021**, *262*, 120090. [CrossRef]
29. Faix, O.; Andersons, B.; Zakis, G. Determination of carbonyl groups of six round robin lignins by modified oximation and FTIR spectroscopyy. *Holzforschung* **1998**, *52*, 268–274. [CrossRef]
30. Black, S.; Ferrell, J.R. Determination of Carbonyl Groups in Pyrolysis Bio-oils Using Potentiometric Titration: Review and Comparison of Methods. *Energy Fuels* **2016**, *30*, 1071–1077. [CrossRef]
31. Constant, S.; Lancefield, C.S.; Weckhuysen, B.M.; Bruijnincx, P.C.A. Quantification and Classification of Carbonyls in Industrial Humins and Lignins by 19F NMR. *ACS Sustain. Chem. Eng.* **2017**, *5*, 965–972. [CrossRef]
32. Delliere, P.; Guigo, N. Monitoring the Degree of Carbonyl-Based Open Structure in a Furanic Macromolecular System. *Macromolecules* **2022**, *55*, 1196–1204. [CrossRef]
33. González, R.; Figueroa, J.M.; González, H. Furfuryl alcohol polymerisation by iodine in methylene chloride. *Eur. Polym. J.* **2002**, *38*, 287–297. [CrossRef]
34. Zavaglia, R.; Guigo, N.; Sbirrazzuoli, N.; Mija, A.; Vincent, L. Complex kinetic pathway of furfuryl alcohol polymerization catalyzed by green montmorillonite clays. *J. Phys. Chem. B* **2012**, *116*, 8259–8268. [CrossRef]
35. Li, W.; Wang, H.; Ren, D.; Yu, Y.S.; Yu, Y. Wood modification with furfuryl alcohol catalysed by a new composite acidic catalyst. *Wood Sci. Technol.* **2015**, *49*, 845–856. [CrossRef]
36. Pranger, L.; Tannenbaum, R. Biobased nanocomposites prepared by in situ polymerization of furfuryl alcohol with cellulose whiskers or montmorillonite clay. *Macromolecules* **2008**, *41*, 8682–8687. [CrossRef]
37. Falco, G.; Guigo, N.; Vincent, L.; Sbirrazzuoli, N. FA polymerization disruption by protic polar solvents. *Polymers* **2018**, *10*, 529. [CrossRef]
38. Cadenato, A.; Salla, J.M.; Ramis, X.; Morancho, J.M.; Marroyo, L.M.; Martin, J.L. Determination of gel and vitrification times of thermoset curing process by means of TMA, DMTA and DSC techniques: TTT diagram. *J. Therm. Anal.* **1997**, *49*, 269–279. [CrossRef]
39. Barsberg, S.; Berg, R.W. Combined Raman Spectroscopic and Theoretical Investigation of Fundamental Vibrational Bands of Furfuryl Alcohol (2-furanmethanol). *J. Phys. Chem. A* **2006**, *110*, 9500–9504. [CrossRef]
40. Socrates, G. The Carbonyl Group: C=O. In *Infrared and Raman Characteristic Group Frequencies. Tables and Charts*, 3rd ed.; John Wiley and Sons Ltd.: Chichester, UK, 2004; pp. 110–154.

Article

Plant Oil-Based Acrylic Latexes towards Multisubstrate Bonding Adhesives Applications

Vasylyna Kirianchuk [1], Bohdan Domnich [2], Zoriana Demchuk [3], Iryna Bon [2], Svitlana Trotsenko [2], Oleh Shevchuk [1], Ghasideh Pourhashem [2] and Andriy Voronov [2,*]

[1] Department of Organic Chemistry, Institute of Chemistry and Chemical Technologies, Lviv Polytechnic National University, 79013 Lviv, Ukraine
[2] Department of Coatings and Polymeric Materials, North Dakota State University, Fargo, ND 58102, USA
[3] Oak Ridge National Laboratory, Chemical Sciences Division, Oak Ridge, TN 37830, USA
* Correspondence: andriy.voronov@ndsu.edu

Abstract: To investigate the utility of acrylic monomers from various plant oils in adhesives manufacturing, 25–45 wt. % of high oleic soybean oil-based monomer (HOSBM) was copolymerized in a miniemulsion with commercially applied butyl acrylate (BA), methyl methacrylate (MMA), or styrene (St). The compositions of the resulting ternary latex copolymers were varied in terms of both "soft" (HOSBM, BA) and "rigid" (MMA or St) macromolecular fragments, while total monomer conversion and molecular weight of copolymers were determined after synthesis. For most latexes, results indicated the presence of lower and higher molecular weight fractions, which is beneficial for the material adhesive performance. To correlate surface properties and adhesive performance of HOSBM-based copolymer latexes, contact angle hysteresis (using water as a contact liquid) for each latex-substrate pair was first determined. The data showed that plant oil-based latexes exhibit a clear ability to spread and adhere once applied on the surface of materials differing by polarities, such as semicrystalline polyethylene terephthalate (PET), polypropylene (PP), bleached paperboard (uncoated), and tops coated with a clay mineral paperboard. The effectiveness of plant oil-based ternary latexes as adhesives was demonstrated on PET to PP and coated to uncoated paperboard substrates. As a result, the latexes with high biobased content developed in this study provide promising adhesive performance, causing substrate failure instead of cohesive/adhesive break in many experiments.

Keywords: plant oils; plant oil-based acrylic monomers; miniemulsion polymerization; biobased latexes; waterborne contact adhesive

1. Introduction

It has become evident that the synthesis of sustainable polymers (based on natural resources) has developed into a booming research area that targets the replacing of petroleum-based counterparts in manufacturing with a broad range of polymeric materials. Renewable raw materials, such as cellulose, lignin, vegetable oils, starches, mono- and di-saccharides have attracted growing attention from both industrial and academic researchers in biobased polymeric material design [1,2]. Among others, plant/vegetable oils have become a prospective natural feedstock for synthesizing various biobased polymers and polymeric materials [3]. Chemically, they are mixtures of triglycerides, glycerol esters and various fatty acids. Depending on the oil composition, chain length, unsaturation degree, and types of fatty acids, moieties differ, essentially thus determining the physico-chemical properties of plant oils, as well as the synthesis possibilities and the prospects of particular oils. In fact, their chemical diversity is exciting, as well as offering various synthetic opportunities. Most chemical reactions of plant oil triglycerides proceed by reactions of the ester group, while some other synthetic pathways include reactions of allyl fragments [4]. When various other synthetic methods were applied, oxy-polymerized oils, polyesters [5], polyurethanes [6], polyamides, acrylates, and epoxy resins based on plant oil triglycerides [7] were successfully synthesized.

In the field of adhesives, polymeric materials from biobased renewable resources are increasingly being considered [8,9]. While using biopolymers, such as natural rubber, proteins, and polysaccharides (cellulose and starch, in particular) has been historically introduced, research interests in synthesizing polymeric adhesives, based on new renewable monomers, from various natural resources have recently grown, due to expansion in the sustainable products market. However, popular renewable feedstocks, like proteins, tend to increase the hydrophilicity of adhesives, due to the many polar groups in their structure, which creates a barrier to their incorporation in adhesive formulations. The use of plant-based polymeric materials can bring advantages to adhesives that were not previously possible [10]. Incorporated in adhesives formulation, structural elements from plant oils can increase adhesive hydrophobicity and, thus, water resistance [11–13]. Aside from hydrophobicity, polymer fragments derived from vegetable/plant oils can bring other advantages, such as plasticizing effects, compatibility between reagents, cross-linking sites, [14,15], etc. While the market for biobased adhesives is still limited, their advantages can be leveraged in different areas [16,17]. Moreover, using plant oil-based adhesives can improve the curing process, and strengthen adhesive bonds, while improving material sustainability by simplification of the recycling process, due to their inherent biodegradability, which saves efforts and costs, in terms of health and safety regulations.

Sustainability, however, is often not the only sufficient decision-making factor for commercializing biobased products. Modern chemical technologies should also enhance the performance compared to petroleum-based materials, especially if additional costs are required for new material implementation.

Synthesized in our group, plant oil-based monomers (POBMs) can be applied directly in the production of biobased polymeric materials that utilize acrylic monomers in free radical polymerization mechanisms, including emulsion/miniemulsion processes to yield latex polymers [18,19]. POBMs undergo free radical polymerization and, at the same time, retain reactive sites for post-polymerization cross-linking to generate materials with advanced properties and long-term performance. Depending on the applied oil(s), the chemical composition properties of POBM-based polymers can be tuned, based on the unsaturation amount of the fatty acid side chains [20–22]. The high functionality of POBM molecules facilitates control of the resulting polymer molecular weight and may enhance adhesion to a variety of substrates. Feasibility of synthesizing up to 70 wt. % plant oil-based cross-linkable latex binary copolymers has been demonstrated by our groups [20,21]. Such cross-linking may improve the mechanical properties and bonding strength of the resulting adhesives.

It is important to note that, despite the fact that plant oils are considered as a promising sustainable feedstock in biobased adhesive manufacturing, direct comparison of such adhesives with materials based on POBMs is not possible without an additional feasibility study which we report on in this publication.

This study focuses on synthesizing and characterizing novel plant oil-based ternary latex copolymers and evaluating their feasibility as adhesives on various substrates. For this purpose, a range of copolymers with high POBM content (25–45 wt. %) were synthesized using miniemulsion polymerization. To investigate the potential of POBMs as petroleum-based counterparts' replacement in adhesives manufacturing, a high oleic soybean oil-based monomer (HOSBM) was chosen for copolymerization with the following substances, common in commercial latex adhesives: butyl acrylate (BA), methyl methacrylate (MMA), and styrene (St). We varied copolymer composition [including plant oil-based content and ratio between "soft" (POBMs, BA) and "rigid" (MMA or St) macromolecular fragments, based on merit] of the latex copolymers, while total monomer conversion, and particle size, as well as thermal characteristics of latexes, were considered as synthetic criteria and monitored after completion of polymerization.

Furthermore, the adhesive properties of the synthesized ternary latexes were evaluated using various substrate materials. As a result, newly developed POBM-based polymeric materials provide equivalent and, in some cases, better adhesive performance, causing substrate failure in many experiments. The utility of POBM-based ternary latexes as adhesives was clearly demonstrated on all substrates.

2. Results and Discussion

2.1. Synthesis and Characterization of Plant Oil-Based Terpolymers

For the synthesis of ternary latex copolymers, the concentration of plant oil-based fragments of HOSBM in monomer feed was varied (up to 45 wt. %) in copolymerization with MMA/St, while the content of the other (soft) counterpart, n-butyl acrylate, was held at 10 wt. %.

Miniemulsion copolymerization was completed within 10 h, during which essentially all monomers were polymerized (total monomer conversion of 90–95%) to yield thermal stable latexes at 30 wt. % solid content (Table 1).

Table 1. Solid content, conversion, and composition of terpolymer latexes.

Latex Formulations (wt. % in Monomer Feed)		Solid, %	Conv., %	Copolymers Composition, wt. %	
				Calculated	^{1}H NMR
HOSBM-BA-MMA	25–10–65	30.3	94.2	0.18–0.05–0.77	0.19–0.06–0.75
	35–10–55	30.0	92.5	0.27–0.05–0.68	0.27–0.09–0.64
	45–10–45	30.4	90.5	0.36–0.05–0.59	0.39–0.08–0.53
HOSBM-BA-St	25–10–65	30.7	94.5	0.2–0.11–0.69	0.24–0.09–0.67
	35–10–55	30.9	93.0	0.28–0.11–0.61	0.26–0.11–0.63
	45–10–45	30.4	93.0	0.35–0.11–0.54	0.42–0.1–0.48

The polymer composition is a significant parameter determining the properties of the terpolymer and, ultimately, its practical applications. The Alfrey–Goldfinger equations were used to estimate the compositions of the biobased terpolymers [23,24]. The resulting copolymers' compositions were determined by ^{1}H NMR spectroscopy, which is routinely used for the characterization of multicomponent polymers. The technique works best when the individual monomers exhibit well-defined signals unique to the specific monomer so that the accuracy of the signal integration is reliable. However, some uncertainty may arise when signals for individual monomers overlap. According to data in the literature for individual monomers ^{1}H NMR spectra integrals, the molar fraction of BA/MMA/St, can be evaluated from ^{1}H NMR-based integrated areas for the signals designated on the spectrum for -O-**CH$_2$**- of BA (at 4.1–3.8 ppm, 2H), for -O-**CH$_3$** of MMA (at 3.75–3.3 ppm, 3H), and for "aromatic hydrogens" of St (spectral area for 6.2–7.5 ppm, 5H) [25].

The ^{1}H NMR spectrum of purified terpolymer HOSBM-BA-MMA (Figure 1) indicates the incorporation of all three monomer fragments through the appearance of characteristic signals for the protons of -**CH=CH**- (HOSBM), -O-**CH$_3$** (MMA), and -O-**CH$_2$**- (BA) groups at 5.37 (peak *a*), 4.15 (peak *b*), and 3.7 ppm (peak *c*), respectively. However, calculations of the composition for HOSBM-BA-MMA terpolymer are complicated, since the -O-**CH$_2$**- signals of BA at 4.15 ppm overlap with HOSBM signals -O-**CH$_2$**-**CH$_2$**- at 4.2 ppm, as well as the -O-**CH$_3$** signals of MMA at 3.75 ppm overlap with HOSBM signals -**CH$_2$**-**CH$_2$**-NH- at 3.6 ppm (Figure S1). To overcome this challenge, the ^{1}H NMR spectrum of HOSBM homopolymer was used for calculations.

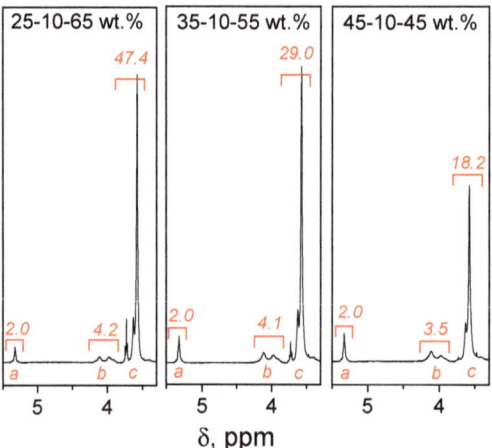

Figure 1. ^{1}H NMR spectra of terpolymers HOSBM-BA-MMA.

The resulting molar composition of terpolymers, based on integrals of characteristic peaks (I_a, I_b, I_c) (Figure 1), was calculated using Equation (1). The amount of BA/MMA was calculated by subtracting the integral value of the signals at HOSBM homopolymer spectrum at ~4.15 and 3.7 ppm (2.2 and 2.3, respectively) from the determined value. The ^{1}H NMR spectrum for HOSBM homopolymer is presented in the supporting information (Figure S2). The conversion calculations from mole fraction to weight fractions have been performed for all compositions presented in the Table 1. Terpolymers HOSBM-BA-MMA were obtained with high conversion (>90%), and their compositions based on ^{1}H NMR measurements were very similar to the theoretical one (Table 1).

$$[HOSBM] : [BA] : [MMA] = \frac{I_a/2}{I_t} : \frac{I_b/2}{I_t} : \frac{I_c/3}{I_t} \quad (1)$$

$$I_t = \frac{I_a}{2} + \frac{I_b}{2} + \frac{I_c}{3} \quad (2)$$

The ^{1}H NMR spectrum of purified terpolymer HOSBM-BA-St (Figure 2) indicated the incorporation of St units through the appearance of characteristic signals for the aromatic hydrogens at 6.2–7.5 ppm (peak c). The characteristic signals for the protons of **-CH=CH-** group (HOSBM) appeared at 5.37 ppm (peak a). Since the -CH$_2$-**CH$_3$** signals of BA at 0.96 ppm (peak b) overlapped with HOSBM signals -CH$_2$-**CH$_3$** at 0.9 ppm the ^{1}H NMR spectrum of the HOSBM homopolymer was used to calculate the composition for HOSBM-BA-St terpolymers. The molar composition of terpolymers was determined via ^{1}H NMR spectra based on integrals of these characteristic peaks (I_a, I_b, I_c) using Equation (3). The weight fractions were calculated from the mole fraction and presented in Table 1. The amount of BA was calculated by subtracting the integral value of the signals at HOSBM homopolymer spectrum at ~0.9 (3.2) from the determined value.

$$[HOSBM] : [BA] : [St] = \frac{I_a/2}{I_t} : \frac{I_b/3}{I_t} : \frac{I_c/5}{I_t} \quad (3)$$

$$I_t = \frac{I_a}{2} + \frac{I_b}{3} + \frac{I_c}{5} \quad (4)$$

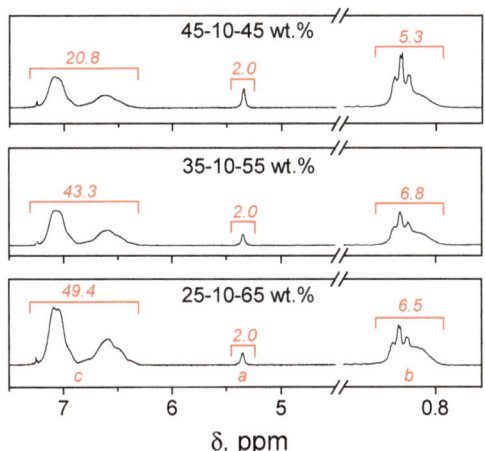

Figure 2. ^{1}H NMR spectra of terpolymers HOSBM-BA-St.

Overall, obtained data on biobased content in the resulting latexes confirmed that the vast majority of plant oil-based ingredients were incorporated into the copolymer macromolecules during miniemulsion polymerization, and the composition of copolymers determined using NMR data coincided well with the composition calculated by the Alfrey–Goldfinger equations.

The mean particle size distribution of the latex polymer particles was determined using dynamic light scattering. Figure 3 shows that the volume-average particle size for all terpolymer latexes ranged between 43 and 82 nm, while particle size distribution increased with increasing HOSBM content in monomer feed (PDI = 0.08–0.52).

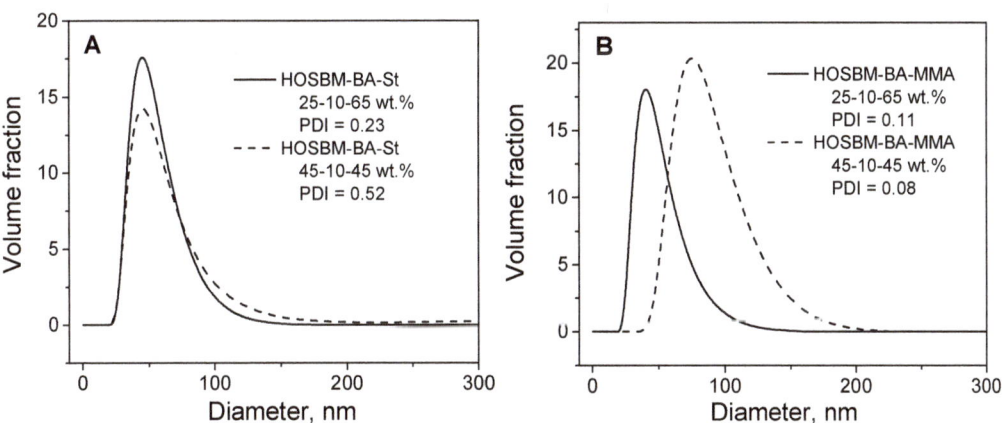

Figure 3. Latex particle size distribution of the biobased terpolymers HOSBM-BA-St (**A**) and HOSBM-BA-MMA (**B**).

Polymer molecular weight and dispersity are important parameters impacting adhesive performance. It is not uncommon that these parameters may have a complex influence on different properties. For example, good shear adhesion requires high molecular weight polymers with high entanglement molecular weight. The segment molecular weight between the crosslink points can be a factor if the formation of the crosslinked adhesive network is induced. On the other hand, to ensure good tackiness, macromolecules with lower molecular weight (and Tg) are required. Thus, polymers with bimodal (lower and higher fractions simultaneously) molecular weight distribution may facilitate enhanced values of both shear adhesion and tackiness of the adhesive [26].

In this work, obtained latexes were characterized by gel permeation chromatography (GPC) to evaluate their molecular weight and dispersity. Figure 4 shows the logarithmic dependence of number average molecular weight on the concentration of HOSBM (ln [HOSBM]) in monomer feed.

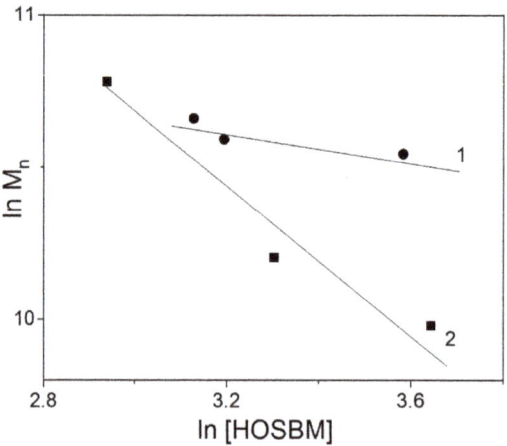

Figure 4. A log-log dependence of number average molecular weight (Mn) and HOSBM-BA-St (1)/HOSBM-BA-MMA (2) weight concentration.

As expected, the molecular weight of the final latex polymers decreased as the fraction of POBMs in the initial monomer mixture (degree of unsaturation) increased. This is explained by the degradative chain transfer effect on HOSBM (allylic termination) provided by allylic hydrogen atoms in the HOSBM molecules, which was reported in our previous studies [19,20]. This observation was in agreement with the previously described more extensive chain transfer effect, caused by a higher unsaturation degree of the monomer feed. Interestingly, bimodal molecular weight distribution has been observed for latexes synthesized in the copolymerization of HOSBM and BA with MMA, while GPC analysis of material synthesized in copolymerization with St yields a single peak, although with broader dispersity (Figure S3). We explain this difference by various termination modes for macroradicals during chain copolymerization. The pathways of termination of MMA and St polymerizations are the most extensively studied [27]. The disproportionation (yielding two shorter macromolecules)/coupling ratio in MMA polymerization was 85/15, while for St polymerization the contribution of the coupling mode could be varied from nearly 60% up to 100%. According to the literature data [27], the GPC analysis of the MMA polymers revealed a bimodal trace which is divided into two components (low and high molecular weight) and confirms the mixed pathway of termination. For both compositions, though, the measurements clearly showed the presence of lower and higher molecular weight fractions, which is clearly beneficial for their adhesive performance.

To evaluate the effect of HOSBM content on the thermal properties of plant oil-based latex adhesives, latex samples underwent differential scanning calorimetry (DSC) measurements. As illustrated in Figure 5, the DSC diagrams show a dependence of glass transition temperature (Tg) for synthesized latexes on different HOSBM content. The plasticizing effect of POBM fragments has been previously reported by our group [20,21]. The obtained results confirmed that Tg decreased when HOSBM content increased, making the resulting materials much softer if compared to polystyrene and polymethyl methacrylate (both having glass transition temperature in a range of 100–110 °C). A variation of oil-based monomer content changed the Tg (and, thus, the thermal properties) of the resulting latex polymers for all synthesized compositions. An increasing fraction of HOSBM in copolymers made the macromolecules more flexible, as indicated by decreasing Tg.

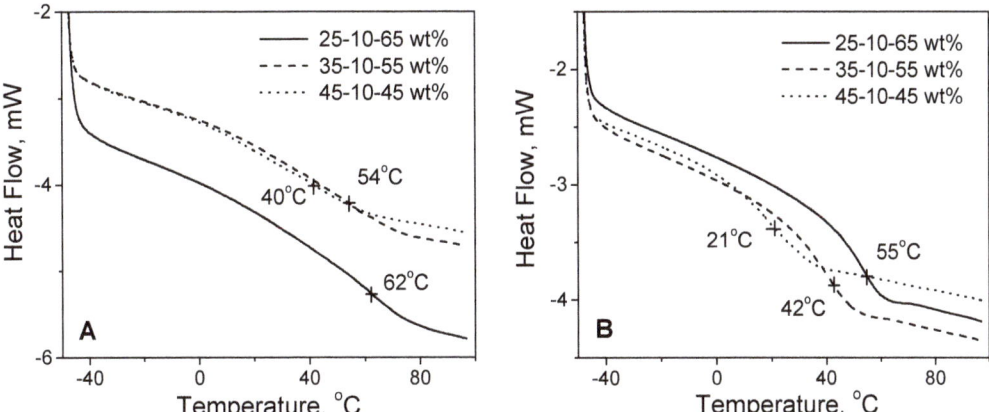

Figure 5. DSC diagrams for terpolymers HOSBM-BA-MMA (**A**) and HOSBM-BA-St (**B**) (heating rate: 10 °C min^{-1}).

2.2. Correlation between Wettability and Adhesive Performance Plant Oil-Based Terpolymers

Good wettability of material surfaces is an important factor for enabling bonding formation between adhesive and substrate. It is critical that the applied adhesive formulation wets and spreads over the surface to achieve close contact between two materials to ensure physical interactions and the developing intermolecular forces.

The substrate surface properties, including surface energy and roughness, have an essential impact on the resulting bonding quality. Another surface characteristic that can be translated into adhesive performance is contact angle hysteresis (CAH). CAH is determined as a difference between advancing contact angle, θ_A of the liquid that spreads over a surface, and receding contact angle, θ_R of a liquid that retreats from a wet surface. Surface roughness, chemical heterogeneity, and surface reorganization upon close contact with a liquid during the measurement are considered to be the most influential factors for CAH values [28,29].

Hence, the next step in this study was to investigate how latex copolymer composition (including POBM content) impacts wettability (ability to maintain contact with substrate surface) of various material surfaces by latexes and if the latter correlates with the plant oil-based latexes adhesive performance. A range of substrates representing semicrystalline polyethylene terephthalate film (PET), polypropylene sheet (PP), bleached paperboard (uncoated), and a top coated with a clay mineral bleached paperboard were sourced for this purpose. Prior to studying the wettability of each material substrate by the synthesized latexes, surface energy for each substrate material was determined using the Zisman method [30,31]. According to this model, the surface free energy splits between dispersive and polar energy components, which can be calculated from the contact angles between the materials' surface and probe liquids of different (known) interfacial properties (for example, water and diiodomethane).

As shown in Table 2, the surface energy of PP was below 35 mN/m, while the polar energy component was very low, which meant that PP was the most non-polar material in the range. Uncoated paperboard, with its polar functional groups of cellulose/hemicellulose, possesses slightly higher surface energy. Among chosen materials, the highest surface energy of PET can be attributed to the polar ester groups. However, the polar energy component of the PET surface was still lower than for the top coated with hydrophilic clay mineral paperboard substrate.

Table 2. Surface free energy parameters of the used materials.

Substrate	Water θ, °	CH$_2$I$_2$ θ, °	Surface Energy, mN/m		
			λ_S^d	λ_S^p	λ_S
PET	73.4 ± 1.7	37.4 ± 1.3	40.9 ± 0.6	5.6 ± 0.7	46.5 ± 1.3
PP	93.1 ± 4.0	53.5 ± 2.9	32.3 ± 1.6	1.2 ± 0.8	33.5 ± 2.5
Paperboard coated	66 ± 3.4	58 ± 3.6	30 ± 2.1	12.8 ± 2.1	42.8 ± 4.2
Paperboard uncoated	83 ± 6.3	51 ± 2.6	33.9 ± 1.5	3.6 ± 2.1	37.5 ± 3.6

Water θ—water contact angle; CH$_2$I$_2$ θ—diiodomethane contact angle; λ_S^d—dispersive component of the surface energy; λ_S^p—polar component of the surface energy; λ_S—surface free energy.

Table 3 presents experimentally determined surface free energy for each HOSBM-based latex synthesized in this work. The obtained values (ranging between 27 and 30 mN/m) do not depend significantly on copolymer composition (including biobased content), revealing that latexes possess rather low surface energy, which can be explained by the presence of remaining surfactant molecules at the liquid–air interface. Taking into account the experimental precision, no difference between latexes with various POBM content was observed.

Table 3. Contact angle hysteresis of HOSBM-based latexes on PP and PET.

Sample (Monomer Feed)		Surface Tension, mN/m	PP			PET		
			θ_A, Deg	θ_R, Deg	θ_A–θ_R	θ_A, Deg	θ_R, Deg	θ_A–θ_R
HOSBM-BA-MMA	25–10–65	27.4	56 ± 1.4	25 ± 2.0	31	60 ± 1.2	29 ± 1.2	31
	45–10–45	27.7	61 ± 1.0	32 ± 1.8	29	60 ± 1.9	25 ± 1.7	35
HOSBM-BA-ST	25–10–65	28.9	65 ± 3.0	43 ± 3.2	22	62 ± 0.7	34 ± 0.6	28
	45–10–45	30.0	60 ± 2.5	34 ± 1.4	26	59 ± 1.0	31 ± 0.9	28

θ_A—advancing contact angle; θ_R—receding contact angle; θ_A–θ_R—contact angle hysteresis.

To correlate surface properties and adhesive performance of HOSBM-based copolymer latexes and ensure latex spreading ability and adherence over the substrate materials differing by polarity, contact angle hysteresis for each latex-substrate pair was next determined using water as a contact liquid.

Tables 3 and 4 collect contact angle hysteresis values for latexes with 25 and 45 wt. % of HOSBM on PP, PET, and coated/uncoated paperboard substrates. Even though the surface free energy of substrate materials differed in a range of 34–47 mN/m (with a polar energy component differing by one order of magnitude), no significant difference appeared for the experimentally determined values of hysteresis with increasing plant oil-based content in the copolymer. Nevertheless, essentially higher values for both advancing and receding contact angles were observed for coated/uncoated paperboard, compared to PP and PET substrates, which may indicate different adherence of latexes to both paperboard substrates and might be explained by more expressed chemical heterogeneity of the substrate surface [32].

Overall, wettability measurements indicated that the synthesized ternary copolymers latexes exhibited an ability to spread and adhere once applied to all chosen substrates, and, thus, could be considered further for adhesive performance evaluation. Therefore, the adhesive performance of latexes was studied next using the T-peel test.

Since PET, PP, and paperboard are widely used in food packaging, the textile industry in manufacturing carpets, woven materials, etc., as well as other consumer goods, T-peel strength testing can be applied for assessing the adhesive joints in laminated or packaging materials [33]. Having this in mind, the synthesized HOSBM-based copolymers were tested on PET to PP and coated to uncoated paperboard substrate pairs to demonstrate the latexes' feasibility in multisubstrate bonding applications.

Table 4. Contact angle hysteresis of HOSBM-based latexes on coated/uncoated paperboard.

Sample	(Monomer Feed)	Coated			Uncoated		
		θ_A, Deg	θ_R, Deg	$\theta_A - \theta_R$	θ_A, Deg	θ_R, Deg	$\theta_A - \theta_R$
HOSBM-BA-MMA	25–10–65	81 ± 1.9	49 ± 0.9	32	79 ± 1.3	49 ± 0.5	30
	45–10–45	80 ± 1.5	47 ± 1.0	33	82 ± 2.5	47 ± 0.5	35
HOSBM-BA-ST	25–10–65	80 ± 1.1	55 ± 1.5	25	83 ± 4.6	55 ± 3.1	28
	45–10–45	80 ± 0.8	49 ± 2.0	31	89 ± 1.6	59 ± 1.7	30

θ_A—advancing contact angle; θ_R—receding contact angle; $\theta_A - \theta_R$—contact angle hysteresis.

Typically, adhesion force is determined by wettability and the resulting physicochemical interactions between adhesive and substrate, including strong covalent bonding as well as weaker physical (van der Waals or hydrogen) interactions [24]. Considering the chemical structure of plant oil-based monomers, it can be assumed that the presence of oil-derived unsaturated fatty acid fragments in latexes may provide cross-linking sites, thus strengthening adhesion forces. Figure 6A shows two selected plots for HOSBM-BA-St latexes with different biobased content tested by T-peel test on paperboard (coated to uncoated) substrates. As seen qualitatively, cohesive failure was observed for latex with lower HOSBM content (Figure 6B), while the presence of 45 wt. % biobased content in the material resulting in higher peel strength and substrate failure during the testing (Figure 6C). In the case of cohesive failure, the peel plot stayed at a plateau during testing, and both substrates contained adhesive residues after the experiment ended. The peeling plot looked more complex for the latexes with higher biobased content. The load initially rose to a maximum point, then dropped to a lower value, corresponding to the noticeable delamination of the substrate (Figure 6C).

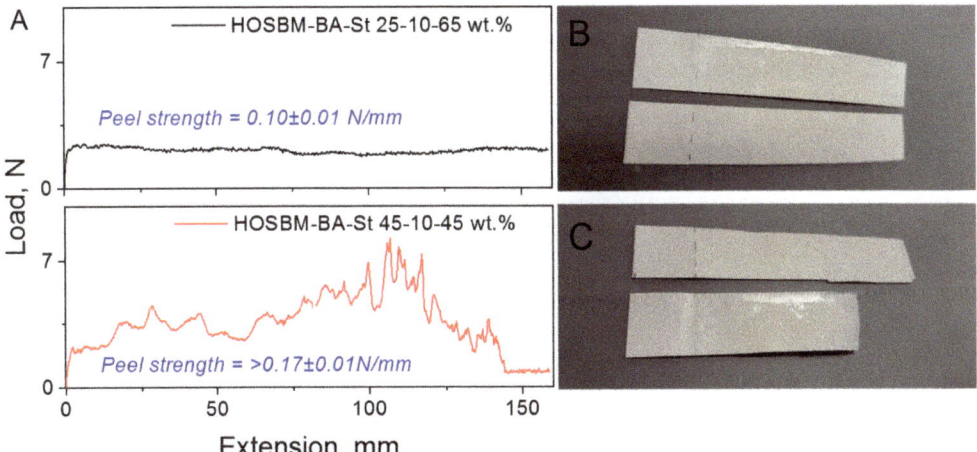

Figure 6. Peel strength of HOSBM-BA-St 25–10–65 wt. % and 45–10–45 wt. % latexes copolymers on paperboard (coated to uncoated) substrates (**A**) and substrate samples after peel strength testing ((**B**,**C**) [substrate failure]).

Peel strength of HOSBM-BA-MMA latexes with various biobased contents were tested on PET-PP substrate pairs (Figure 7). The obtained results showed that the presence of HOSBM fragments in the latex terpolymers (up to 45 wt. % in monomer feed) improved adhesion and increased the peel strength. Similar to the testing on paperboard, using latexes with lower biobased content resulted in cohesive failure, and substrate failure was observed once a higher concentration of biobased fragments was incorporated into latex copolymers. We attributed an increase in peel strength to the stronger chemical bonding formation at the terpolymer-substrate interface. As a result, substrate failure occurred due to the strong adhesion of the terpolymers with higher biobased content to the substrate, which was more substantial than the structural cohesion force of the latex (Figure S4). Therefore, it can be concluded that increasing the concentration of HOSBM fragments in the latex terpolymers led to differences in adhesive joint failure mode during peel strength testing (which was cohesive failure for a lower amount of HOSBM vs. substrate failure when the amount of HOSBM in the terpolymer went up).

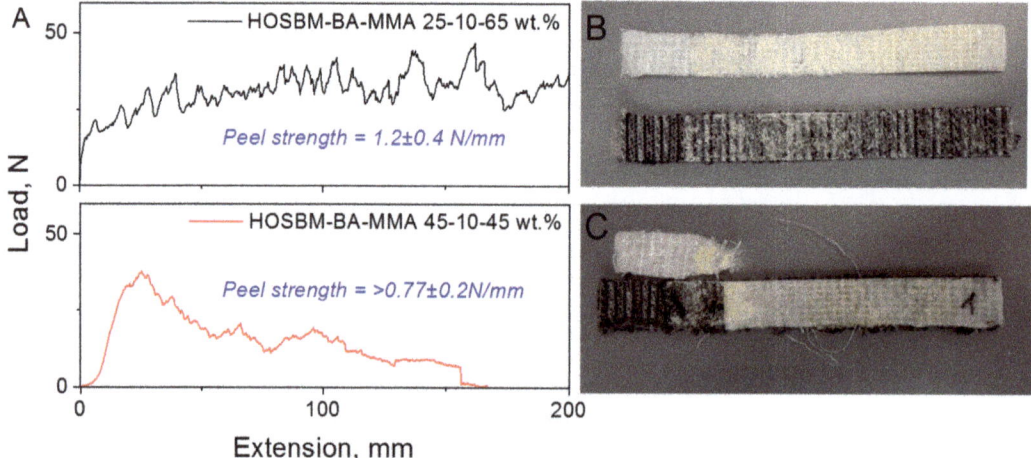

Figure 7. Peel strength of HOSBM-BA-MMA 25–10–65 wt. % and 45–10–45 wt. % latexes copolymers on PET-PP substrates (**A**) and substrates samples after T-peel strength testing ((**B,C**) [substrate failure]).

The obtained results indicated that the adherence level of latexes overall correlated with wettability, thus contact angle hysteresis measurements reflected POBM-based ternary copolymers behavior on substrates with various chemical heterogeneities.

In summary, the performance of the synthesized latexes with up to 45 wt.% of high oleic soybean oil-based monomer showed their utility on multiple substrates, resulting in substrates' failure in adhesive peel strength testing.

3. Materials and Methods

3.1. Materials

High oleic soybean oil (Perdue Agribusiness LLC, Salisbury, MD, US), sodium dodecyl sulfate (SDS, VWR; Solon, OH, US), and sodium chloride (VWR; Solon, OH, US) were used as received. Methyl methacrylate (MMA, Alfa-Aesar; Ward Hill, MA, USA), butyl acrylate (BA, TCI America, Portland, OR, USA), styrene (St, Sigma-Aldrich, St. Louis, MO, US), and acrylic acid (AA, Alfa-Aesar; Ward Hill, MA, USA) were distilled under vacuum to remove the inhibitor and stored in a refrigerator. Azobisisobutyronitrile (AIBN; Sigma-Aldrich, St. Louis, MO, USA) was purified with recrystallization from methanol. All other solvents used were reagent grade or better and used as received. Deionized water was used throughout the study (Millipore water, MilliQ, 18 MΩ).

3.2. Latex Preparation

The high oleic soybean oil-based latexes were synthesized in mini-emulsion copolymerization of respective acrylic monomers from plant oils with BA, St, and MMA. For this purpose, the oil phase (25.5 g) was prepared by mixing HOSBM (25–45 wt. %, 6.4–11.5 g), BA (10 wt. %, 2.55 g) with MMA or St at different ratios (40–55 wt. %, 11.45–16.55 g) in the presence of 1.5 wt. % (0.4 g) oil-soluble initiator per oil phase.

The aqueous phase was formed by dissolving the emulsifier (SDS, 5 wt. % per oil phase, 1.3 g) in Millipore water with added small amounts of NaCl (0.02 mol/L, 0.05 g). The oil phase was added dropwise to the aqueous phase and mixed for 45 min to form a pre-emulsion. The pre-emulsions were sonicated with three pulses for 60 sec each using Q-Sonica (500 W digital sonicator, $\frac{1}{2}$ in. tip, 20 kHz, Newtown, CN, USA) and placed in an ice bath to maintain the temperature at 25 °C. The resulting miniemulsions were purged with nitrogen for 10 min and polymerized at 70 °C for 10 h. The latex solid content was kept at 30 wt. %.

3.3. Plant Oil-Based Latex Characterization

Total monomer conversion was determined by multiple precipitations of latex copolymers in methanol to remove residual unreacted monomers. The purified copolymers were dried in an oven until a constant weight was achieved.

The latex solid content was measured gravimetrically by drying the samples in an oven at an elevated temperature for 45 min.

The latex copolymers composition was analyzed using ^{1}H NMR spectroscopy (JEOL ECA 400 MHz NMR Spectrometer, Akishima, Japan) using $CDCl_3$ as a solvent.

The molecular weight of the latex copolymers was determined using gel permeation chromatography (GPC) (Waters Corporation Modular Chromatograph, which consists of a Waters' 1515 HPLC pump, a Waters' 2410 refractive index detector, and two 10 μm PL-gel mixed-B columns) at 40 °C using tetrahydrofuran as a carrier solvent.

Particle size distribution of the plant oil-based latex particles was measured using dynamic light scattering, Malvern Zetasizer Nano-ZS90 with a fixed scattering angle 90°, and a 633-nm wavelength laser. Samples were prepared by diluting one drop of latex in approximately 7 mL of water.

The glass transition temperature of POBM-based latex copolymers was determined by differential scanning calorimetry (DSC) (TA Instruments Q1000 calorimeter, New Castle, DE, USA) at heat/cool/heat mode (−50 °C/150 °C) with dry nitrogen purging through the sample at 50 mL/min flow rate. Latex sample's heating/cooling rate was 10–20 °C/min.

3.4. Measurement of the Surface Free Energy, Surface Tension, Dynamic Contact Angles, and Contact Angle Hysteresis

The surface free energy of substrates, the surface tension of latexes, dynamic contact angles, and contact angle hysteresis of the latex coatings from plant oil-based adhesives were characterized using KRÜSS contact angle instruments DSA100 with an external tilting device PA3220.

Surface free energy, as well as the disperse and polar fractions of substrates, were determined according to the Zisman method. Water and diiodomethane were used as the standard test liquids. The surface free energy of substrates was measured at room temperature without pretreatment. For each sample, three to five contact angle measurements with water and diiodomethane were made at different locations.

The pendant drop method was used to determine the surface tension of synthesized latex terpolymers. In this method a latex drop of 4 μL volume was suspended from the syringe needle and allowed to stabilize for 5–10 min prior to surface tension measurement. Surface tension was calculated from the shadow image of a pendant drop using drop shape analysis. All measurements were conducted at room temperature.

Dynamic contact angles were measured as a drop moved across a tilted surface. Initially, the drop of the plant oil-based latex was deposited on a level substrate surface, then the table was slowly tilted and the inclination angle increased. For comparative measurements between different samples, the suitable measuring conditions (tilting speed = 60°/min, tilting position = 30°, and drop volume = 18–23 µL) were kept. Contact angle hysteresis was defined as the difference between the advancing (θ_A) and receding (θ_R) contact angles.

3.5. Adhesive Performance Testing

Peel strength of the POBM-based latexes was measured according to the ASTM D1876−08 method using MTS's Insight Electromechanical 5 kN Extended Length Testing System with load cell 5 kN at a test speed of 304.8 mm/min.

3.5.1. Peel Strength of PET-PP Substrates

The range of the synthesized latexes was applied on PET primary backing and polypropylene secondary backing. The test samples had a rectangular shape (length: 229 mm, width: 25 mm). The test panels were dried for 24 h at room temperature under a 4.5 kg weight press, then held at elevated temperature for 24 h, and cooled at standard conditions for 3 h.

The unbonded ends of the test specimen were clamped bent in the test grips of the tension testing machine. The load at a constant head speed of 304.8 mm/min was applied. The autographic recording of load versus distance peeled was made during the T-peel test. The peel resistance over at least a 127-mm length of the bond line after the initial peak was determined. The measurement of peel strength was repeated five times, and the average value was recorded.

3.5.2. Peel Strength of Paperboard Substrates

A range of synthesized latexes was applied on paperboard substrate (bleached paperboard with a clay mineral coated top surface, 12 pt). The test samples had a rectangular shape (length: 152 mm, width: 25 mm). The tested panels were held at elevated temperature for 24 h and cooled at standard conditions for 3 h.

The unbonded ends of the test specimen were clamped bent in the test grips of the tension testing machine. The load at a constant head speed of 304.8 mm/min was applied. The peel resistance over at least a 100-mm length of the bond line after the initial peak was determined. The measurement of peel strength was repeated 5 times, and the average value was recorded.

4. Conclusions

Plant oil-based acrylic latex ternary copolymers with a range of performance capabilities and controlled Tg were developed and exhibited adhesive capabilities on various substrates. Adhesive performance of the latexes could be tailored by using different plant oil-based monomers in copolymerization, which might provide features required for specific applications. In this study, the best-performing latex adhesives containing up to 45 wt. % of high-oleic soybean oil-based monomer fragments demonstrated promising efficiency in the testing of PET to PP and coated to uncoated paperboard substrate pairs, resulting in substrate failure during the adhesive testing.

Since the incorporation of a higher amount of hydrophobic POBM in latex copolymers brings some limitations to the synthetic process, in the future, process parameters (solid content, surfactant amount, monomers feeding) need to be adjusted accordingly for different formulations.

5. Patents

Biobased Acrylic Monomers US 10,315,985 B2, 11 June 2019.
Biobased Acrylic Monomers and Polymers Thereof US 10,584,094 B2, 10 March 2020.

Supplementary Materials: The following supporting information can be downloaded at: https://www.mdpi.com/article/10.3390/molecules27165170/s1, Figure S1: Chemical structure of plant oil-based terpolymer HOSBM-BA-MMA with indicated integral proton shifts, Figure S2: ^{1}H NMR spectrum of poly(HOSBM), Figure S3: Gel permeation chromatography (GPC) analysis diagram of terpolymers HOSBM-BA-MMA (A) and HOSBM-BA-St (B), Figure S4: T-peel adhesion strength testing on paperboard (A) and PP-PET (B) substrates after applying plant oil-based terpolymers.

Author Contributions: Conceptualization, V.K., B.D., Z.D., I.B. and A.V.; validation, A.V.; formal analysis, V.K., B.D., Z.D., I.B., O.S. and S.T.; investigation, V.K., B.D., Z.D., I.B., O.S. and S.T.; data curation, G.P. and A.V.; writing—original draft preparation, V.K. and A.V.; writing—review and editing, V.K., B.D., Z.D., I.B., O.S., S.T., G.P. and A.V.; visualization, V.K. and B.D.; supervision, G.P. and A.V.; project administration G.P. and A.V.; funding acquisition, G.P. and A.V. All authors have read and agreed to the published version of the manuscript.

Funding: This work was funded by the NSF the Center for Bioplastics and Biocomposition; North Dakota Soybean Council, and North Dakota Department of Agriculture.

Institutional Review Board Statement: Not applicable.

Informed Consent Statement: Not applicable.

Data Availability Statement: Data is contained within the article.

Acknowledgments: It is a pleasure to acknowledge fruitful discussions with Kellie Ballew, Candi Hampton (Shaw Industries), Rahul Bhardwaj (WestRock), Hart Haugen (Sherwin-Williams), Jed Randel (NatureWorks), Tina Tosukhowong (GC Innovation America), and Neal Williams (AkzoNobel) on the subject of this paper. We thank the Shaw Industries Group, Inc. (Dalton, GA, USA) and WestRock Company (Richmond, VA, USA) for providing substrate samples. We also thank Frederik Haring for assistance with MTS Insight Electromechanical Testing System.

Conflicts of Interest: The authors declare no conflict of interest. The funders had no role in the design of the study; in the collection, analyses, or interpretation of data; in the writing of the manuscript, or in the decision to publish the results.

Sample Availability: Not applicable.

References

1. Papageorgiou, G.Z. Thinking Green: Sustainable Polymers from Renewable Resources. *Polymers* **2018**, *10*, 952. [CrossRef]
2. Caillol, S. Special Issue "Natural Polymers and Biopolymers II". *Molecules* **2021**, *26*, 112. [CrossRef] [PubMed]
3. Sharmin, E.; Zafar, F.; Akram, D.; Alam, M.; Ahmad, S. Recent advances in vegetable oils-based environment-friendly coatings: A review. *Ind. Crops Prod.* **2015**, *76*, 215–229. [CrossRef]
4. Gunstone, F.D. Chemical reactions of fatty acids with special reference to the carboxyl group. *Eur. J. Lipid Sci. Technol.* **2001**, *103*, 307–314. [CrossRef]
5. Igwe, I.; Ogbobe, O. Studies on the properties of polyester and polyester blends of selected vegetable oils. *J. Appl. Polym. Sci.* **2000**, *75*, 1441–1446. [CrossRef]
6. Gultekin, G.; Atalay-Oral, C.; Erkal, S.; Sahin, F.; Karastova, D.; Tantekin-Ersolmaz, S.B.; Guner, F.S. Fatty acid-based polyurethane films for wound dressing applications. *J. Mater. Sci. Mater. Med.* **2009**, *20*, 421–431. [CrossRef] [PubMed]
7. Guner, F.S.; Yagci, Y.; Erciyes, A.T. Polymers from Triglyceride Oils. *Prog. Polym. Sci.* **2006**, *31*, 633–670. [CrossRef]
8. Heinrich, L.A. Future opportunities for biobased adhesives—Advantages beyond renewability. *Green Chem.* **2019**, *21*, 1866. [CrossRef]
9. Islam, N.; Rahman, F.; Kumar Das, A.; Hiziroglu, S. An overview of different types and potential of biobased adhesives used for wood products. *Int. J Adhes. Adhes.* **2022**, *112*, 102992. [CrossRef]
10. Finlay, M.R. Old Efforts at New Uses: A Brief History of Chemurgy and the American Search for Biobased Materials. *J. Ind. Ecol.* **2003**, *7*, 33–46. [CrossRef]
11. Kong, X.; Liu, G.; Curtis, J.M. Characterization of canola oil based polyurethane wood adhesives. *Int. J. Adhes. Adhes.* **2011**, *31*, 559–564. [CrossRef]
12. Saetung, A.; Rungvichaniwat, A.; Tsupphayakornake, P.; Bannob, P.; Tulyapituk, T.; Saetung, N. Properties of waterborne polyurethane films: Effects of blend formulation with hydroxyl telechelic natural rubber and modified rubber seed oils. *J. Polym. Res.* **2016**, *23*, 264. [CrossRef]
13. Li, C.; Xiao, H.; Wang, X.; Zhao, T. Development of green waterborne UV-curable vegetable oil-based urethane acrylate pigment prints adhesive: Preparation and application. *J. Cleaner Prod.* **2018**, *180*, 272–279. [CrossRef]

14. Li, Y.; Chou, S.H.; Qian, W.; Sung, J.; Chang, S.I.; Sun, X.S. Optimization of Soybean Oil Based Pressure-Sensitive Adhesives Using a Full Factorial Design. *J. Am. Oil Chem. Soc.* **2017**, *94*, 713–721. [CrossRef]
15. Petrović, Z.S. Polyurethanes from Vegetable Oils. *Polym. Rev.* **2008**, *48*, 109–155. [CrossRef]
16. Frost & Sullivan; Krishnan, S. Investment Analysis of European Adhesives and Sealants Market. Available online: https://www.prnewswire.com/news-releases/investment-analysis-of-the-european-adhesives-and-sealants-2015-499563041.html (accessed on 30 June 2022).
17. Frost & Sullivan. North American and European Construction Adhesives and Sealants Market, Forecast to Emphasis on Light Weighting in Building Components to Drive Revenue Growth. Available online: https://store.frost.com/north-american-and-european-construction-adhesives-and-sealants-market-forecast-to-2022.html (accessed on 30 June 2022).
18. Tarnavchyk, I.; Popadyuk, A.; Popadyuk, N.; Voronov, A. Synthesis and Free Radical Copolymerization of a Vinyl Monomer from Soybean Oil. *ACS Sustainable Chem. Eng.* **2015**, *3*, 1618–1622. [CrossRef]
19. Demchuk, Z.; Shevchuk, O.; Tarnavchyk, I.; Kirianchuk, V.; Kohut, A.; Voronov, S.; Voronov, A. Free Radical Polymerization Behavior of the Vinyl Monomers from Plant Oil Triglycerides. *ACS Sustain. Chem. Eng.* **2016**, *4*, 6974–6980. [CrossRef]
20. Demchuk, Z.; Kohut, A.; Voronov, S.; Voronov, A. Versatile Platform for Controlling Properties of Plant Oil-Based Latex Polymer Networks. *ACS Sustain. Chem. Eng.* **2018**, *6*, 2780–2786. [CrossRef]
21. Demchuk, Z.; Kirianchuk, V.; Kingsley, K.; Voronov, S.; Voronov, A. Plasticizing and Hydrophobizing Effect of Plant Oil Based Acrylic Monomers in Latex Copolymers with Styrene and Methyl Methacrylate. *J. Theor. Appl. Nanotechnol.* **2018**, *6*, 29–37. [CrossRef]
22. Kohut, A.; Voronov, S.; Demchuk, Z.; Kirianchuk, V.; Kingsley, K.; Shevchuk, O.; Caillol, S.; Voronov, A. Non-Conventional Features of Plant Oil-Based Acrylic Monomers in Emulsion Polymerization. *Molecules* **2020**, *25*, 2990. [CrossRef]
23. Turner, A.J.; George, G. The mechanism of copolymerization. *J. Chem. Phys.* **1944**, *12*, 205–209. [CrossRef]
24. Demchuk, Z.; Shevchuk, O.; Tarnavchyk, I.; Kirianchuk, V.; Lorenson, M.; Kohut, A.; Voronov, S.; Voronov, A. Free Radical Copolymerization Behavior of Plant Oil-Based Vinyl Monomers and Their Feasibility in Latex Synthesis. *ACS Omega* **2016**, *1*, 1374–1382. [CrossRef]
25. Nguyen, M.N.; Pham, Q.T.; Le, V.D.; Nguyen, T.B.V.; Bressy, C.; Margaillan, A. Synthesis and characterization of random and block-random diblock silylated terpolymers via RAFT polymerization. *Asian J. Chem.* **2018**, *50*, 1125–1130. [CrossRef]
26. Foreman, P.B. Acrylic adhesives. In *Handbook of Pressure-Sensitive Adhesives and Products*; Benedek, I., Feldstein, M.M., Eds.; Taylor & Francis Groups: Boca Raton, FL, USA, 2009; 165p, ISBN 9781420059397.
27. Nakamura, Y.; Yamago, S. Termination Mechanism in the Radical Polymerization of Methyl Methacrylate and Styrene Determined by the Reaction of Structurally Well-Defined Polymer End Radicals. *Macromolecules* **2015**, *48*, 6450–6456. [CrossRef]
28. Johnson, R.E.; Dettre, R.H. Wetting of low-energy surfaces. In *Wettability*; Berg, J.C., Ed.; Marcel Dekker: New York, NY, USA, 1993; p. 1. ISBN 9780824790462.
29. Chan, C.M. *Polymer Surface Modification and Characterization*; Hanser Publisher: Munich, Germany, 1994; ISBN 9783446158702.
30. Fox, H.W.; Zisman, W.A. The spreading of liquids on low energy surfaces. I. polytetrafluoroethylene. *J. Colloid Sci.* **1950**, *5*, 520. [CrossRef]
31. Good, R.J.; Shu, L.K.; Chiu, H.-C.; Yeung, C.K. Adhesion Science and Technology. In Proceedings of the International Adhesion Symposium, Yokohama, Japan, 6–10 November 1994; Mizumachi, H., Ed.; Taylor & Francis Group: Boca Raton, FL, USA, 1994.
32. Bistac, S.; Kunemann, P.; Schultz, J. Tentative Correlation between Contact Angle Hysteresis and Adhesive Performance. *J. Col. Interf. Sci.* **1998**, *201*, 247–249. [CrossRef]
33. Geyer, R.; Jambeck, J.R.; Law, K.L. Production, use, and fate of all plastics ever made. *Sci. Adv.* **2017**, *3*, 19–24. [CrossRef]

Article

Preparation of Biocomposites with Natural Reinforcements: The Effect of Native Starch and Sugarcane Bagasse Fibers

Muriel Józó [1,2], Róbert Várdai [1,2,*], András Bartos [1,2], János Móczó [1,2] and Béla Pukánszky [1,2]

1. Laboratory of Plastics and Rubber Technology, Department of Physical Chemistry and Materials Science, Budapest University of Technology and Economics, Műegyetem rkp. 3, H-1111 Budapest, Hungary
2. Institute of Materials and Environmental Chemistry, Research Centre for Natural Sciences, P.O. Box 286, H-1519 Budapest, Hungary
* Correspondence: vardai.robert@vbk.bme.hu; Tel.: +36-1-463-4337

Abstract: Biocomposites were prepared from poly(lactic acid) and two natural reinforcements, a native starch and sugarcane bagasse fibers. The strength of interfacial adhesion was estimated by model calculations, and local deformation processes were followed by acoustic emission testing. The results showed that the two additives influence properties differently. The strength of interfacial adhesion and thus the extent of reinforcement are similar because of similarities in chemical structure, the large number of OH groups in both reinforcements. Relatively strong interfacial adhesion develops between the components, which renders coupling inefficient. Dissimilar particle characteristics influence local deformation processes considerably. The smaller particle size of starch results in larger debonding stress and thus larger composite strength. The fracture of the bagasse fibers leads to larger energy consumption and to increased impact resistance. Although the environmental benefit of the prepared biocomposites is similar, the overall performance of the bagasse fiber reinforced PLA composites is better than that offered by the PLA/starch composites.

Keywords: biopolymers; biocomposites; reinforcements; adhesion; micro-mechanics; acoustic emission

1. Introduction

The interest in bioplastics has increased enormously in recent years [1,2]. They are used in a wide range of applications including packaging [3,4], agriculture [5,6], and various consumer goods [7], but also as technical parts in the automotive or machine industry [8–11]. The increasing use of biopolymers is adequately justified by the huge amount of plastic waste accumulating each year, but also by other environmental concerns like microplastic pollution [12,13]. Biopolymers have several advantages including the use of renewable resources, compostability, advantageous carbon footprint, etc. On the other hand, these materials also possess some drawbacks, like somewhat inferior properties compared to commodity polymers, frequent complications during their conversion, sensitivity to water during processing [3], and higher price.

Presently, the biopolymer produced and used in the largest quantity is poly(lactic acid) (PLA). Its stiffness and strength compete even with those of engineering plastics, but its physical ageing is fast because of its low glass transition temperature [14] and its impact resistance is also rather small [15]. Accordingly, PLA is frequently modified in various ways including plasticization [16–18], blending [19–23], and the addition of reinforcements [24–31] to further improve stiffness and strength. In order to maintain the inherent advantages of biopolymers of small carbon footprint, natural origin, and compostability, they are often combined with bio-based additives [32–40] such as reinforcements from natural resources. Wood flour [24,25,28] and various natural fibers [26,30,31] are used the most frequently, but microcrystalline [29] or regenerated cellulose and lignin [19] are also added to PLA to modify its properties.

Starch is also a biopolymer produced by plants for energy storage and it is available in large quantities [41]. Because of its large molecular weight and polarity, starch cannot be processed by the usual conversion technologies of plastics [42]. Additionally, heat and water sensitivity further complicate the use of starch for the modification of PLA. Accordingly, it is usually plasticized with water, glycerol, or other polar compounds capable of forming hydrogen bonds with starch [43–45]. However, thermoplastic starch (TPS) has rather poor mechanical properties like small stiffness, strength, and deformability, but the disadvantages like heat and water sensitivity, and larger viscosity, remain practically the same. On the other hand, native starch is a stiff material which can form strong hydrogen bonds with PLA thus reinforcing it in a similar way to inorganic fillers like calcium carbonate, talc, or kaolin. Only a few papers have been published on PLA/starch composites, but most of them deal with the treatment or the modification of starch in order to improve its adhesion to PLA [46–48]. In spite of their goal, unfortunately these works do not provide a detailed analysis of the interactions between the components and on local processes taking place during deformation. According to the best of our knowledge, no one has yet pointed out the differences or similarities between natural fibers and native starch as additives for PLA.

In accordance with the considerations presented above, the goal of this work was to prepare PLA/native starch composites and compare their properties to those modified with a natural reinforcement, sugarcane bagasse fiber. All components are derived from natural resources thus they are biodegradable with advantageous environmental impact. The similarity of their chemical structure also justifies comparison. The majority of publications on PLA/natural fiber composites claim weak interactions between the components, thus an attempt was made to modify them by the addition of a functionalized, maleated PLA polymer. Since both the natural fibers and starch are expected to increase stiffness and strength, the attention is focused mainly on the mechanical properties of the composites, i.e., stiffness, strength, and impact resistance, and efforts were made to explore the mechanism of the deformation and failure of the composites. Some comments on the relevance of the results for practice are also mentioned in the final section of the paper.

2. Results

The following sections summarize the results of the work and include some discussion of the results as well. The chemical and physical structures of the two modifying components are compared in the first section and then their effect on composite properties is shown in the next. The quantitative analysis of the extent of reinforcement is presented in the subsequent section, followed by the discussion of local deformation processes, as well as the consequence of the results for the possible practical application of the composites.

2.1. Structure of the Reinforcements

The chemical and physical structure of the reinforcement used strongly influence its effect on the matrix polymer and on the properties of the resulting composite [49]. The chemical structure of the two reinforcements used in this study shows strong similarities, but some differences as well. Both polymers are polysaccharides consisting of glucose units. The coupling of these units is different, cellulose contains β-glucosides [50], while starch consists of α units [41]. Cellulose molecules are linear, while starch chains are helical. The chain structure of the two polymers is shown in Scheme 1. Starch consists of linear amylose and branched amylopectin chains (only this latter is shown in the scheme). In spite of the differences mentioned, both polymers contain a large number of hydroxyl groups capable of forming relatively strong hydrogen bonds with PLA.

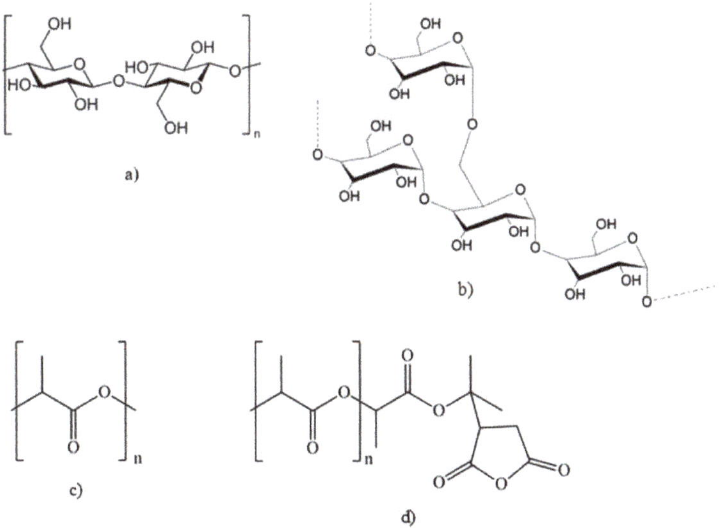

Scheme 1. Chemical structure of the materials used in the study: (**a**) cellulose; (**b**) amylopectin (the structure of amylose is the same but without branching); (**c**) PLA; (**d**) MAPLA.

The phase structure of the two reinforcements also shows some similarities, both are semicrystalline materials containing an ordered, crystalline phase. Both the strong self-interactions and crystallinity result in large stiffness. However, while starch is a relatively pure material, cellulose fibers contain other components as well, mainly hemicellulose, lignin, and waxes [50,51]. The oriented chains located in cellulose crystals result in increased stiffness and strength in the direction of the orientation and smaller strength perpendicularly to it [52]. Such orientation does not exist in starch particles.

Particle morphology is crucial for such materials used as reinforcements in polymer composites. Size, shape, and the extent of anisotropy, i.e., aspect ratio, are extremely important in the determination of deformation processes and final properties [53,54]. The size distribution of the two materials determined by laser light scattering is compared in Figure 1. The average size of starch particles is around 20 μm, while bagasse fibers are much larger, the most frequent size being around 900–1000 μm. However, since sugarcane bagasse fibers have an anisotropic shape, their size was also determined by microscopy. According to these measurements, the average length of the fibers is 2850 μm, their diameter 720 μm and thus their aspect ratio is around 5.1. Moreover, we must call attention to the fact here that the fibers change their size during processing, considerable attrition takes place and thus the final length of the fibers in PLA composites processed by extrusion and injection molding was 730 μm, their diameter 170 μm, and aspect ratio decreased to 3.0 [26]. Although considerable similarities can be found in the chemical structure of the two reinforcements, their physical structure, composition, and particle morphology differ considerably; it remains to be seen which factor determines the extent of reinforcement and composite properties.

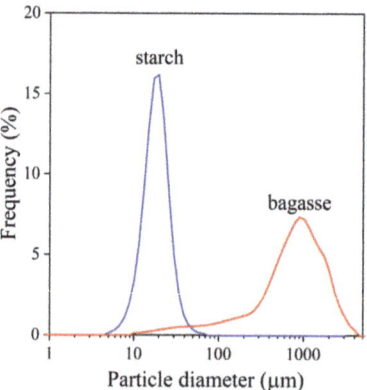

Figure 1. Particle size distribution of the additives as determined by laser light scattering.

2.2. Composite Properties

Since fiber-reinforced materials are mostly used in structural applications, the attention is focused mainly on their mechanical properties. The stiffness of the composites containing various amounts of the two reinforcements is shown in Figure 2. Both reinforcements increase stiffness, but to considerably different extents, bagasse fibers have a much stronger reinforcing effect than starch. The stiffness of natural fibers was shown to cover a wide range from several 10 to several 100 GPa [50,55]; the direct measurement of the stiffness of the fibers used in this study resulted in the value of 25 GPa [51] that is in line with their reinforcing effect. Unfortunately, the direct determination of the stiffness of starch particles is very difficult, if not impossible, and we found only a single paper reporting the modulus value for starch, which was derived indirectly from the study of epoxy resin/starch and polycaprolactone/starch blends [56]. The value given was 2.7 GPa that cannot be correct since the addition of starch clearly increases the stiffness of neat PLA having a modulus of 3.3 GPa.

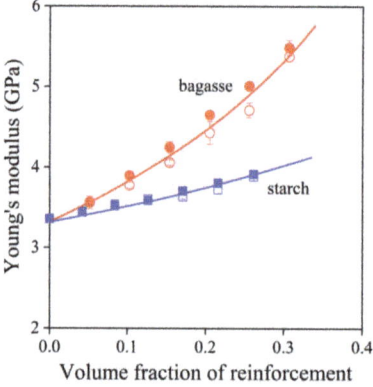

Figure 2. Composition dependence of the stiffness of PLA composites reinforced with starch or bagasse fibers. Effect of coupling. Symbols: (○, ●) sugarcane bagasse fibers, (□, ■) starch; empty symbols without MAPLA, full symbols with MAPLA.

We may assume that the inherent stiffness of starch is smaller than that of the sugarcane bagasse fibers, but the anisotropic particle geometry and the orientation of the fibers must also contribute to the larger stiffness of their composites. Using MAPLA as coupling agent in the composites does not improve or deteriorate stiffness that is not very surprising, since stiffness is not very sensitive to changes in interfacial adhesion [25,57].

The deformability of the composites (not shown) decreases marginally with an increasing amount of the additives independently of their type or coupling. The strength of the composites containing the two reinforcements is presented in Figure 3 as a function of additive content. The results are quite surprising, starch having a stronger reinforcing effect than the bagasse fibers, at least at smaller additive loadings. The orientation of the anisotropic fibers should lead to stronger composites similarly to stiffness. Obviously, other factors, most probably local deformation processes, play a significant role in the determination of composite strength. The much larger size of the fibers, even after attrition, facilitates debonding and leads to smaller strength. Although starch particles of around 20 μm size may also debond, but at a larger stress than the fibers having a size of one order of magnitude larger.

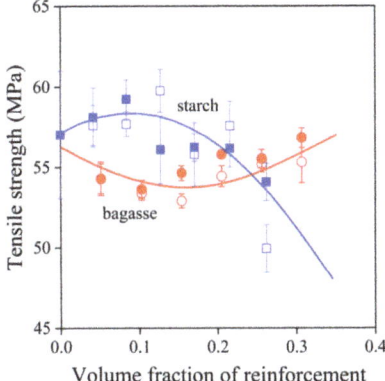

Figure 3. Effect of composition and coupling on the tensile strength of PLA composites. Symbols: (○, ●) sugarcane bagasse fibers, (□, ■) starch; empty symbols without MAPLA, full symbols with MAPLA.

Another surprising result is that the MAPLA coupling agent practically does not affect the tensile strength of the composites either, although properties measured at larger deformations depend on interfacial adhesion quite strongly [57]. In spite of numerous claims published in the literature that the interaction between natural fibers and PLA is weak [28,29,31], we have proved earlier that relatively strong interactions develop in such composites [26,58]. Considering the similarities in the chemical structure of starch and bagasse fibers (large number of OH groups), the lack of any coupling effect is not surprising in the case of starch either. However, the relative influence of particle size and interfacial interactions must be considered in the further evaluation of the results.

Impact resistance is another mechanical property that is important for most structural materials; generally large strength and impact resistance is required from such materials used in practice [59,60]. The impact strength of the composites prepared in this study is presented in Figure 4. The effect of the two reinforcements differs considerably from each other. The addition of bagasse fibers doubles the impact strength of PLA, while the addition of starch decreases it slightly. The opposite effects must be caused by the different failure mechanism of the composites containing the two kinds of additives. The effect of coupling is very slight again. This small effect must be related to the interaction of the components and to the local deformation processes taking place during failure. The main factors determining composite properties are particle characteristics and to some extent interfacial adhesion.

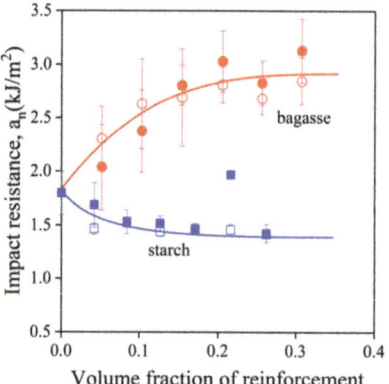

Figure 4. Dependence of the notched Charpy impact resistance of PLA composites containing natural reinforcements. Symbols: (○, ●) sugarcane bagasse fibers, (□, ■) starch; empty symbols without MAPLA, full symbols with MAPLA.

2.3. Reinforcement

The reinforcing effect of additives is difficult to assess based on the primary results of stiffness and especially strength values even though it is almost invariably done so. Reinforcement depends on several factors including the spatial arrangements of the fibers, their size, shape, and orientation as well as on interfacial interactions. Mathematical models give a quantitative estimation of the reinforcing effect of the most various second components. Such a model was developed earlier, which describes the composition dependence of tensile strength [61]. The model can be expressed as (Equation (1))

$$\sigma_T = \sigma_{T0} \lambda^n \frac{1 - \varphi_f}{1 + 2.5\,]\varphi_f} \exp\left(B\,\varphi_f\right) \tag{1}$$

where σ_T shows the true tensile strength of the composite; σ_{T0} is the same for the matrix, φ_f shows the volume fraction of the second component, and B is the load-bearing capacity of the reinforcement. The latter depends on interfacial adhesion. True tensile strength ($\sigma_T = \sigma \lambda$, where $\lambda = L/L_0$, is the relative elongation) expresses the change in the cross-section of the specimen during deformation and λ^n takes into account strain hardening (n can be determined from matrix properties and characterizes the strain hardening tendency. Reduced tensile strength can be expressed by the rearrangement of Equation (1)

$$\sigma_{Tred} = \frac{\sigma_T}{\lambda^n} \frac{1 + 2.5\,\varphi_f}{1 - \varphi_f} = \sigma_{T0} \exp\left(B\,\varphi_f\right) \tag{2}$$

The natural logarithm of tensile strength can be plotted against composition and the plot should result a straight line. The slope of the line expresses the reinforcing effect of the additives quantitatively. The tensile strength of the composites is plotted in the way indicated by Equation (2) in Figure 5. We obtain straight lines as expected verifying the validity of the approach and showing the lack of considerable structural effects.

The analysis indicates that the load-bearing capacity of the two reinforcements is similar in spite of the different primary values shown in Figure 3. The parameters determined by the fitting of the model to the experimental values are compiled in Table 1. The comparison of the data listed in the table shows the very slight overall effect of coupling as well. The similar reinforcing effect of the two additives might be surprising first but can be reasonably explained by the smaller size of starch particles increasing the value of parameter B and the orientation of bagasse fibers, which compensates for the negative effect of large size.

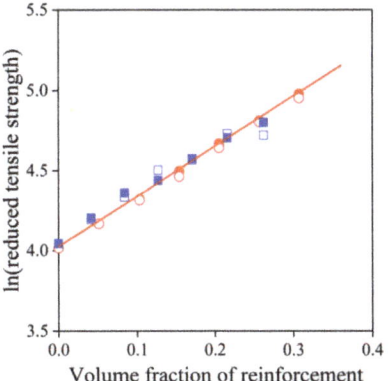

Figure 5. Reinforcing effect of the natural additives used in PLA composites. Tensile strength plotted in the linear representation of Equation (2). Effect of coupling. Symbols: (○, ●) sugarcane bagasse fibers, (□, ■) starch; empty symbols without MAPLA, full symbols with MAPLA.

Table 1. Parameter (B) characterizing the load-bearing capacity of the reinforcements studied (see Equations (1) and (2)).

Reinforcement	Coupling	σ_{T0} (MPa)	Parameter B	R^2 [a]
Starch	−	58.2	3.11 ± 0.16	0.9896
	+	59.2	2.84 ± 0.12	0.9910
Bagasse	−	55.0	3.07 ± 0.04	0.9992
	+	55.1	3.14 ± 0.03	0.9994

[a] determination coefficient indicating the goodness of the fit.

The strength of interfacial adhesion is difficult to predict from the model because of the different size and shape of the particles, but it can be estimated by the reversible work of adhesion. This latter can be calculated from the surface tension of the components. The matrix of the composites was the same, and the surface tension of the two reinforcements is quite similar, 35 mJ/m^2 was measured for starch, while 38 mJ/m^2 for bagasse fibers by inverse gas chromatography leading to similar work of adhesions (99.4 mJ/m^2 and 100.2 mJ/m^2 respectively), i.e., strength of interfacial adhesion.

2.4. Deformation and Failure

The results presented above indicate that considerable differences exist in the properties of the composites prepared with the two kinds of reinforcements. Stiffness and impact resistance was larger for the PLA/bagasse fiber composites that is a slight contradiction in itself since larger stiffness is usually accompanied by smaller impact strength [62], while the strength of the PLA/starch composites exceeded that of the other set of materials. Since the strength of interfacial adhesion is similar, the main reason for the differences must lay in the particle characteristics of the additives and the local deformation processes initiated by them.

Composites are heterogeneous materials containing components with dissimilar elastic properties. External load results in the development of local stress maxima, which initiate local deformation processes. These processes like the separation of the interface between the matrix and the reinforcement (debonding), and the fracture of the particles can be followed by various techniques including volume strain measurements, or by acoustic emission tests. The results of the latter measurements are presented in Figure 6 for two composites, one containing starch (Figure 6a) and the other prepared with bagasse fibers (Figure 6b). The small circles in the figure indicate individual acoustic events, while the two continuous lines show the stress vs. strain curve for reference (left axis), as well as the cumulative number of signal vs. deformation trace (right axis). This latter shows the total number of signals detected up to a certain deformation and its shape offers information about the process itself. Debonding is often accompanied by a saturation-like correlation [59], but the shape in itself does not allow the unambiguous identification of the dominating local deformation process [63].

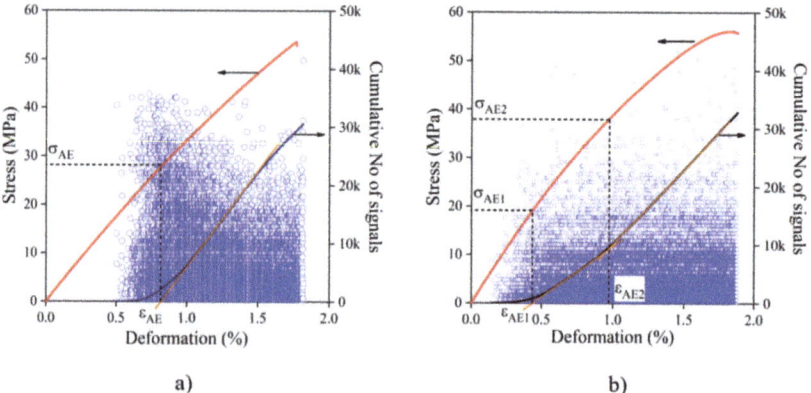

Figure 6. Results of the acoustic emission testing of PLA composites containing starch or bagasse fibers. Additive content: 20 vol%. Symbols: (○) individual acoustic events, full lines: stress vs. deformation (left axis, red) and cumulative number of signal vs. deformation (right axis, navy) correlations: (**a**) starch, (**b**) sugarcane bagasse fiber.

The comparison of the two plots presented in Figure 6a,b reveals a number of similarities, but also differences. The amplitude, i.e., strength, of the signals indicated by the height of the points on the vertical axis is approximately the same, although signals with higher energy seem to be located in larger numbers at the first part of the measurement in the starch composite (Figure 6a). Both plots show that signals start to appear after a certain, critical deformation, which is larger for the starch composite than for the one containing bagasse. If debonding is the main local deformation process, this difference can be clearly explained by the smaller size of the starch particles, since debonding stress is inversely proportional to particle size [64]. The signals in Figure 6a seem to form only one group but their high amplitude and the shape of the cumulative number of signals trace indicate the occurrence of a second process besides debonding which may be the fracture of starch particles, although one would not expect this to happen and it needs further proof.

The plotting of the cumulative number of signal traces allows the determination of characteristic stresses which can be assigned to the local processes taking place during the deformation of the specimen. The determination of these values from the plots is shown in Figure 6. Characteristic stresses determined for the processes detected, σ_{AE1} and σ_{AE2}, are plotted against composition in Figure 7. Only one characteristic stress could be determined for the starch composites with any reliability. The first process in bagasse fiber composites is initiated at smaller stress and is unambiguously assigned to debonding. The larger stress obtained for starch is consistent with the explanation given above, i.e., debonding needs larger stress for smaller particles. In the case of the fiber-reinforced polymer the second process is mostly related to the fracture of the fibers [52,57,59].

Fracture is difficult to imagine for starch particles, but it is not impossible. The stress belonging to fiber fracture is much larger and can be clearly separated from the first process for the bagasse composites. The existence of a second, competitive process cannot be identified definitely for starch-filled composites but SEM micrographs may provide further evidence about the processes occurring during failure.

SEM micrographs recorded on the fracture surface of specimens broken during tensile tests are presented in Figure 8 for samples containing starch or bagasse fibers with or without coupling. Considerable debonding can be observed in Figure 8a in the composite containing starch without coupling, while less debonding and the fracture of some particles are observed in Figure 8b in the presence of the coupling agent (MAPLA). One could draw the conclusion from these results that coupling is very efficient that would contradict all earlier conclusions. However, we must call the attention here to the fact that SEM micrographs cover only a very small area of the fractured surface of the specimen. Some micrographs prepared showed less debonding than in Figure 8a and even some broken particles in the absence of the coupling agent, and, on the other hand, larger number of debonded particles in its presence. Nevertheless, the micrographs verify the occurrence of two processes in these composites and identify them as debonding and particle fracture. Very similar micrographs were also recorded on composites containing the bagasse fibers, as shown by Figure 8c,d; the two processes, debonding and fiber fracture, can be clearly identified also in these micrographs.

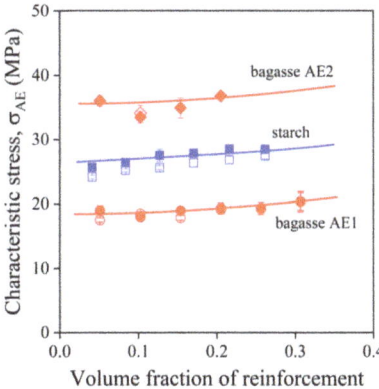

Figure 7. Composition dependence of characteristic stresses derived from acoustic emission testing. Symbols: (○, ●) sugarcane bagasse fibers, σ_{AE1} (□, ■) starch, σ_{AE}, (◇, ◆) bagasse, σ_{AE2}; empty symbols without MAPLA, full symbols with MAPLA.

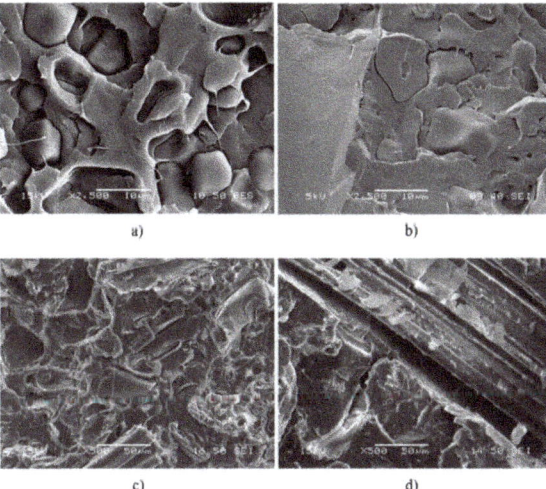

Figure 8. SEM micrographs recorded on the fractured surfaces of specimens broken during tensile testing: (**a**) starch, 20 wt%; (**b**) starch, 20 wt%, MAPLA; (**c**) sugarcane bagasse fiber, 20 wt%; and (**d**) bagasse fiber, 20 wt%, MAPLA.

3. Discussion

The results presented in the previous sections clearly show that the main factors determining composite properties are particle characteristics and the inherent properties of the additives. Bagasse fibers increase stiffness to a larger extent than starch because of the larger stiffness of the fibers and their anisotropic particle characteristics. Strength is strongly influenced by local processes taking place during deformation. The smaller particle size of starch results in larger debonding stress (see also Figure 7, σ_{AE1}) and thus larger composite strength compared to composites containing bagasse fibers. The presumably larger inherent strength of the fibers leads to considerably larger impact resistance for the PLA/bagasse fiber composites. Because of the similar interfacial interactions of the two reinforcements, interfacial adhesion is also very similar for the two (see Figure 5 and Table 1) and coupling has a very slight influence on properties. The limited efficiency of coupling can be explained also by the small number of functional groups and smaller molecular weight of the modified polymer as well as the small number of entanglements in PLA. The results are consistent and help to identify the main factors determining composite properties, but also the advantages and drawbacks of the two reinforcements.

One of the main requirements towards composites used as structural materials is large stiffness and impact resistance. The requirement is quite difficult to satisfy because they are inversely proportional in most structural materials, i.e., larger stiffness is usually accompanied by smaller impact strength. The two quantities are plotted against each other in Figure 9. The general tendency is indeed valid for the PLA/starch composites, but shows the opposite direction for the polymer reinforced with the bagasse fibers. The reason for the discrepancy is the relatively large inherent strength of the bagasse fibers and the fact that considerable fiber fracture takes place during the deformation and failure of its composites; in fact, fiber fracture might be the dominating local deformation process over debonding and it consumes considerable energy during failure. We must call attention here, though, that the absolute value of fracture resistance is not very large in any of the composites, and that much larger fracture strengths are required in certain applications. Nevertheless, bagasse fibers offer a reasonable overall performance over starch.

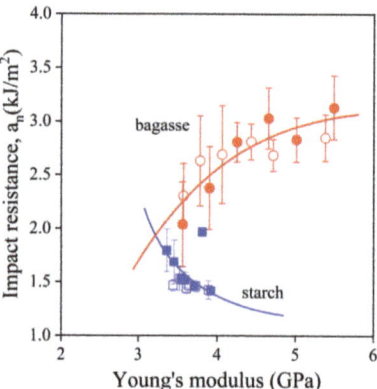

Figure 9. Correlation of the stiffness and impact strength of PLA composites reinforced with starch and sugarcane bagasse fibers. Opposing effects as a result of different local processes. Symbols: (○, ●) sugarcane bagasse fibers, (□, ■) starch; empty symbols without MAPLA, full symbols with MAPLA.

4. Materials and Methods

4.1. Materials

The matrix, poly(lactic acid) (Ingeo 4032D, M_n = 88,500 g/mol and M_w/M_n = 1.8) was provided by NatureWorks LLC (Minnetonka, MN, USA). The extrusion grade polymer (<2% D isomer) has a density of 1.24 g/cm³, and its MFR is 3.9 g/10 min (190 °C and 2.16 kg). The production method of the maleic anhydride grafted PLA (MAPLA) coupling agent was described earlier in detail [25]. The Ingeo 3251D grade PLA (NatureWorks LLC, Minnetonka, MN, USA; MFR: 35 g/10 min at 190 °C and 2.16 kg load) was used for the grafting reaction. The reactive extrusion was carried out in a Brabender LabStation (Brabender GmbH, Duisburg, Germany) single screw extruder with the temperature profile of 175–180–185–190 °C. The screw speed was set to 12 rpm. 2 wt% maleic anhydride and 2 wt% Luperox 101 peroxide were added to the reaction mixture. MAPLA was characterized by NMR (Varian NMR System, Agilent Technologies, Inc., Santa Clara, CA, USA), however, it was not purified; it was applied as produced in the reactive extrusion process.

The bagasse fibers used as reinforcement were obtained from a sugar mill in Sidoarjo, Indonesia. The chemical composition of the fibers was determined by the van Soest method. The detailed description of the method can be found in the paper of van Soest [65]. According to the method, hemicellulose content is determined after treatment with an acidic detergent solution, the amount of cellulose by treating the fibers with sulfuric acid of 72 wt% concentration and lignin by burning the sample in an oven. The fibers contained 47% cellulose, 35% hemicellulose, 15% lignin, and 3% ash. The fibers were dried and sieved before extrusion. Fiber characteristics were determined by digital optical microscopy (Keyence VHX 5000, Keyence Co., Osaka, Japan). The native starch was derived from corn and was purchased from Hungrana Kft., Hungary. The amount of the additives in the polymer changed from 0 to 30 wt% in 5 wt% steps. The coupling agent (MAPLA) was added to the matrix polymer in 0.1:1 ratio calculated for the weight of starch or sugarcane bagasse fibers.

4.2. Sample Preparation

The modifying components were dried in a Memmert UF450 type oven (Memmert GmbH, Schwabach, Germany) before extrusion (bagasse fibers for 4 h, and starch for 24 h at 105 °C) to eliminate their residual moisture content. PLA and MAPLA were dried together in a vacuum oven at 100 °C and 150 mbar pressure. A Brabender DSK 42/7 (Brabender GmbH, Duisburg, Germany) twin-screw compounder was employed for the appropriate homogenization of the components at the set temperatures of 170–180–185–190 °C and the rate of homogenization of 40 rpm. The granules were injection molded into tensile bars according to the ISO 527 1A standard using a Demag IntElect 50/330-100 electric injection molding machine. The temperature profile of the barrel was the same as in extrusion, while injection pressure (800–1200 bar) depended on the amount of the reinforcement added. The value of the most important processing parameters were: injection speed: 50 mm/s; holding pressure: 650–800 bar; holding time: 15 s; cooling time: 45 s. The temperature of the mold was set to 20 °C. The specimens were kept under ambient conditions (23 °C, 50% RH) until further testing.

4.3. Characterization

An Instron 5566 universal testing machine (Instron, Norwood, MA, USA) was used for the characterization of tensile properties with a gauge length of 115 mm and 5 mm/min crosshead speed. An acoustic emission (AE) equipment was used simultaneously with tensile testing to detect local deformation processes. The specific equipment used was a Sensophone AED 404 apparatus manufactured by Geréb és Társa Ltd. (Budapest, Hungary). A single piezoelectric sensor (150 kHz resonance frequency) was attached to the center of the specimen. The threshold level of detection was set to 25 dB. A Ceast Resil 5.5 impact tester (Ceast spa, Pianezza, Italy) was used for the determination of Charpy impact strength (ISO 179 standard at 23 °C with 2 mm notch depth) of the specimens. Fracture surfaces were studied by scanning electron microscopy (Jeol JSM 6380 LA, Jeol Ltd., Tokyo, Japan). Micrographs were recorded on surfaces created during tensile and fracture testing, respectively. Accelerating voltage (5 and 15 kV) and distance to the sample changed according to magnification and the quality of the image. Detectors were used both for secondary and back-scattered electrons. Surfaces were sputtered with gold before the recording of micrographs from the composites. The particle size and size distribution of the modifying components were determined by laser light scattering using a Horiba LA 950 A2 (Horiba, Kyoto, Japan) analyzer. The size, size distribution, and aspect ratio of the bagasse fibers were also determined after processing by digital optical microscopy (Keyence VHX 5000, Keyence Co., Osaka, Japan). The polymer was dissolved in tetrahydrofuran (Molar Chemicals Kft., Hungary) to extract the fibers from the composites.

5. Conclusions

The comparison of the properties of PLA composites reinforced with native starch and sugarcane bagasse fibers shows that the two additives influence properties differently. The analysis of their chemical and physical structure indicates that the former is quite similar, while considerable differences exist in the latter. Particle characteristics differ even more and are shown to influence properties considerably. The strength of interfacial adhesion, and thus the extent of reinforcement, are similar because of the similarities in the chemical structure of the reinforcements. Relatively strong interfacial adhesion develops between the components which renders coupling inefficient. Dissimilar particle characteristics, on the other hand, influence local deformation processes considerably. The smaller particle size of starch results in larger debonding stress and thus larger composite strength. Besides debonding, considerable fiber or particle fracture also takes place during the failure of the composites. The larger inherent strength of the bagasse fibers leads to larger energy consumption during fracture and increased impact resistance. Although the environmental benefit of the prepared biocomposites is very similar, the overall performance of the bagasse fibers reinforced PLA composites is better than that offered by the PLA/starch composites.

Author Contributions: Investigation, M.J. and R.V.; data curation, M.J., R.V., A.B. and J.M.; writing—review and editing, M.J., R.V., A.B. and J.M.; visualization, R.V.; methodology, R.V., A.B., J.M. and B.P.; conceptualization, J.M. and B.P.; writing—original draft, B.P.; supervision, B.P. All authors have read and agreed to the published version of the manuscript.

Funding: The research reported in this paper is part of project no. TKP2021-EGA-02, implemented with the support provided by the Ministry for Innovation and Technology of Hungary from the National Research, Development and Innovation Fund, financed under the TKP2021 funding scheme. The research was partly funded by the National Research, Development and Innovation Fund of Hungary (FK 129270).

Institutional Review Board Statement: Not applicable.

Informed Consent Statement: Not applicable.

Data Availability Statement: The raw/processed data required to reproduce these findings cannot be shared at this time due to legal or ethical reasons.

Acknowledgments: The authors thank Bence Csótai and Kristóf Nagy for their technical support in sample preparation and measurement.

Conflicts of Interest: The authors declare no conflict of interest. The funders had no role in the design of the study; in the collection, analyses, or interpretation of data; in the writing of the manuscript; or in the decision to publish the results.

References

1. Ulonska, K.; König, A.; Klatt, M.; Mitsos, A.; Viell, J. Optimization of Multiproduct Biorefinery Processes under Consideration of Biomass Supply Chain Management and Market Developments. *Ind. Eng. Chem. Res.* **2018**, *57*, 6980–6991. [CrossRef]
2. Miller, S.A. Sustainable Polymers: Opportunities for the Next Decade. *ACS Macro Lett.* **2013**, *2*, 550–554. [CrossRef] [PubMed]
3. Auras, R.; Harte, B.; Selke, S. An overview of polylactides as packaging materials. *Macromol. Biosci.* **2004**, *4*, 835–864. [CrossRef] [PubMed]
4. Arrieta, M.P.; López, J.; Hernández, A.; Rayón, E. Ternary PLA–PHB–Limonene blends intended for biodegradable food packaging applications. *Eur. Polym. J.* **2014**, *50*, 255–270. [CrossRef]
5. Tomadoni, B.; Casalongué, C.; Alvarez, V.A. Biopolymer-Based Hydrogels for Agriculture Applications: Swelling Behavior and Slow Release of Agrochemicals. In *Polymers for Agri-Food Applications*; Gutiérrez, T.J., Ed.; Springer International Publishing: Cham, Switzerland, 2019; pp. 99–125.

6. George, A.; Sanjay, M.R.; Srisuk, R.; Parameswaranpillai, J.; Siengchin, S. A comprehensive review on chemical properties and applications of biopolymers and their composites. *Int. J. Biol. Macromol.* **2020**, *154*, 329–338. [CrossRef] [PubMed]
7. Gigante, V.; Seggiani, M.; Cinelli, P.; Signori, F.; Vania, A.; Navarini, L.; Amato, G.; Lazzeri, A. Utilization of coffee silverskin in the production of Poly(3-hydroxybutyrate-co-3-hydroxyvalerate) biopolymer-based thermoplastic biocomposites for food contact applications. *Compos. Part A Appl. Sci. Manuf.* **2021**, *140*, 106172. [CrossRef]
8. Niaounakis, M. 6-Automotive Applications. In *Biopolymers: Applications and Trends*; William Andrew Publishing: Oxford, UK, 2015; pp. 257–289.
9. Verma, D.; Dogra, V.; Chaudhary, A.K.; Mordia, R. 5-Advanced biopolymer-based composites: Construction and structural applications. In *Sustainable Biopolymer Composites*; Verma, D., Sharma, M., Goh, K.L., Jain, S., Sharma, H., Eds.; Woodhead Publishing: Sawston, UK, 2022; pp. 113–128.
10. Alam, M.A.; Sapuan, S.M.; Ya, H.H.; Hussain, P.B.; Azeem, M.; Ilyas, R.A. Chapter 1-Application of biocomposites in automotive components: A review. In *Biocomposite and Synthetic Composites for Automotive Applications*; Sapuan, S.M., Ilyas, R.A., Eds.; Woodhead Publishing: Duxford, UK, 2021; pp. 1–17.
11. Rusu, D.; Boyer, S.A.E.; Lacrampe, M.F.; Krawczak, P. Bioplastics and vegetal fiber reinforced bioplastics for automotive applications. In *Handbook of Bioplastics and Biocomposites Engineering Applications*; Pilla, S., Ed.; Scrivener Publishing LLC.: Beverly, MA, USA, 2011; pp. 397–449.
12. Ivar do Sul, J.A.; Costa, M.F. The present and future of microplastic pollution in the marine environment. *Environ. Pollut.* **2014**, *185*, 352–364. [CrossRef]
13. Pabortsava, K.; Lampitt, R.S. High concentrations of plastic hidden beneath the surface of the Atlantic Ocean. *Nat. Commun.* **2020**, *11*, 4073. [CrossRef]
14. Cui, L.; Imre, B.; Tátraaljai, D.; Pukánszky, B. Physical ageing of poly(lactic acid): Factors and consequences for practice. *Polymer* **2020**, *186*, 122014. [CrossRef]
15. Coltelli, M.-B.; Mallegni, N.; Rizzo, S.; Cinelli, P.; Lazzeri, A. Improved impact properties in poly(lactic acid) (PLA) blends containing cellulose acetate (CA) prepared by reactive extrusion. *Materials* **2019**, *12*, 270. [CrossRef]
16. Hassouna, F.; Raquez, J.-M.; Addiego, F.; Toniazzo, V.; Dubois, P.; Ruch, D. New development on plasticized poly(lactide): Chemical grafting of citrate on PLA by reactive extrusion. *Eur. Polym. J.* **2012**, *48*, 404–415. [CrossRef]
17. Garcia-Garcia, D.; Carbonell-Verdu, A.; Arrieta, M.P.; López-Martínez, J.; Samper, M.D. Improvement of PLA film ductility by plasticization with epoxidized karanja oil. *Polym. Degrad. Stab.* **2020**, *179*, 109259. [CrossRef]
18. Boyacioglu, S.; Kodal, M.; Ozkoc, G. A comprehensive study on shape memory behavior of PEG plasticized PLA/TPU bio-blends. *Eur. Polym. J.* **2020**, *122*, 109372. [CrossRef]
19. Li, X.; Hegyesi, N.; Zhang, Y.; Mao, Z.; Feng, X.; Wang, B.; Pukánszky, B.; Sui, X. Poly(lactic acid)/lignin blends prepared with the Pickering emulsion template method. *Eur. Polym. J.* **2019**, *110*, 378–384. [CrossRef]
20. Dasan, Y.K.; Bhat, A.H.; Ahmad, F. Polymer blend of PLA/PHBV based bionanocomposites reinforced with nanocrystalline cellulose for potential application as packaging material. *Carbohydr. Polym.* **2017**, *157*, 1323–1332. [CrossRef]
21. Siparsky, G.L.; Voorhees, K.J.; Dorgan, J.R.; Schilling, K. Water transport in polylactic acid (PLA), PLA/polycaprolactone copolymers, and PLA/polyethylene glycol blends. *J. Environ. Polym. Degrad.* **1997**, *5*, 125–136. [CrossRef]
22. Gao, H.; Hu, S.; Su, F.; Zhang, J.; Tang, G. Mechanical, thermal, and biodegradability properties of PLA/modified starch blends. *Polym. Compos.* **2011**, *32*, 2093–2100. [CrossRef]
23. Yao, M.; Deng, H.; Mai, F.; Wang, K.; Zhang, Q.; Chen, F.; Fu, Q. Modification of poly(lactic acid)/poly(propylene carbonate) blends through melt compounding with maleic anhydride. *Express Polym. Lett.* **2011**, *5*, 937–949. [CrossRef]
24. Bledzki, A.K.; Franciszczak, P.; Meljon, A. High performance hybrid PP and PLA biocomposites reinforced with short man-made cellulose fibres and softwood flour. *Compos. Part A Appl. Sci. Manuf.* **2015**, *74*, 132–139. [CrossRef]
25. Csikós, Á.; Faludi, G.; Domján, A.; Renner, K.; Móczó, J.; Pukánszky, B. Modification of interfacial adhesion with a functionalized polymer in PLA/wood composites. *Eur. Polym. J.* **2015**, *68*, 592–600. [CrossRef]
26. Bartos, A.; Nagy, K.; Anggono, J.; Purwaningsih, H.; Móczó, J.; Pukánszky, B. Biobased PLA/sugarcane bagasse fiber composites: Effect of fiber characteristics and interfacial adhesion on properties. *Compos. Part A Appl. Sci. Manuf.* **2021**, *143*, 106273. [CrossRef]
27. Lila, M.K.; Shukla, K.; Komal, U.K.; Singh, I. Accelerated thermal ageing behaviour of bagasse fibers reinforced poly(lactic acid) based biocomposites. *Compos. Part B Eng.* **2019**, *156*, 121–127. [CrossRef]
28. Huda, M.S.; Drzal, L.T.; Misra, M.; Mohanty, A.K. Wood-fiber-reinforced poly(lactic acid) composites: Evaluation of the physicomechanical and morphological properties. *J. Appl. Polym. Sci.* **2006**, *102*, 4856–4869. [CrossRef]
29. Mathew, A.P.; Oksman, K.; Sain, M. Mechanical properties of biodegradable composites from poly lactic acid (PLA) and microcrystalline cellulose (MCC). *J. Appl. Polym. Sci.* **2005**, *97*, 2014–2025. [CrossRef]
30. Bax, B.; Müssig, J. Impact and tensile properties of PLA/Cordenka and PLA/flax composites. *Compos. Sci. Technol.* **2008**, *68*, 1601–1607. [CrossRef]
31. Oksman, K.; Skrifvars, M.; Selin, J.-F. Natural fibres as reinforcement in polylactic acid (PLA) composites. *Compos. Sci. Technol.* **2003**, *63*, 1317–1324. [CrossRef]
32. Yusuf, M.; Shabbir, M.; Mohammad, F. Natural Colorants: Historical, Processing and Sustainable Prospects. *Nat. Prod. Bioprospecting* **2017**, *7*, 123–145. [CrossRef]

33. Van Den Oever, M.J.A.; Boeriu, C.G.; Blaauw, R.; Van Haveren, J. Colorants based on renewable resources and food-grade colorants for application in thermoplastics. *J. Appl. Polym. Sci.* **2004**, *92*, 2961–2969. [CrossRef]
34. Atarés, L.; Chiralt, A. Essential oils as additives in biodegradable films and coatings for active food packaging. *Trends Food Sci. Technol.* **2016**, *48*, 51–62. [CrossRef]
35. Kmiotek, M.; Bieliński, D.; Piotrowska, M. Propolis as an antidegradant and biocidal agent for natural rubber. *J. Appl. Polym. Sci.* **2018**, *135*, 45911. [CrossRef]
36. Liu, L.; Qian, M.; Song, P.; Huang, G.; Yu, Y.; Fu, S. Fabrication of Green Lignin-based Flame Retardants for Enhancing the Thermal and Fire Retardancy Properties of Polypropylene/Wood Composites. *ACS Sustain. Chem. Eng.* **2016**, *4*, 2422–2431. [CrossRef]
37. Costes, L.; Laoutid, F.; Brohez, S.; Dubois, P. Bio-based flame retardants: When nature meets fire protection. *Mater. Sci. Eng. R Rep.* **2017**, *117*, 1–25. [CrossRef]
38. Strandberg, C.; Albertsson, A.C. Process efficiency and long-term performance of α-tocopherol in film-blown linear low-density polyethylene. *J. Appl. Polym. Sci.* **2005**, *98*, 2427–2439. [CrossRef]
39. Dabbaghi, A.; Jahandideh, A.; Kabiri, K.; Ramazani, A.; Zohuriaan-Mehr, M.J. The synthesis and incorporation of a star-shaped bio-based modifier in the acrylic acid based superabsorbent: A strategy to enhance the absorbency under load. *Polym. Plast. Technol. Mater.* **2019**, *58*, 1678–1690. [CrossRef]
40. Wypych, G. 3-Impact Modifiers. In *Databook of Impact Modifiers*; ChemTec Publishing: Scarborough, ON, Canada, 2022; pp. 17–448.
41. Carvalho, A.J.F. Starch: Major sources, properties and applications as thermoplastic materials. In *Handbook of Biopolymers and Biodegradable Plastics*; Ebnesajjad, S., Ed.; William Andrew Publishing: Boston, MA, USA, 2013; pp. 129–152.
42. Jiang, T.; Duan, Q.; Zhu, J.; Liu, H.; Yu, L. Starch-based biodegradable materials: Challenges and opportunities. *Adv. Ind. Eng. Polym. Res.* **2020**, *3*, 8–18. [CrossRef]
43. Forssell, P.M.; Mikkilä, J.M.; Moates, G.K.; Parker, R. Phase and glass transition behaviour of concentrated barley starch-glycerol-water mixtures, a model for thermoplastic starch. *Carbohydr. Polym.* **1997**, *34*, 275–282. [CrossRef]
44. Basiak, E.; Lenart, A.; Debeaufort, F. How glycerol and water contents affect the structural and functional properties of starch-based edible films. *Polymers* **2018**, *10*, 412. [CrossRef]
45. Li, H.B.; Huneault, M.A. Comparison of Sorbitol and Glycerol as Plasticizers for Thermoplastic Starch in TPS/PLA Blends. *J. Appl. Polym. Sci.* **2011**, *119*, 2439–2448. [CrossRef]
46. Wokadala, O.C.; Emmambux, N.M.; Ray, S.S. Inducing PLA/starch compatibility through butyl-etherification of waxy and high amylose starch. *Carbohydr. Polym.* **2014**, *112*, 216–224. [CrossRef]
47. Yang, X.; Finne-Wistrand, A.; Hakkarainen, M. Improved dispersion of grafted starch granules leads to lower water resistance for starch-g-PLA/PLA composites. *Compos. Sci. Technol.* **2013**, *86*, 149–156. [CrossRef]
48. Zhang, J.-F.; Sun, X. Mechanical Properties of Poly(lactic acid)/Starch Composites Compatibilized by Maleic Anhydride. *Biomacromolecules* **2004**, *5*, 1446–1451. [CrossRef] [PubMed]
49. Schlumpf, H.P. Fillers and reinforcing materials in plastics-physicochemical aspects for the processor. *Kunstst.-Ger. Plast.* **1983**, *73*, 511–515.
50. Bledzki, A.K.; Gassan, J. Composites reinforced with cellulose based fibres. *Prog. Polym. Sci.* **1999**, *24*, 221–274. [CrossRef]
51. Bartos, A.; Anggono, J.; Farkas, Á.E.; Kun, D.; Soetaredjo, F.E.; Móczó, J.; Purwaningsih, H.; Pukánszky, B. Alkali treatment of lignocellulosic fibers extracted from sugarcane bagasse: Composition, structure, properties. *Polym. Test.* **2020**, *88*, 106549. [CrossRef]
52. Bartos, A.; Utomo, B.P.; Kanyar, B.; Anggono, J.; Soetaredjo, F.E.; Móczó, J.; Pukánszky, B. Reinforcement of polypropylene with alkali-treated sugarcane bagasse fibers: Mechanism and consequences. *Compos. Sci. Technol.* **2020**, *200*, 108428. [CrossRef]
53. Kwon, H.-J.; Sunthornvarabhas, J.; Park, J.-W.; Lee, J.-H.; Kim, H.-J.; Piyachomkwan, K.; Sriroth, K.; Cho, D. Tensile properties of kenaf fiber and corn husk flour reinforced poly(lactic acid) hybrid bio-composites: Role of aspect ratio of natural fibers. *Compos. Part B Eng.* **2014**, *56*, 232–237. [CrossRef]
54. Pukánszky, B.; Belina, K.; Rockenbauer, A.; Maurer, F.H.J. Effect of nucleation, filler anisotropy and orientation on the properties of PP composites. *Composites* **1994**, *25*, 205–214. [CrossRef]
55. Rowell, R.M. Natural fibres: Types and properties. In *Properties and Performance of Natural-Fibre Composites*; Pickering, K.L., Ed.; Woodhead Publishing: Boca Raton, FL, USA, 2008; pp. 3–66.
56. Schroeter, J.; Hobelsberger, M. On the Mechanical Properties of Native Starch Granules. *Starch* **1992**, *44*, 247–252. [CrossRef]
57. Dányádi, L.; Renner, K.; Szabó, Z.; Nagy, G.; Móczó, J.; Pukánszky, B. Wood flour filled PP composites: Adhesion, deformation, failure. *Polym. Adv. Technol.* **2006**, *17*, 967–974. [CrossRef]
58. Faludi, G.; Dora, G.; Imre, B.; Renner, K.; Móczó, J.; Pukánszky, B. PLA/lignocellulosic fiber composites: Particle characteristics, interfacial adhesion, and failure mechanism. *J. Appl. Polym. Sci.* **2014**, *131*, 39902. [CrossRef]
59. Várdai, R.; Ferdinánd, M.; Lummerstorfer, T.; Pretschuh, C.; Jerabek, M.; Gahleitner, M.; Faludi, G.; Móczó, J.; Pukánszky, B. Effect of various organic fibers on the stiffness, strength and impact resistance of polypropylene; a comparison. *Polym. Int.* **2021**, *70*, 145–153. [CrossRef]
60. Várdai, R.; Lummerstorfer, T.; Pretschuh, C.; Jerabek, M.; Gahleitner, M.; Pukánszky, B.; Renner, K. Impact modification of PP/wood composites: A new approach using hybrid fibers. *Express Polym. Lett.* **2019**, *13*, 223–234. [CrossRef]

61. Pukánszky, B. Influence of interface interaction on the ultimate tensile properties of polymer composites. *Composites* **1990**, *21*, 255–262. [CrossRef]
62. Callister, W.D.; Rethwisch, D.G. *Materials Science and Engineering: An Introduction*; John Wiley & Son: New York, NY, USA, 2018.
63. Ferdinánd, M.; Várdai, R.; Móczó, J.; Pukánszky, B. Deformation and Failure Mechanism of Particulate Filled and Short Fiber Reinforced Thermoplastics: Detection and Analysis by Acoustic Emission Testing. *Polymers* **2021**, *13*, 3931. [CrossRef]
64. Pukánszky, B.; Vörös, G. Mechanism of interfacial interactions in particulate filled composites. *Compos. Interfaces* **1993**, *1*, 411–427. [CrossRef]
65. Soest, P.J.V. Use of Detergents in the Analysis of Fibrous Feeds. II. A Rapid Method for the Determination of Fiber and Lignin. *J. Assoc. Off. Agric. Chem.* **1963**, *46*, 829–835. [CrossRef]

Article

HPLC Enantioseparation of Rigid Chiral Probes with Central, Axial, Helical, and Planar Stereogenicity on an Amylose (3,5-Dimethylphenylcarbamate) Chiral Stationary Phase

Simona Rizzo [1], Tiziana Benincori [2], Francesca Fontana [3,4], Dario Pasini [5] and Roberto Cirilli [6,*]

[1] CNR Istituto di Scienze e Tecnologie Chimiche "Giulio Natta", Via C. Golgi 19, 20133 Milano, Italy
[2] Dipartimento di Scienza e Alta Tecnologia, Università degli Studi dell'Insubria, Via Valleggio 11, 22100 Como, Italy
[3] Dipartimento di Ingegneria e Scienze Applicate, Università di Bergamo, Viale Marconi 5, 24044 Dalmine, Italy
[4] CSGI Bergamo R.U., Viale Marconi 5, 24044 Dalmine, Italy
[5] Department of Chemistry and INSTM Research Unit, University of Pavia, 27100 Pavia, Italy
[6] Centro Nazionale per il Controllo e la Valutazione dei Farmaci, Istituto Superiore di Sanità, Viale Regina Elena 299, 00161 Rome, Italy
* Correspondence: roberto.cirilli@iss.it

Citation: Rizzo, S.; Benincori, T.; Fontana, F.; Pasini, D.; Cirilli, R. HPLC Enantioseparation of Rigid Chiral Probes with Central, Axial, Helical, and Planar Stereogenicity on an Amylose (3,5-Dimethylphenylcarbamate) Chiral Stationary Phase. *Molecules* 2022, 27, 8527. https://doi.org/10.3390/molecules27238527

Academic Editors: Victor Mamane and Sylvain Caillol

Received: 13 October 2022
Accepted: 25 November 2022
Published: 3 December 2022

Publisher's Note: MDPI stays neutral with regard to jurisdictional claims in published maps and institutional affiliations.

Copyright: © 2022 by the authors. Licensee MDPI, Basel, Switzerland. This article is an open access article distributed under the terms and conditions of the Creative Commons Attribution (CC BY) license (https://creativecommons.org/licenses/by/4.0/).

Abstract: The chiral resolving ability of the commercially available amylose (3,5-dimethylphenylcarbamate)-based chiral stationary phase (CSP) toward four chiral probes representative of four kinds of stereogenicity (central, axial, helical, and planar) was investigated. Besides chirality, the evident structural feature of selectands is an extremely limited conformational freedom. The chiral rigid analytes were analyzed by using pure short alcohols as mobile phases at different column temperatures. The enantioselectivity was found to be suitable for all compounds investigated. This evidence confirms that the use of the amylose-based CSP in HPLC is an effective strategy for obtaining the resolution of chiral compounds containing any kind of stereogenic element. In addition, the experimental retention and enantioselectivity behavior, as well as the established enantiomer elution order of the investigated chiral analytes, may be used as key information to track essential details on the enantiorecognition mechanism of the amylose-based chiral stationary phase.

Keywords: planar stereogenicity; axial stereogenicity; trypticene; helicene; HPLC enantioseparation; amylose (3,5-dimethylphenylcarbamate); Chiralpak AD-3

1. Introduction

The polysaccharide-based chiral stationary phases (CSPs) are now routinely used in ultra-high-performance liquid chromatography (UHPLC) [1–3], high-performance liquid chromatography (HPLC) [4], as well as supercritical fluid chromatography (SFC) [5] enantiomeric separations. Their application allows the resolution of a broad range of chiral compounds from nano to preparative scale under multimodal elution conditions.

Despite their success, at present, there is no reliable way to predict whether or not an enantioselective separation will be achieved on a given polysaccharide-based CSP. Although in silico molecular models capable of mimicking the behavior of the polysaccharide selectors have been developed and, in some circumstances, successfully applied to actual enantioseparations [6], the research of enantioselective conditions is still carried out through the evaluation of literature data or the screening of commercially available columns. The importance of predictive tools in the enantioselective HPLC analysis of chiral compounds has attracted interest from many researchers and stimulated the study of the interactions involved in the enantiorecognition process promoted by polysaccharide derivatives through NMR, IR, HPLC, and computational techniques [7–11].

One strategy for the development of new pieces of knowledge on the mechanistic aspect of the enantioseparation process is the design of tailored chiral probes that display a chromatographic behavior traceable to specific structural elements [12–20]. This approach allows the construction of reliable structure–enantioselectivity relationships and, indirectly, the identification of the key portions of the selector involved in the enantioseparation.

In this context, this work reports on the enantioseparation of a small set of chiral compounds (compounds **1–4**, Figure 1) on the commercially available amylose (3,5-dimethylphenylcarbamate) (ADMPC)-based Chiralpak AD-3 CSP. The amylose derivative is considered to be one of the most effective selectors for achieving chiral resolution, and it is used in the preparation of commercially available chiral packing materials [4,10]. As shown in Figure 1, for each kind of stereogenicity (central, axial, helical, and planar), a representative chiral analyte was chosen.

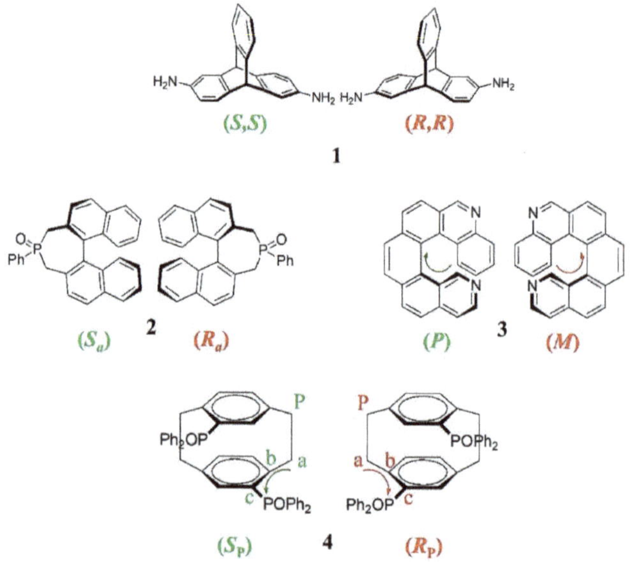

Figure 1. Chemical structures and stereochemical descriptors of enantiomers of chiral compounds **1–4** representative of the four stereogenic elements: **1** central stereogenicity, **2** axial stereogenicity, **3** helical stereogenicity, **4** planar stereogenicity.

The triptycene **1** has a paddle-wheel-shaped structure consisting of three benzene rings fused to a bicyclo [2.2.2]octatriene bridgehead system. The unique three-dimensional rigid structure of **1** and the ample possibilities for the installment of reactive positions are attractive points for this class of molecules in order to either modify/tune the π-scaffold and introduce reactive handles for further manipulation and integration into nanostructures. Compared to other C_2-symmetrical chiral synthons, such as trans-1,2-disubstituted cycloalkanes and 1,1,2,2-tetrasubstituted ethane-based scaffolds, triptycenes exhibit outstanding features that are attractive for the development of new functional molecular design, including a robust chiral backbone and extremely limited conformational structure. Recent advances in the synthesis of chiral triptycenes and in their introduction as molecular scaffolds for the assembly of functional supramolecular materials have been recently reviewed [21].

Compound **2** is prepared by oxidation of the enantiomers of the corresponding phosphane, the 4-phenyl-4,5- dihydro-3H-dinaphtho [2,1-c;10,20-e]phosphepine (Binepine), which is a versatile monodentate ligand of transition metals, Rh in particular, employed as a mediator in a wide variety of successful homogeneous stereoselective reactions. The 2,7-dihydrophosphepine oxide **2** does not display any catalytic activity. It can be prepared by oxidation of the enantiomers of Binepine and employed as an intermediate for the synthesis of 3,5-dialkyl-Binepines [22].

The diaza [6]helicene (compound 3) is characterized by an extensively conjugated, inherently chiral structure capable of fluorescence and phosphorescence emission and endowed with interesting optical properties [23]; besides, the presence of the nitrogen atoms allows functionalization, quaternarization, or complexation with metal ions [24], thus opening the way to the preparation of active materials for chiral sensing and optoelectronics [25–27].

The planar chiral diphosphane oxide with a *p*-cyclophane scaffold (compound 4) is the key compound in the resolution process of phanephos, a very popular C_2-symmetric diphosphane, employed as a ligand of Ru(II)- and Rh(I) in asymmetric hydrogenation of stereogenic C=O [28] and C=C double bonds [29].

Although compounds **1–4** are profoundly different from a structural point of view, they share a specific characteristic, namely that of being rigid and bulky molecules with extremely limited conformational freedom because their structure is fixed by a ring system (compounds **1, 2**, and **4**) or formed by an extended aromatic system with non-coplanar extremities (compound **3**). One challenge in selector–selectand docking is the treatment of molecular flexibility and changes in the conformational states. Any change in the selectand conformation can lead to a large difference in the resulting docked poses. Thus, the use of rigid chiral molecules such as **1–4** is expected to be an intriguing strategy to facilitate enantiorecognition process investigations by docking studies.

To support this hitherto untapped application and to verify the versatility of the Chiralpak AD-3 CSP, this work aimed to investigate the chromatographic behavior of **1–4** on the amylose-type CSP by (i) using different polar organic eluents, (ii) changing the column temperature, and (iii) determining the elution order of enantiomers in all conditions investigated.

2. Results and Discussion

2.1. HPLC Enantioseparation under Polar Organic Mode

Before discussing the chromatographic results on the enantioseparation of **1–4**, it is useful to remember the stereogenic elements of such unusual chiral molecules. The chirality of triptycene **1** is due to the NH_2 substituents at the 2,6-positions that make two bridgehead carbon atoms as stereogenic centers. The 2,7-dihydrophosphepine **2** is characterized by a 1,1′-binaphthalene scaffold and, consequently, displays an atropoisomeric framework as the stereogenic element. The third chiral compound studied in this work is an inherently chiral diaza [6]helicene (compound **3**) with a nonplanar screw-shaped structure formed from fused benzene and pyridine rings. The last term of the series is a planar chiral diphosphane oxide with a *p*-cyclophane scaffold (compound **4**). According to the CIP rules concerning the attribution of the configurational descriptors, (S_P) and (R_P), to stereogenic planar molecules, (i) the stereogenic plane of **4** is indifferently one of the two planes containing the aromatic ring, two phosphorous atoms, and two methylene groups; (ii) the first priority atom located out of the plan (indicated as P in Figure 1) is indifferently one of the two equivalent methylene carbons of the cyclophane bridge; (iii) from the pilot atom P, starting from the atom directly connected to it (indicated as *a* in Figure 1), the sequence of atoms is that along with the *b* and *c* atoms, which have the highest CIP priority; (iv) the sequence is clockwise for the (R_P)-enantiomer and counterclockwise for the (S_P)-enantiomer.

Compounds **1–4** are, in all cases, enantiomerically stable at room temperature and thus resolvable.

Figure 2 and Table S1 of Supplementary Materials resume the retention (*k*) and enantioseparation (*α*) factors obtained by: (i) setting the column temperature at 25 °C, (ii) using the 100 mm × 4.6 mm Chiralpak AD-3 column packed with 3-μm ADMPC-based particles, and (iii) selecting neat methanol, ethanol, 1-propanol, and 2-propanol as mobile phases.

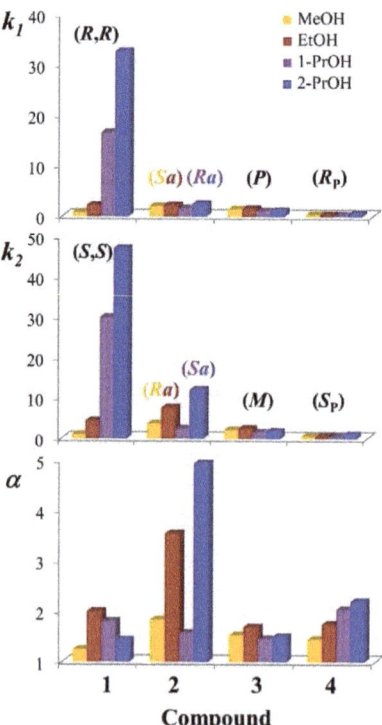

Figure 2. Effect of mobile phase on the retention (k_1 and k_2) and enantioseparation (α) factors of **1–4**. Chromatographic conditions: column, Chiralpak AD-3 (100 mm × 4.6 mm, 3 μm); flow rate, 1 mL/min; temperature, 25 °C; detection, UV and CD at 241 nm. 2-PrOH: 2-propanol; 1-PrOH: 1-propanol, EtOH: ethanol; MeOH: methanol.

The most marked dependence of enantioselectivity from mobile phase composition was recorded for compound **2**. As the less retained enantiomer elutes at close retention times, this variability is attributed essentially to the different retention of the more retained enantiomer. The maximum value of the enantioseparation factor was recorded in 2-propanol (α = 4.94) and the minimum in 1-propanol (α = 1.58). In both elution modes, the enantiomer (R_a)-**2** was eluted before (S_a)-**2**. Passing to methanol and ethanol, the enantiomer elution reversed, and the enantioselectivity became almost double. As visible in Figure 3, the evaluation of the sign of the ellipticity recorded during chromatography leads to a solid interpretation of the elution order of enantiomers.

The (S_a)-**2** enantiomer was eluted as the first species from the Chiralpak AD-3 column and showed a positive circular dichroism (CD) peak at 241 nm. The offline CD signal assumed the same positive sign at the wavelength of 241 nm (see the CD spectrum depicted in Figure 3). Passing from ethanol to 2-propanol, the sign of the online CD peaks reversed, and the (R_a)-**2** enantiomer became the less retained species.

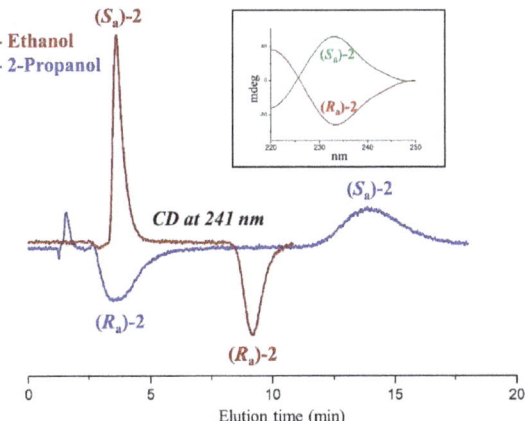

Figure 3. Typical chromatograms illustrating the differences in enantiomer elution order of **2** using pure ethanol and 2-propanol as mobile phases. Chromatographic conditions: column, Chiralpak AD-3 (100 mm × 4.6 mm, 3 µm); flow rate, 1 mL/min; temperature, 25 °C; detection, CD at 241 nm. 2-PrOH: 2-propanol; 1-PrOH: 1-propanol, EtOH: ethanol; MeOH: methanol. Inset: CD spectra of enantiomers of **2**.

The extreme variability of enantioselectivity and the inversion of enantiomer elution order can be attributed to the ability of the molecules of alcohol used in the mobile phase to alter the conformation of the polymeric ADMPC selector as well as the size and shape of the chiral periodical helical grooves occurring along the polymeric backbone [10,29,30]. The incorporation of molecules of alcohol within the nano-sized chiral cavities in the ADMPC structure leads to a fine modulation of their steric environment. Thus, depending on the type of molecules incorporated, the wetted chiral cavities can differently select the portions of the enantiomers of chiral analyte **2** that can interact with the carbamate sites of selector through hydrogen bond and dipole-dipole interactions [12,29,30].

Contrary to what was observed in the case of compound **2**, the enantiomeric separation of **3** was weakly influenced by the nature of alcoholic eluent. Notably, the retention factor of the more retained enantiomer was unusually low in the polar organic conditions used and ranged from 1.57 (with 1-propanol) to 2.43 (with ethanol).

For compound **4**, the enantioselectivity increased in parallel to the chain-lengthening from methanol to 1-propanol. A further increase in the enantioseparation factor was recorded using 2-propanol. Under these eluent conditions, the enantioseparation factor values changed from 1.45 to 2.2. As in the case of **3**, the retention of both enantiomers was very weak, and the retention factor values were lower than 1, irrespective of the nature of the mobile phase employed. The poor retentive properties of the selector can be attributed to the strong competitive interactions established by alcohol molecules with the active sites of the Chiralpak AD-3 CSP. This evidence leads to the hypothesis that P=O groups of **4** are involved in hydrogen bonds with the carbamate groups of the amylose derivative [10], and this type of interaction is the driving force of the enantiorecognition process.

To conclude this section, let us look at the chromatographic data obtained for compound **1**.

Both the retention factor values for the first and second eluted enantiomers dramatically and progressively increased using the following series of solvents: methanol < ethanol < 1-propanol < 2-propanol. Using 2-propanol, the enantioseparation factor of the second eluted (*R,R*)-enantiomer reached the remarkable value of 47.11. For the same enantiomer, the value collapsed to 0.97 with methanol. Despite this trend, the highest enantioselectivity was observed in ethanol (α = 1.58), while using 2-propanol, the enantioseparation factor was only 1.44. A possible explanation for interpreting the remarkable retention of **1** employing 2-propanol as a mobile phase is the involvement of the NH_2 group in one or more strong and poorly selective hydrogen bonds with the carbamate moieties of the (ADMPC)-based CSP.

2.2. Thermodynamic Aspects of Enantioseparation

The thermodynamic parameters associated with the enantioseparation of compounds **1–4** on the Chiralpak AD-3 CSP were determined by correlating the enantioseparations factors recorded between 25 and 45 °C and the column temperature. According to the following Equation:

$$\ln \alpha = -\Delta\Delta H°/RT + \Delta\Delta S°/R \qquad (1)$$

the differences between the two enantiomers in enthalpy ($\Delta\Delta H°$) and entropy ($\Delta\Delta S°$) of adsorption onto stationary phase were calculated from the slope and intercept, respectively, of $\ln \alpha$ vs. $1/T$ plots (van't Hoff plots).

Table 1 shows the enantioseparation factors recorded at 25 °C and the thermodynamic parameters obtained by van't Hoff analysis.

Table 1. Absolute configuration (AC) of the first eluted enantiomer, enantioseparation factors at 25 °C, and thermodynamic data of **1–4**. Chromatographic conditions: column, Chiralpak AD-3 (100 mm × 4.6 mm, 3 μm); flow rate, 1 mL/min; temperature, 25 °C; detection, CD at 241 nm. 2-PrOH: 2-propanol; 1-PrOH: 1-propanol, EtOH: ethanol; MeOH: methanol. NA: not applicable.

Entry	Compound	Mobile Phase	AC First Eluted/CD Sign at 241 nm	α (25 °C)	$\Delta\Delta H°$ (kcal/mol)	$\Delta\Delta S°$ (e.u.)	T_{ISO} (°C)
1	1	MeOH	(R,R)/(−)	1.23	−0.62	−1.65	100
2		EtOH	(R,R)/(−)	2.00	−1.40	−3.32	148
3		1-PrOH	(R,R)/(−)	1.81	−1.06	−2.38	172
4		2-PrOH	(R,R)/(−)	1.47	0.33	1.89	−97
5	2	MeOH	(Sa)/(+)	1.84	−2.72	−7.90	71
6		EtOH	(Sa)/(+)	3.54	−2.45	−5.72	156
7		1-PrOH	(Ra)/(−)	1.58	1.38	5.53	−24
8		2-PrOH	(Ra)/(−)	4.94	3.98	16.52	−32
9	3	MeOH	(P)/(−)	1.53	−0.52	−0.90	304
10		EtOH	(P)/(−)	1.69	−0.55	−0.80	412
11		1-PrOH	(P)/(−)	1.46	−0.22	0.02	NA
12		2-PrOH	(P)/(−)	1.50	−2.62	−7.97	56
13	4	MeOH	(Rp)/(+)	1.55	−0.13	0.43	NA
14		EtOH	(Rp)/(+)	1.75	0.84	3.92	−60
15		1-PrOH	(Rp)/(+)	2.16	0.12	1.95	−209
16		2-PrOH	(Rp)/(+)	2.22	6.51	23.42	5

It follows from Equation (1) that, for the chiral separations in which both the terms $\Delta\Delta H°$ and $\Delta\Delta S°$ were characterized by equal signs (positive or negative), it was possible to calculate the isoenantioselective temperature, namely the temperature at which enthalpy-entropy compensation occurs ($|T_{ISO}\Delta\Delta S°| = |\Delta\Delta H°|$) and the enantiomers coelute ($\alpha = 1$).

An inspection of the T_{ISO} values shown in Table 1 reveals that when the column temperatures were higher than the computed T_{ISO} (entries 4, 7, 8, 14, 15, 16), $\Delta\Delta H°$ and $\Delta\Delta S°$ assumed positive signs, and the enantioseparation process occurred within the entropy-controlled domain ($|T\Delta\Delta S°| > |\Delta\Delta H°|$). Accordingly, the separation factors recorded at 45 °C were higher than those recorded at 25 °C. As an example, the separation factor of **2** in the presence of 2-propanol changed from 4.94 to 6.82 as a result of increased temperature. Vice versa, when the column temperatures were lower than the T_{ISO} values, the enantiodiscrimination was enthalpy-driven ($|T\Delta\Delta S°| < |\Delta\Delta H°|$), and the separation factors lowered as temperature increased. In no case, by increasing the temperature in the range of values explored, a change in enantiomer elution order was observed. As representative examples of enantioseparations occurring within entropy and enthalpy domains, Figure 4 shows the van't Hoff plots obtained by variable-temperature analysis of **1** and **2**.

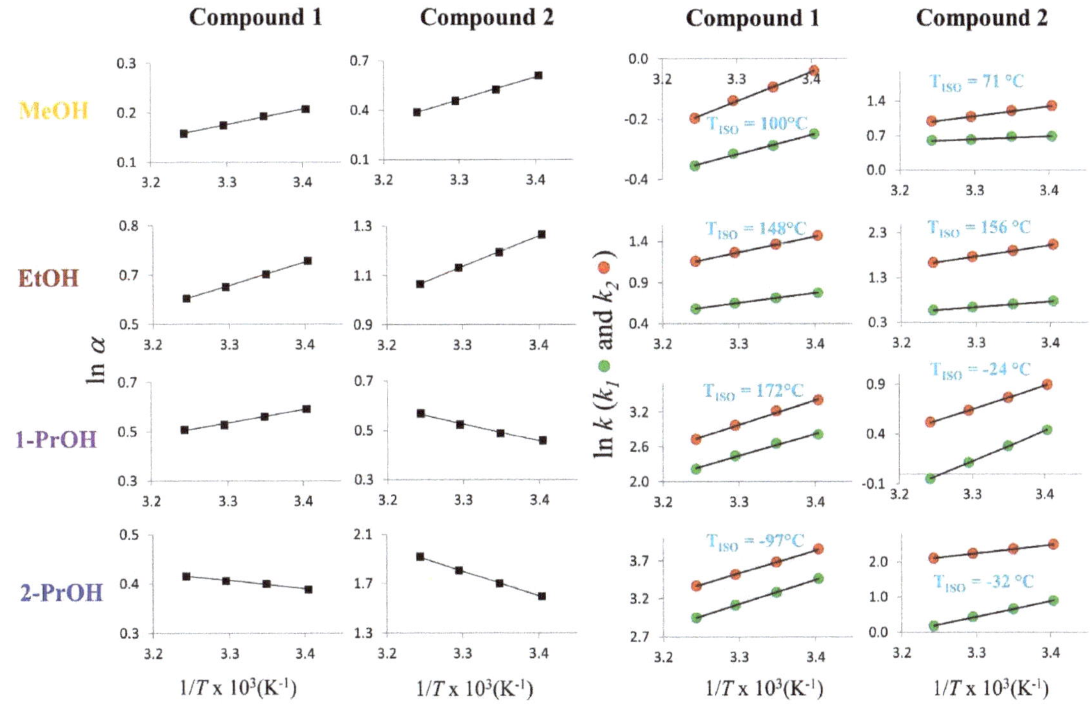

Figure 4. Plots of ln k vs. $1/T \times 10^3$ and ln α vs. $1/T \times 10^3$ for **1** and **2**. Chromatographic conditions: column, Chiralpak AD-3 (100 mm × 4.6 mm, 3 µm); flow rate, 1 mL/min; temperature, 25 °C; detection, CD at 241 nm. 2-PrOH: 2-propanol; 1-PrOH: 1-propanol, EtOH: ethanol; MeOH: methanol.

It is interesting to note that by changing the eluent conditions from pure methanol and ethanol (entries 5 and 6 of Table 1 and Figure 4) to 1-propanol and 2-propanol (entries 7 and 8 of Table 1 and Figure 4), T_{ISO} of **2** becomes lower than room temperature and the elution order of the enantiomers reverses. This once again proves that the temperature of isoelution is a critical parameter to be considered for controlling enantioseparation. Finally, although the enantioseparation of **1** was entropy-controlled with 2-propanol and enthalpy-driven in the other elution conditions, the enantiomer elution order was the same with the (R,R)-eluted before than (S,S)-enantiomer.

3. Materials and Methods

3.1. Reagents and Chemicals

HPLC-grade solvents were obtained from Aldrich (Milan, Italy) and filtered (0.22 µm filter) before use. HPLC analyses were carried out on a Chiralpak AD-3 (100 mm × 4.6 mm, 3 µm) column (Chiral Technologies Europe, Illkirch Graffenstaden, France).

Compounds **1-4** were synthesized according to previously reported procedures [31–35].

3.2. Instruments

HPLC apparatus consisted of a Perkin-Elmer (Norwalk, CT, USA) 200 LC pump equipped with a Rheodyne (Cotati, CA, USA) injector, a 50-µL sample loop, an HPLC Perkin-Elmer oven, and a Jasco (Jasco, Tokyo, Japan) Model CD 2095Plus UV/CD detector. The signal was acquired and processed by Clarity software (DataApex, Prague, Czech Republic).

The CD spectra of the enantiomers of **2** were recorded at 25 °C by using a Jasco Model J-700 spectropolarimeter. The optical path was 1 mm. The spectra are averagely computed over four instrumental scans, and the intensities are presented in terms of ellipticity values (mdeg).

The absolute configuration of the enantiomers was determined in previous works [30–33].

3.3. HPLC Operating Conditions

Fresh standard solutions of chiral samples were prepared by dissolving the analytes in ethanol solution (concentration about 0.5 mg mL^{-1}). The injection volume was 10–30 µL. Solvents and samples were filtered through 0.22-µm filters. The flow rate was set at 1.0 mL min^{-1}. The hold-up time was estimated by using 1,3,5-tri-tert-butylbenzene as a marker and pure ethanol as a mobile phase.

4. Conclusions

The ADMPC-based Chiralpak AD-3 CSP has been tested in HPLC enantioseparation of four chiral compounds, each of which represents a kind of stereogenicity. The outcomes of the enantioselective HPLC analysis carried out in polar organic conditions indicate that Chiralpak AD-3 CSP can discriminate the enantiomers of all compounds investigated, regardless of their rigid stereogenic element, but its performance is significantly influenced by column temperature and nature of the polar organic mobile phase, which impact the conformation of the polymeric ADMPC selector and the size and shape of its chiral cavities. In particular, in the case of compound **2**, an inversion of the enantiomer elution order occurred, passing from methanol and ethanol to 1-propanol and 2-propanol. Thus, besides proposing an effective approach to the HPLC isolation of enantiopure forms of **1–4**, this work adds new information on the resolving capability of the amylose-based Chiralpak AD-3 CSP and highlights the importance of considering mobile phase composition and temperature as key parameters to effectively establish the chiral recognition mechanism of this versatile chromatographic packing material. Finally, tuning molecular rigidity can be an effective strategy for controlling and studying the interactions of the chiral analytes with the active sites of the polysaccharide-based CSPs.

Supplementary Materials: The following supporting information can be downloaded at: https://www.mdpi.com/article/10.3390/molecules27238527/s1, Table S1: Effect of mobile phase on the retention of the first eluted enantiomer (k_1) and enantioseparation (α) factors of **1–4**.

Author Contributions: Synthesis, S.R., D.P., T.B. and F.F.; writing—review and editing, S.R., D.P., T.B., F.F. and R.C.; conceptualization, supervision, HPLC and CD analysis, writing—original draft preparation, R.C. All authors have read and agreed to the published version of the manuscript.

Funding: This research received no external funding.

Institutional Review Board Statement: Not applicable.

Informed Consent Statement: Not applicable.

Data Availability Statement: Data sharing not applicable.

Conflicts of Interest: The authors declare no conflict of interest.

Sample Availability: Samples of the compounds **1** and **3** can be available from the authors.

References

1. Malgorzata Ceboa, M.; Fua, X.; Gawaz, M.; Chatterjee, M.; Lämmerhofer, M. Enantioselective ultra-high performance liquid chromatography-tandem mass spectrometry method based on sub-2µm particle polysaccharide column for chiral separation of oxylipins and its application for the analysis of autoxidized fatty acids and platelet releasates. *J. Chromat. A* **2020**, *1624*, 461206.
2. Peluso, P.; Chankvetadze, B. Recognition in the domain of molecular chirality: From noncovalent interactions to separation of enantiomers. *Chem. Rev.* **2022**, *122*, 13235–13400. [CrossRef] [PubMed]

3. Ibrahim, D.; Ghanem, A. On the Enantioselective HPLC separation ability of sub-2 μm columns: Chiralpak®IG-U and ID-U. *Molecules* **2019**, *24*, 1287. [CrossRef] [PubMed]
4. Chankvetadze, B. Recent trends in preparation, investigation and application of polysaccharide-based chiral stationary phases for separation of enantiomers in high-performance liquid chromatography. *Trends Anal Chem.* **2020**, *122*, 115709. [CrossRef]
5. Kalíková, K.; Slechtová, T.; Vozka, J.; Tesarová, E. Supercritical fluid chromatography as a tool for enantioselective separation; A review. *Anal. Chim. Acta* **2014**, *821*, 1–33. [CrossRef]
6. Alcaro, S.; Bolasco, A.; Cirilli, R.; Ferretti, R.; Fioravanti, R.; Ortuso, F. Computeraided molecular design of asymmetric pyrazole derivatives with exceptional enantioselective recognition toward the Chiralcel OJ-H stationary phase. *J. Chem. Inf. Model.* **2012**, *52*, 649–654. [CrossRef]
7. Peluso, P.; Mamane, V. Stereoselective Processes Based on σ-Hole Interactions. *Molecules* **2022**, *27*, 4625. [CrossRef]
8. Peluso, P.; Chankvetadze, B. The molecular bases of chiral recognition in 2-(benzylsulfinyl)benzamide enantioseparation. *Anal. Chim. Acta* **2021**, *1141*, 194e205. [CrossRef]
9. Uccello-Barretta, G.; Vanni, L.; Balzano, F. Nuclear magnetic resonance approaches to the rationalization of chromatographic enantiorecognition processes. *J. Chromat. A* **2010**, *1217*, 928–940. [CrossRef]
10. Yamamoto, C.; Yashima, E.; Okamoto, Y. Structural Analysis of Amylose Tris(3,5-dimethylphenylcarbamate) by NMR Relevant to Its Chiral Recognition Mechanism in HPLC. *J. Am. Chem. Soc.* **2002**, *124*, 12583–12589. [CrossRef]
11. Kasat, R.B.; Zvinevich, Y.; Hillhouse, H.W.; Thomson, K.T.; Wang, N.-H.L.; Franses, E.I. Direct probing of sorbent-solvent interactions for amylose tris(3,5-dimethylphenylcarbamate) using infrared spectroscopy, X-ray diffraction, solid-state NMR, and DFT modeling. *J. Phys. Chem. B* **2006**, *110*, 14114–14122. [CrossRef] [PubMed]
12. Cantatore, C.; Korb, M.; Lang, H.; Cirilli, R. ON/OFF receptor-like enantioseparation of planar chiral 1,2-ferrocenes on an amylose-based chiral stationary phase: The role played by 2-propanol. *Anal. Chim. Acta* **2022**, *1211*, 339880. [CrossRef]
13. Carradori, S.; Secci, D.; Guglielmi, P.; Pierini, M.; Cirilli, R. High-performance liquid chromatography enantioseparation of chiral 2-(benzylsulfinyl)benzamide derivatives on cellulose tris(3,5-dichlorophenylcarbamate) chiral stationary phase. *J. Chromat. A* **2020**, *1610*, 460572. [CrossRef] [PubMed]
14. Chankvetadze, B.; Yamamoto, C.; Okamoto, Y. Extremely high enantiomer recog- nition in hplc separation of racemic 2-(benzylsulfinyl)benzamide using cellu- lose tris(3,5-dichlorophenylcarbamate) as a chiral stationary phase. *Chem. Lett.* **2000**, *29*, 1176–1177. [CrossRef]
15. Carradori, S.; Secci, D.; Faggi, C.; Cirilli, R. A chromatographic study on the exceptional chiral recognition of 2-(benzylsulfinyl)benzamide by an immobilized-type chiral stationary phase based on cellulose tris(3,5-dichlorophenylcarbamate). *J. Chromatogr. A* **2018**, *1531*, 151–156. [CrossRef] [PubMed]
16. Cirilli, R.; Alcaro, S.; Fioravanti, R.; Secci, D.; Fiore, S.; La Torre, F.; Ortuso, F. Unusually high enantioselectivity in high-performance liquid chromatography using cellulose tris(4-methylbenzoate) as a chiral stationary phase. *J. Chromatogr. A* **2009**, *1216*, 4673–4678. [CrossRef] [PubMed]
17. Cirilli, R.; Alcaro, S.; Fioravanti, R.; Ferretti, R.; Bolasco, A.; Gallinella, B.; Faggi, C. A chromatographic study on the exceptional enantioselectivity of cellulose tris(4-methylbenzoate) towards C5-chiral 4,5-dihydro-(1H)-pyrazole deriva- tives. *J. Chromatogr. A* **2011**, *1218*, 5653–5657. [CrossRef]
18. Ortuso, F.; Alcaro, S.; Menta, S.; Fioravanti, R.; Cirilli, R. A chro- matographic and computational study on the driving force operating in the exceptionally large enantioseparation of N-thiocar- bamoyl-3-(4′-biphenyl)-5-phenyl-4,5-dihydro-(1H) pyrazole on a 4-methylben- zoate cellulose-based chiral stationary phase. *J. Chromatogr. A* **2014**, *1324*, 71–77.
19. Pierini, M.; Carradori, S.; Menta, S.; Secci, D.; Cirilli, R. 3-(Phenyl-4-oxy)-5-phenyl-4,5-dihydro-(1H)-pyrazole: A fascinating molec- ular framework to study the enantioseparation ability of the amylose (3,5-dimethylphenylcarbamate) chiral stationary phase. part II. solvophobic effects in enantiorecognition process. *J. Chromatogr. A* **2017**, *1499*, 140–148. [CrossRef]
20. Pisani, L.; Rullo, M.; Catto, M.; de Candia, M.; Carrieri, A.; Cellamare, S.; Altomare, C.D. Structure–property relationship study of the HPLC enantioselective retention of neuroprotective 7-[(1-alkylpiperidin-3-yl)methoxy]coumarin derivatives on an amylose-based chiral stationary phase. *J. Sep. Sci.* **2018**, *41*, 1376–1384. [CrossRef]
21. Preda, G.; Nitti, A.; Pasini, D. Chiral triptycenes in supramolecular and materials chemistry. *ChemistryOpen* **2020**, *9*, 719–727. [CrossRef] [PubMed]
22. Gladiali, S.; Alberico, E.; Jungec, K.; Beller, M. BINEPINES: Chiral binaphthalene-core monophosphepine ligands for multipurpose asymmetric catalysis. *Chem. Soc. Rev.* **2011**, *40*, 3744–3763. [CrossRef] [PubMed]
23. Abbate, S.; Longhi, G.; Lebon, F.; Castiglioni, E.; Superchi, S.; Pisani, L.; Fontana, F.; Torricelli, F.; Caronna, T.; Villani, C.; et al. Helical sense-responsive and substituent-sensitive features in vibrational and electronic circular dichroism, in circularly polarized luminescence and in Raman spectra of some simple optically active hexahelicenes. *J. Phys. Chem. C* **2014**, *118*, 1682–1695. [CrossRef]
24. Mendola, D.; Saleh, N.; Hellou, N.; Vanthuyne, N.; Roussel, C.; Toupet, L.; Castiglione, F.; Melone, F.; Caronna, T.; Marti-Rujas, J.; et al. Synthesis and structural properties of Aza[n]helicene platinum complexes: Control of cis and trans stereochemistry. *Inorg. Chem.* **2016**, *55*, 2009–2017. [CrossRef] [PubMed]
25. Zanchi, C.; Lucotti, A.; Cancogni, D.; Fontana, F.; Trusso, S.; Ossi, P.M.; Tommasini, M. Functionalization of nanostructured gold substrates with chiral chromophores for SERS applications: The case of 5-Aza[5]helicene. *Chirality* **2018**, *30*, 875–882. [CrossRef]

26. Fontana, F.; Carminati, G.; Bertolotti, B.; Mussini, P.R.; Arnaboldi, S.; Grecchi, S.; Cirilli, R.; Micheli, L.; Rizzo, S. Helicity: A non-conventional stereogenic element for designing inherently chiral ionic liquids for electrochemical enantiodifferentiation. *Molecules* **2021**, *26*, 311. [CrossRef]
27. Pye, P.J.; Rossen, K.; Reomer, R.A.; Volante, R.P.; Reider, P.J. [2.2] Phanephos-Ruthenium(II) complexes: Highly active asymmetric catalysts for the hydrogenation of β-ketoesters. *Tetrahedron Lett.* **1998**, *39*, 4441–4444. [CrossRef]
28. Pye, P.J.; Rossen, K.; Reomer, R.A.; Tsou, N.N.; Volante, R.P.; Reider, P.J. A new planar chiral bisphosphine ligand for asymmetric catalysis: Highly enantioselective hydrogenations under mild conditions. *J. Am. Chem. Soc.* **1997**, *119*, 6207–6208. [CrossRef]
29. Fontana, F.; Melone, F.; Iannazzo, D.; Leonardi, S.G.; Neri, G. Synthesis, characterization and electrochemical properties of 5-aza[5]helicene-CH$_2$O-CO-MWCNTs nanocomposite. *Nanotechnology* **2017**, *28*, 135501. [CrossRef]
30. Wang, T.; Wenslow, R.M. Effects of alcohol mobile-phase modifiers on the structure and chiral selectivity of amylose tris(3,5-dimethylphenylcarbamate) chiral stationary phase. *J. Chromatogr. A* **2003**, *1015*, 99–110. [CrossRef]
31. Wenslow, R.M.; Wang, T. Solid-State NMR characterization of amylose tris(3,5-dimethylphenylcarbamate) chiral stationary-phase structure as a function of mobile-phase composition. *Anal. Chem.* **2001**, *73*, 4190–4195. [CrossRef] [PubMed]
32. Zhang, Q.P.; Wang, Z.; Zhang, Z.W.; Zhai, T.L.; Chen, J.J.; Ma, H.; Tan, B.; Zhang, C. Triptycene-based chiral porous polyimides for enantioselective membrane separation. *Angew. Chem. Int. Ed.* **2021**, *60*, 12781–12785. [CrossRef] [PubMed]
33. Vaghi, L.; Cirilli, R.; Pierini, M.; Rizzo, S.; Terraneo, G.; Benincori, T. PHANE-TetraPHOS, the first D$_2$ symmetric chiral tetraphosphane. Synthesis, metal complexation, and application in homogeneous stereoselective hydrogenation. *Eur. J. Org. Chem.* **2021**, *2021*, 2367–2374. [CrossRef]
34. Fontana, F.; Bertolotti, B.; Grecchi, S.; Mussini, P.R.; Micheli, L.; Cirilli, R.; Tommasini, M.; Rizzo, S. 2,12-Diaza[6]helicene: An efficient non-conventional stereogenic scaffold for enantioselective electrochemical interphases. *Chemosensors* **2021**, *9*, 216. [CrossRef]
35. Vaghi, L.; Benincori, T.; Cirilli, R.; Alberico, E.; Mussini, P.R.; Pierini, M.; Pilati, T.; Rizzo, S.; Sannicolò, F. Ph-tetraMe-bithienine, the first member of the class of chiral heterophosphepines: Synthesis, electronic and steric properties, metal complexes and catalytic activity. *Eur. J. Org. Chem.* **2013**, *2013*, 8174–8184. [CrossRef]

Article

Synthesis, Characterization, and Optimization Studies of Starch/Chicken Gelatin Composites for Food-Packaging Applications

Jorge Iván Castro [1,†], Diana Paola Navia-Porras [2,†], Jaime Andrés Arbeláez Cortés [2], José Herminsul Mina Hernández [3] and Carlos David Grande-Tovar [4,*]

1. Grupo de Investigación SIMERQO, Departamento de Química, Universidad del Valle, Calle 13 No. 100-00, Santiago de Cali 76001, Colombia; jorge.castro@correounivalle.edu.co
2. Grupo de Investigación Biotecnología, Facultad de Ingeniería, Universidad de San Buenaventura Cali, Carrera 122 # 6-65, Santiago de Cali 76001, Colombia; dpnavia@usbcali.edu.co (D.P.N.-P.); valencia.mayra@correounivalle.edu.co (J.A.A.C.)
3. Grupo de Materiales Compuestos, Escuela de Ingeniería de Materiales, Facultad de Ingeniería, Universidad del Valle, Calle 13 No. 100-00, Santiago de Cali 76001, Colombia; jose.mina@correounivalle.edu.co
4. Grupo de Investigación de Fotoquímica y Fotobiología, Facultad de Ciencias, Universidad del Atlántico, Carrera 30 Número 8-49, Puerto Colombia 081008, Colombia
* Correspondence: carlosgrande@mail.uniatlantico.edu.co; Tel.: +57-53-599-484
† These authors contributed equally to this work.

Citation: Castro, J.I.; Navia-Porras, D.P.; Arbeláez Cortés, J.A.; Mina Hernández, J.H.; Grande-Tovar, C.D. Synthesis, Characterization, and Optimization Studies of Starch/Chicken Gelatin Composites for Food-Packaging Applications. *Molecules* 2022, 27, 2264. https://doi.org/10.3390/molecules27072264

Academic Editor: Sylvain Caillol

Received: 10 March 2022
Accepted: 28 March 2022
Published: 31 March 2022

Publisher's Note: MDPI stays neutral with regard to jurisdictional claims in published maps and institutional affiliations.

Copyright: © 2022 by the authors. Licensee MDPI, Basel, Switzerland. This article is an open access article distributed under the terms and conditions of the Creative Commons Attribution (CC BY) license (https://creativecommons.org/licenses/by/4.0/).

Abstract: The indiscriminate use of plastic in food packaging contributes significantly to environmental pollution, promoting the search for more eco-friendly alternatives for the food industry. This work studied five formulations (T1–T5) of biodegradable cassava starch/gelatin films. The results showed the presence of the starch/gelatin functional groups by FT-IR spectroscopy. Differential scanning calorimetry (DSC) showed a thermal reinforcement after increasing the amount of gelatin in the formulations, which increased the crystallization temperature (Tc) from 190 °C for the starch-only film (T1) to 206 °C for the film with 50/50 starch/gelatin (T3). It also exhibited a homogeneous surface morphology, as evidenced by scanning electron microscopy (SEM). However, an excess of gelatin showed low compatibility with starch in the 25/75 starch/gelatin film (T4), evidenced by the low Tc definition and very rough and fractured surface morphology. Increasing gelatin ratio also significantly increased the strain (from 2.9 ± 0.5% for T1 to 285.1 ± 10.0% for T5) while decreasing the tensile strength (from 14.6 ± 0.5 MPa for T1 to 1.5 ± 0.3 MPa for T5). Water vapor permeability (WVP) increased, and water solubility (WS) also decreased with gelatin mass rising in the composites. On the other hand, opacity did not vary significantly due to the films' cassava starch and gelatin ratio. Finally, optimizing the mechanical and water barrier properties resulted in a mass ratio of 53/47 cassava starch/gelatin as the most appropriate for their application in food packaging, indicating their usefulness in the food-packaging industry.

Keywords: composites; food packaging; films; gelatin; starch

1. Introduction

The food industry is one of the most significant users of single-use plastic from petroleum, contributing to solid waste [1]. This plastic can even degrade into finer particles (micro- and nano-plastics) that can enter the food chain and affect different trophic levels, including humans [2]. By 2050, it is projected that more than 33 million tons of plastic will end up accumulating in various ecosystems, especially in the ocean [3]. Therefore, searching for more environmentally friendly alternatives is imperative to avoid using these polymers [4]. In this sense, several investigations have been carried out into the production of films containing biodegradable components from extracts of *Pseudomonas oleovorans* bacteria [5], chicken feathers [6], soy protein [7], chicken breast protein [8], sunflower protein [9], spinach flour [10], corn starch–chitosan blends [11], among others. However, many biodegradable films have disadvantages in mechanical properties and high production costs. Therefore, a viable option is to produce cassava starch films due to their low price, availability, and oxygen-barrier properties [12,13].

Cassava or tapioca (*Manihot esculenta* Crantz) is a low-cost tuber native to Thailand, and is used for starch production. Starch is a polymer composed mainly of amylose and amylopectin. Both components consist of chains of D-glucose residues with bonds α-(1,4) that are interconnected through α-(1,6)-glycosidic bonds forming branches in polymers [14]. In addition, amylose is responsible for the isotropic, odorless, tasteless, colorless, non-toxic, and biodegradable film-forming characteristics [15,16]. However, the mechanical properties of cassava starch films are insufficient for food packaging due to their brittle and weak behavior when subjected to tensile stress. In addition, films produced from polysaccharides are susceptible to moisture [17]. The above have promoted the synthesis of films, including plasticizing additives such as sorbitol, glycerol, and polyethylene glycol, to improve flexibility [18].

Gelatin is a polymer obtained from collagen and is used to manufacture bioplastics. Its miscibility with water allows for the production of different matrices in the food-packaging industry [19,20]. There are two types of gelatins, depending on the preparation methodology. Gelatin A is obtained by acid extraction, and gelatin B is obtained by alkaline extraction [21]. The primary sources of gelatin are pig skins, pig bones, and cows' skin [22]. Gelatin's triple-helix protein structure and amino acid composition provide physical barrier capacity and protection against ultraviolet light, protecting foods from physical damage and oxidation caused by reactive oxygen species (ROS) [23–25].

The amino acid content in gelatin (mainly proline, hydroxyproline, and glycine) makes chicken gelatin a potential component for developing new materials applicable to the food industry. Chicken gelatin contains considerably higher gel strength in Bloom value (355 ± 1.48 g) than bovine gelatin (229 ± 0.71 g). However, there are few studies related to films incorporating cassava starch with gelatin in an optimized formulation to maximize the composite's properties [26,27].

Veiga-Santos et al. found that the addition of gelatin did not significantly affect the mechanical properties of tapioca starch sheets plasticized with sucrose [27]. However, the addition of gelatin increased the water vapor permeability of the glycerol-plasticized cassava starch film [26]. There are no studies where optimized films from chicken gelatin and cassava starch ratio are formulated. Despite the reported applications of cassava starch bioplastics with chicken-based gelatin, there is a need to optimize the cassava starch–gelatin mixtures to maximize the packaging properties of the films obtained. In this study, five (T1–T5) formulations of cassava starch and chicken-waste gelatin-based films were developed and characterized in a simplex-lattice mixture design, optimized to determine the starch/gelatin mixture with desirable functional properties in food packaging. Therefore, this study aimed to obtain bioplastic films from mixtures of cassava starch and chicken gelatin, characterize them, and find an optimal blend of the components to generate a bioplastic film whose properties are helpful in food packaging.

2. Results and Discussion

2.1. FT-IR Spectroscopy Analysis of the Cassava Starch/Gelatin Films

Figure 1 shows the FT-IR characterization of the cassava starch/gelatin films. All samples showed higher intensity bands in the amide region for the films with high gelatin content (3287 a 3308 cm^{-1}) due to the -OH groups from starch, gelatin, and water-adsorbed molecules [28].

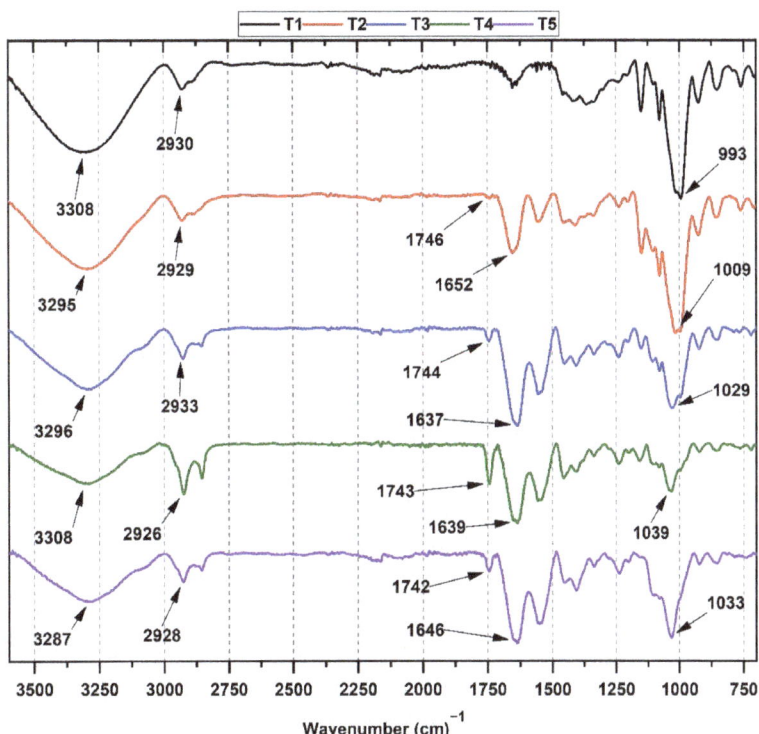

Figure 1. FTIR spectra of cassava starch/gelatin films. T1 gelatin/cassava starch 0/100; T2 gelatin/cassava starch 25/75; T3 gelatin/cassava starch 50/50; T4 gelatin/cassava starch 75/25; T5, gelatin/cassava starch 100/0.

The characteristic bands of cassava starch are present in T1. The symmetrical stress band corresponding to the -C-H group is observed around 2928–2930 cm^{-1}. Additionally, stress bands associated with amide groups are overlapped by -OH stress bands. Bands at 3308 cm^{-1} and 2930 cm^{-1} for O-H and -C-H stretching vibrations are observed. At lower vibration frequencies, the bands associated with in-plane bending vibration for the CH$_2$ group and bending of the C-OH group are also observed between 1413 and 1337 cm^{-1}. Finally, the band corresponding to the stretching of the antisymmetric bridge of the C-O-C group rose at 1149 cm^{-1} [29].

Strong hydrogen bonds between starch and gelatin might generate smooth and homogeneous films, especially in the T3 formulation (50/50 starch/gelatin), as discussed in the morphological section. On the other hand, shifting at higher vibrational frequencies is observed for T2–T4 (starch/gelatin) films, especially for the strain band at 3295 cm^{-1} associated with the -OH group. In addition, asymmetric stretching bands of the amide I group were observed between 1637–1652 cm^{-1}, increasing intensity with higher gelatin content [30]. Additionally, due to the presence of amino groups from the gelatin, the band at 3200 cm^{-1} is shifted. Additionally, gelatin's -C=O group (1743–1746 cm^{-1}) increased, and the CH-O-CH$_2$ groups (1009–1033 cm^{-1}) decreased as the gelatin concentration within the polymeric matrix increased (formulations T2–T4). The shifting of the -OH band and the increase in the amide I band is probably due to hydrogen bonding of the C=O and N-H groups with the -OH group, which demonstrates strong bonding between starch and gelatin components. The exact position of the amide I band is determined by the most stable backbone conformation and hydrogen-bonding pattern associated with the stretching vibrations of the C=O (70–85%) and C-N (10–20%) group [31]. On the other hand, formulation T5 showed the main band at 1742 cm^{-1}, corresponding to the C=O group of amides-I. Likewise, the band located at 1544 cm^{-1} is attributed to amides-II rising from the stretching vibrations of the N-H groups and the stretching vibrations of the C-N group. Finally, the bands located at 3287 and 2928 cm^{-1} correspond to the stretching vibrations of the N-H group for an amide-I and the symmetric tensions of the aliphatic carbons CH$_2$, respectively [32].

2.2. Thermal Analysis of Cassava Starch/Gelatin Films

Thermogravimetric analysis (TGA) aims to evaluate the degradation of any material with increasing temperature. Figure 2 shows the degradation profiles of the different films. The cassava starch/chicken gelatin films (T1, T2, T4, and T5) presented three stages of degradation, while T3 presented four stages of degradation. The first stage of degradation corresponds to weight loss between 25 and 150 °C, attributed to the loss of absorbed bound water in the films and low-molecular-weight materials [33]. Additionally, this degradation step is attributed to the reaction between the carbonyl groups of starch and the amino groups of gelatins [34]. This behavior indicates that thermoplastic processing is not recommendable for starch and gelatin films due to side reactions.

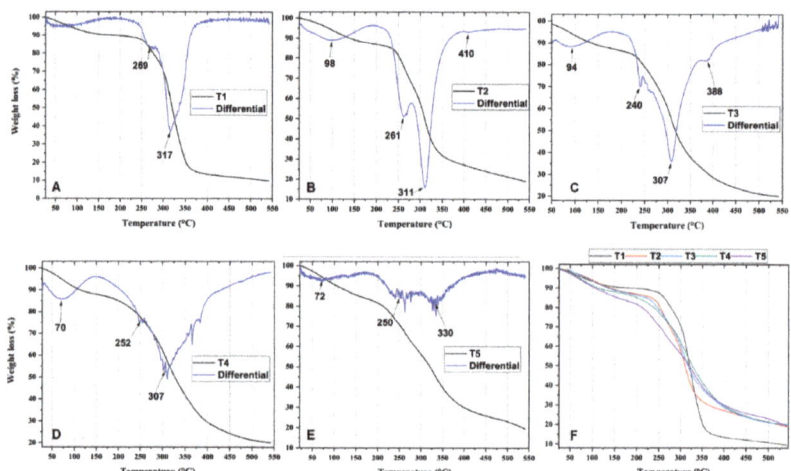

Figure 2. Thermogravimetric analysis for T1, (**A**) gelatin/cassava starch 0/100; T2, (**B**) gelatin/cassava starch 25/75; T3, (**C**) gelatin/cassava starch 50/50; T4, (**D**) gelatin/cassava starch 75/25; T5, (**E**) gelatin/cassava starch 100/0; (**F**) All the samples.

The second stage occurs between 150 and 270 °C, corresponding to the decomposition of organic materials of low molecular weight attributed to cassava starch [33,35]. In this sense, it is highlighted that the gelatin-containing films exhibited lower values in this degradation stage compared to T1 (decreasing from 317 to 311 °C), which did not contain gelatin. According to Figure 2, it is likely that crosslinking was generated during the first heating stage through intramolecular reactions as the temperature increased. Therefore, films containing a higher amount of gelatin facilitated the formation of a higher degree of crosslinking, as observed in the DTGA in Figure 2 (undefined degradations) [34]. Likewise, a continuous thermal degradation was observed for the films containing a higher concentration of starch (T2 and T3), reflected in the decrease in the degradation temperatures compared to T1 [34]. Finally, the third stage of degradation occurs between 270 and 600 °C, attributed to the total decomposition of the material (polysaccharides, protein, and glycerol volatilization), including some contained gases (CO$_2$, CO, H$_2$O) and carbon compounds present in the material [11]. On the other hand, formulation T3 shows a small shoulder of the fourth degradation stage centered at 388 °C. This may correspond to the elimination of hydrogen functional groups, degradation, and depolymerization of the starch–gelatin interactions between the polymer chains, which can be formed due to the higher gelatin amount present in the film that at higher temperatures causes crosslinking and strong interactions with starch, requiring more energy for bond-breaking [36].

The low-resolution degradations observed in the DTGA of samples T4 and T5 are due to crosslinking reactions between carbonyl and amino groups, degrading at low speed and high temperatures. Consequently, the processing of these sheets has limitations related to possible uncontrolled responses at high temperatures [34,37]. On the other hand, the thermal analysis of the films containing a higher amount of gelatin shows that they are susceptible when the temperature exceeds 100 °C, as indicated by the thermogravimetric derivative analysis (DTGA, blue line, Figure 2).

Differential scanning calorimetry (DSC) studies the phase transitions of the material as the temperature increases [38]. DSC analysis of cassava starch/chicken gelatin films (Figure 3) exhibited two endothermic peaks associated with the melting of starch crystalline regions during retrogradation [39]. The two endothermic peaks are attributed to each film's two melting temperatures (Tm). For T1, the Tm was observed at 190 °C. However, the Tm increased to 208 °C and 206 °C for T2 and T3. The increase in the Tm might be due to the above-discussed crosslinking reactions between the C=O and N-H functional groups from starch and gelatin, reducing the mobility of the polymer chains and generating thermostable films [40].

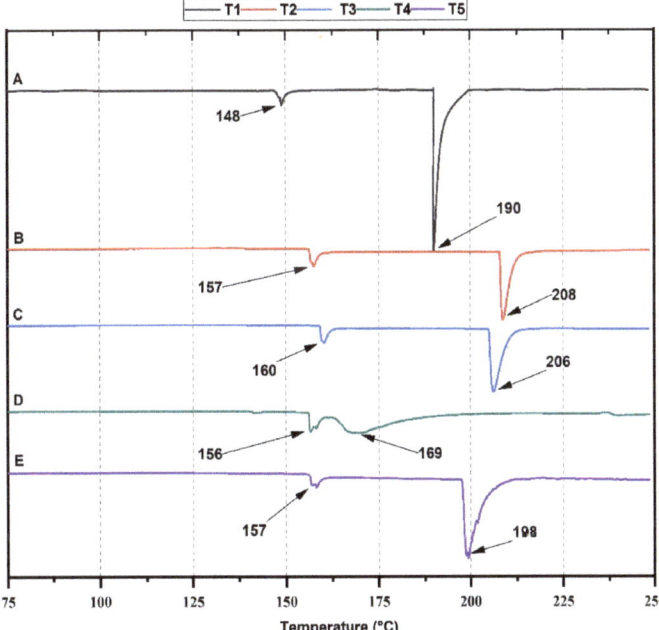

Figure 3. DSC thermogram of films: T1, (**A**) gelatin/cassava starch 0:100; T2, (**B**) gelatin/cassava starch 25/75; T3, (**C**) gelatin/cassava starch 50/50; T4, (**D**) gelatin/cassava starch 75:25; T5, (**E**) gelatin/cassava starch 100/0.

On the other hand, higher amounts of gelatin introduce a higher degree of flexibility, which affects the observation of the well-defined endothermic melting peak of the material [41]. For T4, the two peaks are poorly defined due to the higher amount of gelatin (75%) compared to starch (25%) in the formulation. These observations are consistent with previous studies using chicken gelatin as a polymeric matrix [40,42]. Additionally, formulation T5 shows a decrease in structural stability, because its melting peaks (157 and 198 °C) are at a lower temperature with an approximate difference of 8 °C compared to formulations T2 and T3. Therefore, the collagen structure in chicken gelatin involves breaking hydrogen bonds and helix coil transition between polypeptide chains adjacent to collagen molecules in the denaturation process [43].

2.3. Scanning Electron Microscopy (SEM) of Cassava Starch/Gelatin Films

Figure 4 shows the scanning electron micrographs for the cassava starch/gelatin films. A homogeneous, continuous, and smooth surface appearance (characteristic of polysaccharide films such as starch) was observed for T1 (starch film) due to the excellent molecular packing. For the T5 gelatin film, a rough, heterogeneous, and discontinuous appearance is observed, because low (<4 wt.%)- or high (>16 wt.%)-protein films present an inadequate film formation. A low concentration of proteins generates lower macromolecule interactions, while high-globular-protein concentration reduces the water activity, creating films with greater elongations and tear resistance [44].

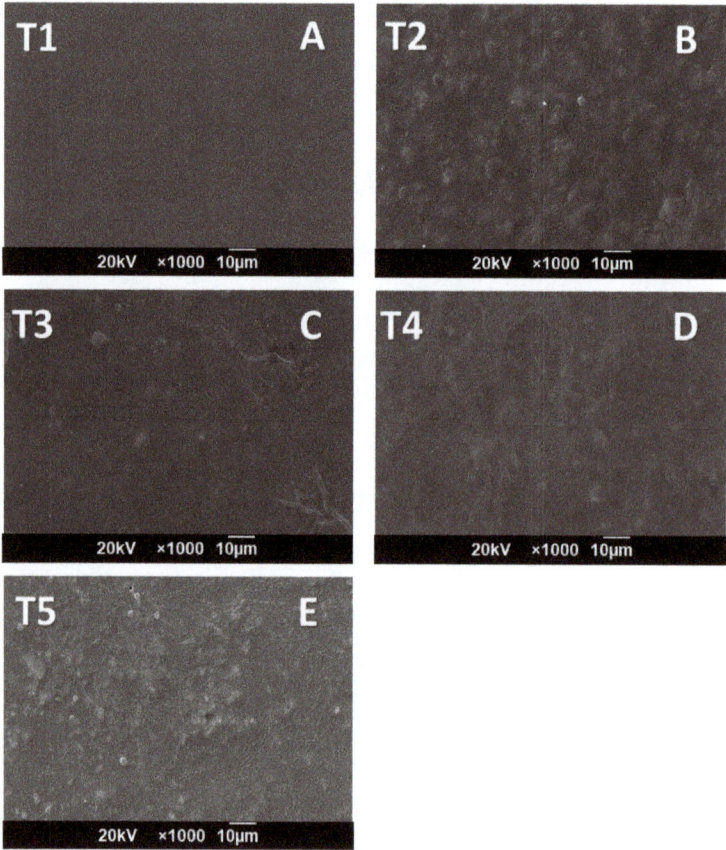

Figure 4. Morphological analysis of slides by scanning electron microscopy (SEM): T1, (**A**) gelatin/cassava starch 0:100; T2, (**B**) gelatin/cassava starch 25/75; T3, (**C**) gelatin/cassava starch 50/50; T4, (**D**) gelatin/cassava starch 75/25; T5, (**E**) gelatin/cassava starch 100/0.

In formulations T2–T4, rough surfaces with pores could be observed. The presence of starch reduced the heterogeneous and irregular appearance of the gelatin film (T5). However, with the introduction of gelatin, a roughness and heterogeneous surface are observed because of the spherical protein presence that intercalates between the polysaccharide chains, forming a rounded appearance. When the gelatin content increases, more intense molecular interaction between starch and gelatin occurs, decreasing the rounded appearance with a heterogeneous aspect.

Formulation T3 (50/50 starch/gelatin) showed the slightest rough appearance, demonstrating good compatibility with this formulation. However, there were differences in the appearance of the starch/gelatin films. Apparent differences are likely due to hydrogen bonds between the components [11,37]. These differences in the microstructure can be associated with the different organization of the polymeric chains during composite formation [40]. Similarly, the roughness of cassava starch/bovine gelatin type B films was observed when the gelatin concentration was increased [45].

2.4. The Tensile Strength of the Cassava Starch/Gelatin Films

Table 1 shows the values of Young's modulus, tensile strength, and deformation percentage of the cassava starch/gelatin films. Significant differences ($p < 0.05$) in the characterization of the mechanical properties between the samples are evident. These results indicate that the sample's flexibility significantly increased with the increase in the amount of gelatin, decreasing, in turn, the stiffness of the material, as observed in the Young's modulus values. The stiffness of the films, shown in Table 1, significantly reduced with increasing gelatin ($p < 0.05$) between T2, T3, and T4 formulations. Young's modulus for the films ranged from 2.2 ± 1.1 MPa for T4 to 440.4 ± 46.1 MPa for T2, a behavior previously observed [40,46], attributed to the formation of new hydrogen bonds between the functional groups of cassava starch and gelatin, which in turn decrease the number of hydrogen bonds between starch chains. During film drying, the formation of intra- and inter-molecular hydrogen bonds and chemical crosslinking is facilitated by the supply of thermal energy, and increases with temperature. As the polymeric network encounters a higher amount of starch, the stiffness of the films also significantly increases [17,45]. Despite that, a higher proportion of starch needs a more elevated amount of glycerol in the formulation, which increases the film's flexibility. The addition of gelatin in the matrix promotes the formation of new hydrogen, increasing the material's ductility.

Table 1. Mechanical properties of cassava starch/gelatin films: T1 (gelatin/cassava starch 0/100; T2 gelatin/cassava starch 25/75; T3 gelatin/cassava starch 50/50; T4 gelatin/cassava starch 75/25; T5 gelatin/cassava starch 100/0.

Formulation	Young's Modulus (MPa) *	Tensile Strength (MPa) *	Deformation (%) *
T1	1116.0 [a] ± 57.6	14.6 [a] ± 0.5	2.9 [a] ± 0.5
T2	440.4 [b] ± 46.1	6.2 [b] ± 0.1	4.6 [a] ± 0.3
T3	246.5 [c] ± 21.9	3.7 [c] ± 0.3	9.0 [a] ± 0.8
T4	2.2 [d] ± 1.1	1.4 [d] ± 0.2	157.8 [b] ± 16.2
T5	0.4 [d] ± 0.0	1.5 [d] ± 0.3	285.1 [c] ± 10.0

* Different letters in the same column indicate significant differences ($p < 0.05$).

On the other hand, tensile strength decreased significantly ($p < 0.05$) as gelatin concentration increased, from 6.2 ± 0.1 MPa for T2 (cassava starch/gelatin 75/25 wt.%) to 1.4 ± 0.2 MPa for T4 (cassava starch/gelatin 25/75 wt.%). These results agree with the significantly increased ($p < 0.05$) flexibility, from 4.6 ± 0.3% for T2 to 157.8 ± 16.2% for T4. From these observations, it is evident that the increase in the amount of gelatin in the composite films significantly decreases the stiffness of the composite. The presence of gelatin proteins with globular characteristics distorts the homogeneous molecular arrangement of the starch chains [40].

2.5. Water Vapor Permeability (WVP) of the Cassava Starch/Gelatin Films

The water vapor permeability (WVP) values of cassava starch/gelatin films are described in Table 2. Statistically, there were significant differences ($p < 0.05$) between the treatments evaluated. The treatment without cassava starch, containing only gelatin (T5), presented the lowest value, while the treatment with 100% cassava starch (T1) showed the highest value. It can be inferred that the variation in gelatin/starch concentration in the films affects water vapor permeability. The increase in cassava starch concentration significantly ($p < 0.05$) increased *WVP* values. A higher proportion of starch increases the active binding points with water molecules, favoring water vapor transport through the films. In addition, the higher the ratio of starch, the higher the proportion of plasticizer, which attracts more water molecules due to its polar character and its contribution to the formation of hydrogen bonds. These results are consistent with recent studies reporting an increase in *WVP* values with increasing cassava starch proportion in films obtained from chicken gelatin and cassava starch [40], and fish gelatin-based films, where the increase in the proportion of rice flour increased the *WVP* values [21]. On the other hand, treatments T2, T3, and T4, which had starch/gelatin ratios of 75/25, 50/50, and 25/75, respectively, were significantly different. This behavior could be related to the interactions formed between gelatin and starch that occupy both polymers' active sites, affecting the passage of water vapor through the crosslinked network.

Table 2. Water vapor permeability, water solubility, and opacity properties of cassava starch/gelatin films.

Formulation	WVP (g mm/ kPa h m^2) *	Water Solubility (%) *	Opacity (%) *
T1	0.437 [a] ± 0.11	4.20 [a] ± 2.12	21.20 [a] ± 0.62
T2	0.395 [b] ± 0.03	5.15 [a] ± 1.29	22.11 [a] ± 0.82
T3	0.409 [b] ± 0.04	3.67 [a] ± 1.31	21.21 [a] ± 0.92
T4	0.408 [b] ± 0.04	21.26 [b] ± 0.97	22.5 [a] ± 0.84
T5	0.365 [c] ± 0.03	13.45 [c] ± 1.61	22.37 [a] ± 0.98

* Different letters in the same column indicate significant differences ($p < 0.05$).

Food packaging has different functions to prevent or reduce the moisture gain of the packaged food inside, so low *WVP* values are expected. Table 2 shows that the *WVP* of the films ranged from 0.365 to 0.437 g mm kPa^{-1} h^{-1} m^{-2}. These values are similar to those reported for films made with cassava starch and gelatin between 0.16 and 0.318 g mm kPa^{-1} h^{-1} m^{-2} [47] and lower than those written by other authors in chicken gelatin films with cassava starch of 1.9 to 5.8 g mm kPa^{-1} h^{-1} m^{-2} [40], and bovine gelatin and cassava starch films of 2.7 a 6.4 g mm kPa^{-1} h^{-1} m^{-2} [48]. However, conventional materials with a low water vapor barrier include polyamide 0.027 to 0.03 g mm kPa^{-1} h^{-1} m^{-2} and low-density polyethylene 0.003 to 0.0035 g mm kPa^{-1} h^{-1} m^{-2}. The film's values obtained in this study are higher, indicating a high water vapor permeability.

2.6. Water Solubility of the Cassava Starch/Gelatin Films

The water solubility values of the cassava starch/gelatin films are shown in Table 2. A significant effect of the cassava starch-gelatin ratio on the water solubility of the films was evident ($p < 0.05$). An increasing trend in solubility was observed with increasing gelatin concentration in the films. The treatment without gelatin (T1) presented a significantly low value, while 100% gelatin film (T5) showed a significant increase. However, treatment T4 (75/25 gelatin/cassava starch) showed the highest solubility value, indicating that the rise in gelatin concentration increases the interactions with water, possibly due to an increase in the formation of hydrogen bridges that favor the bonding between the bioplastic and the water molecules. Likewise, the plasticizer increases with increased starch proportion, incrementing interactions with water given its polar character. However, these interactions are higher when gelatin increases in the formulation, which is associated with the creation of new hydrogen bonds.

The water solubility results of the cassava starch/gelatin films ranged from 3.67 to 5.15% for treatments T1 to T3, similar to those reported in a study of cassava starch and fish gelatin-based films [49]. Treatments T4 and T5 showed values between 13.45 and 21.26%, like those found in sheets made with corn starch and gelatin [47] and in films made with corn starch, gelatin, and glycerol or sorbitol [50]. The behavior of the water solubility of the cassava starch/gelatin films may be related to the fact that the increase in the amount of gelatin in the mixture increases the gelation and compaction rates of the polymeric matrix, which decreases the solubility, as evidenced in the T3 treatment. Treatments T1 to T3 with starch/gelatin proportions 100/0, 75/25, and 50/50 differed significantly from T4 with 25/75. This difference may be because the hydroxyl group (O-H) and the amide group (-C=O-N-H) of gelatin enable the formation of strong hydrogen bonds with the O-H groups of starch [45]. According to Zhang et al. [51], the hydrogen bonds reduce the ability of gelatin to interact with water molecules, decreasing solubility. However, when the proportion of gelatin is unbalanced concerning the proportion of starch, solubility increases, probably because the excess polymer promotes solubility since it has the possibility of forming more bonds with water through available non-cross-linked polar groups.

2.7. Opacity of the Cassava Starch/Gelatin Films

The opacity values of the cassava starch/gelatin films are shown in Table 2. The variation in starch/gelatin concentration did not significantly ($p > 0.05$) affect the opacity values of the cassava starch/gelatin films. This indicates that the interactions between the main suspension components, cassava starch and gelatin, within the polymer matrix did not significantly affect the penetration of light through the films. The opacity values for the five treatments were between 21.20 and 22.50%. There have been previous reports with similar values for potato starch and gelatin films [52], corn starch and gelatin [53], and cassava starch and gelatin [47].

Sensory characteristics play a fundamental role in the selection of packaging materials. Opacity is a property that, in addition to impacting appearance, influences the shelf life of the packaged product. For some packaged foods, films with good opacity provide high lightfastness and therefore help to improve the shelf life of photosensitive foods. However, high levels of opacity are not always required. According to the opacity values found, the films were slightly opaque due to aromatic amino acids in the gelatin acting as a barrier for light [21] and the crystallinity of the starch molecules [52].

2.8. Optimization of the Cassava Starch/Gelatin Films

The mechanical property values of a conventional low-density polyethylene film [54] were used as a target to optimize the properties of the cassava/gelatin starch films in this study. Considering the application in food packaging, low water vapor permeability and water solubility values are required since the film will directly contact food. Consequently, WVP and water solubility (WS) values were minimized. Table 3 shows the maximum and minimum values used in optimizing each response variable (y). For each variable, a goal was established.

Table 3. Optimization conditions.

Response Variable	Minimum	Maximum	Goal
WVP (g mm/ kPa h m^2)	0.365	0.437	Minimum
WS (%)	3.67	21.26	Minimum
Young's modulus (MPa)	0.4	1116	200
Tensile strength (MPa)	1.5	14.68	8
Deformation (%)	2.9	285.1	100

The optimized predictions for each response variable are shown in Figure 5. The optimization shows a prediction percentage that is considered acceptable (91.44%). The predicted values for the variables WVP, WS, Young's modulus, tensile strength, and deformation were 0.4 g mm/kPa h m^2, 8.98%, 203.7 MPa, 3.4 MPa, and 28.4%, respectively, with predictability between $0.64 < d < 0.96$. The above is obtained in films with a cassava starch/gelatin proportion of 53/47.

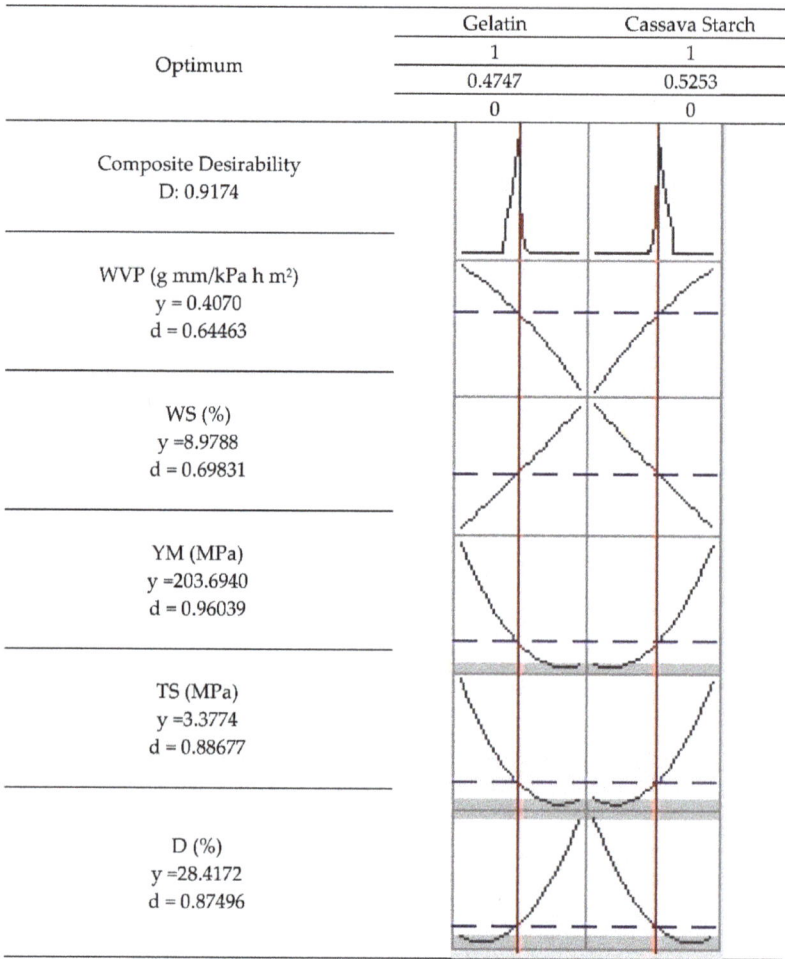

Figure 5. Simultaneous optimization of the variables in the cassava starch/gelatin films.

3. Materials and Methods

3.1. Materials

Cassava starch (*Manihot esculenta* Crantz) was purchased from Tecnas S.A. (Cali, Colombia). Sodium hydroxide and glacial acetic acid were purchased from Merck (Burlington, MA, USA). Extraction of gelatin from chickens (from chicken feet supplied by a local farmer) was reported elsewhere [39]. Food-grade glycerol was purchased from Merck (Burlington, MA, USA). All chemicals were analytical, and no further purification was performed unless otherwise indicated.

3.2. Methods

3.2.1. Preparation of the Sample of Chicken Gelatin

Chicken gelatin was prepared according to the methodology previously reported [39], with slight modifications. The chicken feet were washed, disinfected, and immersed in 400 mL of a 0.25 M NaOH solution for six hours. Subsequently, the sample was transferred to a 4% (v/v) glacial acetic acid solution for three hours. Then, a heat treatment was conducted at 80 °C for eight hours in a ratio of 60 g of chicken feet in 40 mL of water.

Finally, it was filtered, and the resulting solid was dehydrated in an oven at 55 °C until the material was obtained.

3.2.2. Preparation of Cassava Starch/Gelatin Films

Composite films' preparation followed Podshivalov's methodology [55]. The cassava starch suspension was initially heated at 85 °C with constant stirring for one hour. Later, the cassava starch was mixed with 25% w/w glycerol (concerning the weight of cassava starch) to room temperature for 10 min. Subsequently, the suspension was heated to 80 °C for 15 min, cooled to 60 °C, and then we slowly added the gelatin. Finally, the rest of the suspension was sonicated at 60 °C for 50 min using an ultrasonic bath (Branson, Madrid, Spain). All the blends and the simplex-lattice design are reported in Table 4.

Table 4. Simplex-lattice experimental design for the formation of the cassava starch/gelatin composite films.

Sample	Gelatin (wt.%)	Cassava Starch (wt.%)
T1	0	100
T2	25	75
T3	50	50
T4	75	25
T5	100	0

The suspensions in plastic molds were environmentally cured for 24 h and placed for another 24 h in an oven at 40 ± 0.2 °C. After that, the solid film of cassava starch/gelatin was pilled off. The samples were placed in an airtight glass container at 50% relative humidity (RH) until the test time. The samples were cut for mechanical testing according to ASTM D6287, ASTM D618, and ASTM D882. All the thicknesses were measured in a digital micrometer. Film thicknesses ranged from 1.16 to 1.33 mm (±0.17).

3.2.3. Film Characterization

Thermal Analysis

Thermal gravimetric analysis was run on a TA Instruments TGA Q50 V20.13 Build 39 (TA instrument, New Castle, DE, USA) between 30 and 700 ± 2 °C using film samples (5~10 mg) under nitrogen at a flow rate of 60 mL/min of and a heating rate of 10 °C/min. The fusion (Tm) temperatures were determined using a DSC2A-00181 system (TA Instruments) with a heating/cooling rate of 5 °C/min from −25 °C to 250 °C and cooling again to −25 °C using the differential scanning calorimetry (DSC) technique. TGA and DSC results were analyzed using the TA Instruments' Universal Analysis software.

Functional Group Characterization of the Films

The functional group of the films was studied using FTIR spectroscopy on an IR Affinity-1 spectrometer (Shimadzu, Kyoto, Japan). The spectra were obtained using a 4 cm^{-1} resolution and 32 scans in the range of 500 to 4000 cm^{-1}. A diamond tip and ATR (attenuated total reflectance) mode were used for the test.

Morphology Analysis

The surface morphologies were scanned with an electron microscope (JSM-6490LA, JEOL) at 20 kV acceleration voltage, using a copper coating with the mode of secondary backscattered electrons. A gold layer was used for electron conduction on the samples.

Mechanical Properties

A universal SHIMADZU EZ-LZ test machine (Shimadzu, Tokyo, Japan) was used for the tensile test, following the ASTM D882 standard. A 500 N load cell was used. The samples were tested with a gap of 100 mm between jaws at a speed of 50 mm/min. At least

10 samples per formulation were used. The length and width of the tested samples were 100 mm and 20 mm, respectively.

Water Vapor Permeability

Water vapor permeability (*WVP*) determination followed the ASTM-E96. The diameter of the tested samples was 80 mm. The samples were glued to the mouth of a permeation cell containing silica gel. The permeation cells were stored inside the desiccator with 73 ± 2% RH at 25 °C. The slope of the permeation cell mass change as a function of time was obtained by linear regression. *WVP* (g mm/kPa h m^2) was calculated as follows in Equation (1):

$$WVP = \left[\frac{WVTR}{PR \times H}\right] \times l \qquad (1)$$

where *WVTR* is the water vapor transmission rate obtained from the ratio between the water gained as a function of time (g/h) and the permeation area of the sample (m^2). *P* is the saturation vapor pressure of water (kPa); *RH* is the relative humidity in the desiccator, and *l* is the average sample thickness (m).

Solubility in Water

Solubility in water (*WS*) was determined by following the method described by Cheng et al. [56]. Dried films (105 °C for 24 h) of 2 cm × 2 cm were weighted (*Wi*). Then they were hydrated with 50 mL of distilled water at room temperature for 24 h with random agitation. Finally, samples were filtered and dried at 105 °C to obtain their final dry weight (*Wf*). *WS* (%) was calculated as follows in Equation (2):

$$WS = \frac{Wi - Wf}{Wi} \times 100 \qquad (2)$$

Opacity

The opacity test was developed using a colorimeter (Konica Minolta, Japan). Data were processed by Spectra Magic NX software. Film samples with 2 cm × 10 cm were placed under aperture for opacity measurement, determined by using black (P_b) and white patron (P_w) as references. Opacity (*Op*) was calculated as the ratio between the opacity (P_b) and opacity (P_w). Both P_b and P_w were determined in five films with black and white standards (%) using Equation (3).

$$Op(\%) = \frac{P_b}{P_w} \times 100 \qquad (3)$$

3.2.4. Experimental Design and Optimization

A simplex-lattice design was used with two components and four lattices (*m*). The design points are given by *m* + 1. The five design points correspond to:

$x_i = 0, 1/m, 2/m \ldots m/m$, where *i* is the number of components, leaving the following mixtures:

(x_1, x_2): (0, 1); (0.25, 0.75); (0.5, 0.5); (0.75, 0.25); (1, 0).

The mixtures are shown in Table 4.

The optimization was performed by setting the mechanical properties of a commercial polyethylene film as a target and minimizing the values of the *WVP*, and *WS* found in the blends. The desirability function (D: global; d: individual) that converts the functions to a scale between 0 and 1 was used, combining them using the geometric mean and optimizing the general metric means.

The experimental design and the optimization were performed using Minitab (2019) software.

Statistical Analysis

We used an analysis of variance (ANOVA) to establish statistical differences between treatments. Fisher's test was used as a post-ANOVA analysis to compare the means between treatments. ANOVA and post-ANOVA were obtained using Minitab (2019) software.

4. Conclusions

Cassava starch/gelatin composite films were obtained, with good mechanical properties comparable with reported values of commercial polyethylene films. The analysis of the FT-IR spectra for the composites showed the presence of hydroxyl and amide groups characteristic of starch and gelatin and with shifts of their bands at various ratios due to hydrogen bonds. Thermogravimetric analysis of the films showed that with a high amount of gelatin (T4 and T5), degradation profiles were poorly defined, characteristic of films with high flexibility. In the DSC analysis, it was evidenced that the endothermic peak Tc increased due to the presence of gelatin. Still, for T4 and T5, there was a notable decrease in its value, probably due to molecular disorder between the chains and greater flexibility.

Additionally, the morphological analysis of T2 and T3 showed minor roughness, discontinuity, and heterogeneity, and a decrease in the flexibility of gelatin compared to the heterogeneous appearance of the gelatin film (T5). The results of material stiffness through Young's modulus suggest that the flexibility of the films increased with the increase in the amount of gelatin, decreasing, in turn, the stiffness of the material (from 2.9 ± 0.5 to 285.1 ± 10.0 MPa for formulations T1 and T5, respectively). On the other hand, the water solubility results of the cassava starch/gelatin films were between 3.67 and 21.26% for treatments T1 and T5, respectively, similar to those found for cassava starch- or cornstarch-based films with gelatin.

With all these characterization results for the composites, an optimal formulation was obtained to develop cassava starch/gelatin-based films in a 53/47 ratio, plasticized with glycerol using the casting method, that would meet the expectations of the model polyethylene film for food-packaging applications. Young's modulus, tensile strength, deformation, thermal, *WVP*, and water solubility variables were affected by the cassava starch and gelatin mixtures evaluated in the treatments. Based on the predicted values for each response in the optimization, it can be inferred that these films could be used as food-packaging material, as their mechanical properties are close to those of low-density polyethylene. Future studies could incorporate other additives to improve moisture stability properties such as *WVP*.

Author Contributions: Conceptualization, D.P.N.-P. and C.D.G.-T.; Funding acquisition, C.D.G.-T.; Investigation, J.I.C., D.P.N.-P., J.A.A.C. and C.D.G.-T.; Methodology, J.I.C., D.P.N.-P., J.A.A.C., J.H.M.H. and C.D.G.-T.; Resources, J.H.M.H.; Supervision, C.D.G.-T.; Writing—original draft, J.I.C., D.P.N.-P. and C.D.G.-T.; Writing—review and editing, D.P.N.-P., J.H.M.H. and C.D.G.-T. All authors have read and agreed to the published version of the manuscript.

Funding: The APC was funded by Universidad del Atlántico.

Institutional Review Board Statement: Not applicable.

Informed Consent Statement: Not applicable.

Data Availability Statement: Data are available under request to the corresponding author.

Acknowledgments. The authors thank Universidad del Atlántico for the APC payment.

Conflicts of Interest: The authors declare no conflict of interest.

Sample Availability: Samples of the compounds are available from the authors.

References

1. Walker, T.R.; McGuinty, E.; Charlebois, S.; Music, J. Single-Use Plastic Packaging in the Canadian Food Industry: Consumer Behavior and Perceptions. *Humanit. Soc. Sci. Commun.* **2021**, *8*, 80. [CrossRef]
2. Wang, Y.-L.; Lee, Y.-H.; Chiu, I.-J.; Lin, Y.-F.; Chiu, H.-W. Potent Impact of Plastic Nanomaterials and Micromaterials on the Food Chain and Human Health. *Int. J. Mol. Sci.* **2020**, *21*, 1727. [CrossRef] [PubMed]
3. Lebreton, L.; Slat, B.; Ferrari, F.; Sainte-Rose, B.; Aitken, J.; Marthouse, R.; Hajbane, S.; Cunsolo, S.; Schwarz, A.; Levivier, A. Evidence that the Great Pacific Garbage Patch is rapidly accumulating plastic. *Sci. Rep.* **2018**, *8*, 4666. [CrossRef] [PubMed]
4. Basiak, E. Food industry: Use of plastics of the twenty-first century. In *Food Technology*; Apple Academic Press: Palm Bay, FL, USA, 2017; pp. 53–62. ISBN 1315365650.
5. Alves, V.D.; Ferreira, A.R.; Costa, N.; Freitas, F.; Reis, M.A.M.; Coelhoso, I.M. Characterization of biodegradable films from the extracellular polysaccharide produced by Pseudomonas oleovorans grown on glycerol byproduct. *Carbohydr. Polym.* **2011**, *83*, 1582–1590. [CrossRef]
6. Reddy, N.; Chen, L.; Yang, Y. Biothermoplastics from hydrolyzed and citric acid crosslinked chicken feathers. *Mater. Sci. Eng. C* **2013**, *33*, 1203–1208. [CrossRef]
7. González, A.; Igarzabal, C.I.A. Soy protein–Poly (lactic acid) bilayer films as biodegradable material for active food packaging. *Food Hydrocoll.* **2013**, *33*, 289–296. [CrossRef]
8. Cercel, F.; Stroiu, M.; Alexe, P.; Ianițchi, D. Characterization of myofibrillar chicken breast proteins for obtain protein films and biodegradable coatings generation. *Agric. Agric. Sci. Procedia* **2015**, *6*, 197–205. [CrossRef]
9. Salgado, P.R.; Ortiz, S.E.M.; Petruccelli, S.; Mauri, A.N. Biodegradable sunflower protein films naturally activated with antioxidant compounds. *Food Hydrocoll.* **2010**, *24*, 525–533. [CrossRef]
10. Tapia-Blácido, D.; Sobral, P.J.; Menegalli, F.C. Development and characterization of biofilms based on Amaranth flour (Amaranthus caudatus). *J. Food Eng.* **2005**, *67*, 215–223. [CrossRef]
11. Mendes, J.F.; Paschoalin, R.T.; Carmona, V.B.; Neto, A.R.S.; Marques, A.C.P.; Marconcini, J.M.; Mattoso, L.H.C.; Medeiros, E.S.; Oliveira, J.E. Biodegradable Polymer Blends Based on Corn Starch and Thermoplastic Chitosan Processed by Extrusion. *Carbohydr. Polym.* **2016**, *137*, 452–458. [CrossRef]
12. Santayanon, R.; Wootthikanokkhan, J. Modification of cassava starch by using propionic anhydride and properties of the starch-blended polyester polyurethane. *Carbohydr. Polym.* **2003**, *51*, 17–24. [CrossRef]
13. Souza, A.C.; Benze, R.; Ferrão, E.S.; Ditchfield, C.; Coelho, A.C.V.; Tadini, C.C. Cassava starch biodegradable films: Influence of glycerol and clay nanoparticles content on tensile and barrier properties and glass transition temperature. *LWT—Food Sci. Technol.* **2012**, *46*, 110–117. [CrossRef]
14. Bertoft, E. Understanding Starch Structure: Recent Progress. *Agronomy* **2017**, *7*, 56. [CrossRef]
15. Flores, S.; Famá, L.; Rojas, A.M.; Goyanes, S.; Gerschenson, L. Physical properties of tapioca-starch edible films: Influence of filmmaking and potassium sorbate. *Food Res. Int.* **2007**, *40*, 257–265. [CrossRef]
16. Bangyekan, C.; Aht-Ong, D.; Srikulkit, K. Preparation and properties evaluation of chitosan-coated cassava starch films. *Carbohydr. Polym.* **2006**, *63*, 61–71. [CrossRef]
17. Parra, D.F.; Tadini, C.C.; Ponce, P.; Lugão, A.B. Mechanical properties and water vapor transmission in some blends of cassava starch edible films. *Carbohydr. Polym.* **2004**, *58*, 475–481. [CrossRef]
18. Vanin, F.M.; Sobral, P.J.A.; Menegalli, F.C.; Carvalho, R.A.; Habitante, A. Effects of plasticizers and their concentrations on thermal and functional properties of gelatin-based films. *Food Hydrocoll.* **2005**, *19*, 899–907. [CrossRef]
19. Luo, Q.; Hossen, M.A.; Zeng, Y.; Dai, J.; Li, S.; Qin, W.; Liu, Y. Gelatin-based composite films and their application in food packaging: A review. *J. Food Eng.* **2022**, *313*, 110762. [CrossRef]
20. Ahmad, T.; Ismail, A.; Ahmad, S.A.; Khalil, K.A.; Kumar, Y.; Adeyemi, K.D.; Sazili, A.Q. Recent advances on the role of process variables affecting gelatin yield and characteristics with special reference to enzymatic extraction: A review. *Food Hydrocoll.* **2017**, *63*, 85–96. [CrossRef]
21. Ahmad, M.; Hani, N.M.; Nirmal, N.P.; Fazial, F.F.; Mohtar, N.F.; Romli, S.R. Optical and thermo-mechanical properties of composite films based on fish gelatin/rice flour fabricated by casting technique. *Prog. Org. Coatings* **2015**, *84*, 115–127. [CrossRef]
22. Duconseille, A.; Astruc, T.; Quintana, N.; Meersman, F.; Sante-Lhoutellier, V. Gelatin structure and composition linked to hard capsule dissolution: A review. *Food Hydrocoll.* **2015**, *43*, 360–376. [CrossRef]
23. Mohammadi, R.; Mohammadifar, M.A.; Rouhi, M.; Kariminejad, M.; Mortazavian, A.M.; Sadeghi, E.; Hasanvand, S. Physico-mechanical and structural properties of eggshell membrane gelatin- chitosan blend edible films. *Int. J. Biol. Macromol.* **2018**, *107*, 406–412. [CrossRef] [PubMed]
24. Li, K.; Jin, S.; Chen, H.; Li, J. Bioinspired interface engineering of gelatin/cellulose nanofibrils nanocomposites with high mechanical performance and antibacterial properties for active packaging. *Compos. Part B Eng.* **2019**, *171*, 222–234. [CrossRef]
25. Boughriba, S.; Souissi, N.; Jridi, M.; Li, S.; Nasri, M. Thermal, mechanical and microstructural characterization and antioxidant potential of Rhinobatos cemiculus gelatin films supplemented by titanium dioxide doped silver nanoparticles. *Food Hydrocoll.* **2020**, *103*, 105695. [CrossRef]
26. Tongdeesoontorn, W.; Mauer, L.J.; Wongruong, S.; Rachtanapun, P. Water Vapour Permeability and Sorption Isotherms of Cassava Starch Based Films Blended with Gelatin and Carboxymethyl Cellulose. *Asian J. Food Agro-Ind.* **2009**, *2*, 501–514.

27. Veiga-Santos, P.; Oliveira, L.M.; Cereda, M.P.; Scamparini, A.R.P. Sucrose and Inverted Sugar as Plasticizer. Effect on Cassava Starch–Gelatin Film Mechanical Properties, Hydrophilicity and Water Activity. *Food Chem.* **2007**, *103*, 255–262. [CrossRef]
28. Biswal, D.R.; Singh, R.P. Characterisation of carboxymethyl cellulose and polyacrylamide graft copolymer. *Carbohydr. Polym.* **2004**, *57*, 379–387. [CrossRef]
29. Vicentini, N.M.; Dupuy, N.; Leitzelman, M.; Cereda, M.P.; Sobral, P.J.A. Prediction of cassava starch edible film properties by chemometric analysis of infrared spectra. *Spectrosc. Lett.* **2005**, *38*, 749–767. [CrossRef]
30. Tongdeesoontorn, W.; Mauer, L.J.; Wongruong, S.; Sriburi, P.; Rachtanapun, P. Effect of carboxymethyl cellulose concentration on physical properties of biodegradable cassava starch-based films. *Chem. Cent. J.* **2011**, *5*, 6. [CrossRef]
31. Hanani, Z.N.; Roos, Y.; Kerry, J. Fourier Transform Infrared (FTIR) spectroscopic analysis of biodegradable gelatin films immersed in water. In Proceedings of the 11th International Congress on Engineering and Food, ICEF11, Athens, Greece, 22–26 May 2011.
32. Ahmadi, A.; Ahmadi, P.; Ehsani, A. Development of an active packaging system containing zinc oxide nanoparticles for the extension of chicken fillet shelf life. *Food Sci. Nutr.* **2020**, *8*, 5461–5473. [CrossRef]
33. Cyras, V.P.; Manfredi, L.B.; Ton-That, M.-T.; Vázquez, A. Physical and Mechanical Properties of Thermoplastic Starch/Montmorillonite Nanocomposite Films. *Carbohydr. Polym.* **2008**, *73*, 55–63. [CrossRef]
34. Moreno, O.; Cárdenas, J.; Atarés, L.; Chiralt, A. Influence of Starch Oxidation on the Functionality of Starch-Gelatin Based Active Films. *Carbohydr. Polym.* **2017**, *178*, 147–158. [CrossRef] [PubMed]
35. Lee, J.-H.; Lee, J.; Song, K. Bin Development of a chicken feet protein film containing essential oils. *Food Hydrocoll.* **2015**, *46*, 208–215. [CrossRef]
36. Tarique, J.; Sapuan, S.M.; Khalina, A. Effect of glycerol plasticizer loading on the physical, mechanical, thermal, and barrier properties of arrowroot (*Maranta arundinacea*) starch biopolymers. *Sci. Rep.* **2021**, *11*, 13900. [CrossRef]
37. Moreno, O.; Díaz, R.; Atarés, L.; Chiralt, A. Influence of the Processing Method and Antimicrobial Agents on Properties of Starch-gelatin Biodegradable Films. *Polym. Int.* **2016**, *65*, 905–914. [CrossRef]
38. Verdonck, E.; Schaap, K.; Thomas, L.C. A discussion of the principles and applications of Modulated Temperature DSC (MTDSC). *Int. J. Pharm.* **1999**, *192*, 3–20. [CrossRef]
39. De Almeida, P.F. *Análise Da Qualidade de Gelatina Obtida de Tarsos de Frango e Aspectos Envolvidos No Processo Produtivo*; Uninove: São Paulo, Brazil, 2012.
40. Loo, C.P.Y.; Sarbon, N.M. Chicken skin gelatin films with tapioca starch. *Food Biosci.* **2020**, *35*, 100589. [CrossRef]
41. Tongdeesoontorn, W.; Mauer, L.J.; Wongruong, S.; Sriburi, P.; Reungsang, A.; Rachtanapun, P. Antioxidant films from cassava starch/gelatin biocomposite fortified with quercetin and TBHQ and their applications in food models. *Polymers* **2021**, *13*, 1117. [CrossRef]
42. Soo, P.Y.; Sarbon, N.M. Preparation and characterization of edible chicken skin gelatin film incorporated with rice flour. *Food Packag. Shelf Life* **2018**, *15*, 1–8. [CrossRef]
43. Sarbon, N.M.; Badii, F.; Howell, N.K. Preparation and characterisation of chicken skin gelatin as an alternative to mammalian gelatin. *Food Hydrocoll.* **2013**, *30*, 143–151. [CrossRef]
44. Wang, L.Z.; Liu, L.; Holmes, J.; Kerry, J.F.; Kerry, J.P. Assessment of film-forming potential and properties of protein and polysaccharide-based biopolymer films. *Int. J. Food Sci. Technol.* **2007**, *42*, 1128–1138. [CrossRef]
45. Tongdeesoontorn, W.; Mauer, L.J.; Wongruong, S.; Sriburi, P.; Rachtanapun, P. Mechanical and Physical Properties of Cassava Starch-Gelatin Composite Films. *Int. J. Polym. Mater.* **2012**, *61*, 778–792. [CrossRef]
46. Al-Hassan, A.; Norziah, M. Starch–gelatin edible films: Water vapor permeability and mechanical properties as affected by plasticizers. *Food Hydrocoll.* **2012**, *26*, 108–117. [CrossRef]
47. Fakhouri, F.M.; Martelli, S.M.; Bertan, L.C.; Yamashita, F.; Mei, L.H.I.; Queiroz, F.P.C. Edible films made from blends of manioc starch and gelatin–Influence of different types of plasticizer and different levels of macromolecules on their properties. *Postharvest Biol. Technol.* **2012**, *49*, 149–154. [CrossRef]
48. Acosta, S.; Jiménez, A.; Cháfer, M.; González-Martínez, C.; Chiralt, A. Physical properties and stability of starch-gelatin based films as affected by the addition of esters of fatty acids. *Food Hydrocoll.* **2015**, *49*, 135–143. [CrossRef]
49. Najwa, I.S.N.A.; Guerrero, P.; de la Caba, K.; Hanani, Z.A.N. Physical and antioxidant properties of starch/gelatin films incorporated with Garcinia atroviridis leaves. *Food Packag. Shelf Life* **2020**, *26*, 100583. [CrossRef]
50. Fakhouri, F.M.; Costa, D.; Yamashita, F.; Martelli, S.M.; Jesus, R.C.; Alganer, K.; Collares-queiroz, F.P.; Innocentini-mei, L.H. Comparative study of processing methods for starch/gelatin films. *Carbohydr. Polym.* **2013**, *95*, 681–689. [CrossRef]
51. Zhang, Y.; Du, Z.; Xia, X.; Guo, Q.; Wu, H.; Yu, W. Evaluation of the migration of UV-ink photoinitiators from polyethylene food packaging by supercritical fluid chromatography combined with photodiode array detector and tandem mass spectrometry. *Polym. Test.* **2016**, *53*, 276–282. [CrossRef]
52. Kumar, R.; Goyal, G.G.M. Synthesis and functional properties of gelatin/CA–starch composite film: Excellent food packaging material. *J. Food Sci. Technol.* **2019**, *56*, 1954–1965. [CrossRef]
53. Kumar, R. Biodegradable composite films/coatings of modified corn starch/gelatin for shelf life improvement of cucumber. *J. Food Sci. Technol.* **2021**, *58*, 1227–1237. [CrossRef]
54. Stark, N.M.; Matuana, L.M. Trends in sustainable biobased packaging materials: A mini review. *Mater. Today Sustain.* **2021**, *15*, 100084. [CrossRef]

55. Podshivalov, A.; Zakharova, M.; Glazacheva, E.; Uspenskaya, M. Gelatin/potato starch edible biocomposite films: Correlation between morphology and physical properties. *Carbohydr. Polym.* **2017**, *157*, 1162–1172. [CrossRef] [PubMed]
56. Cheng, Y.; Wang, W.; Zhang, R.; Zhai, X.; Hou, H. Effect of gelatin bloom values on the physicochemical properties of starch/gelatin–beeswax composite films fabricated by extrusion blowing. *Food Hydrocoll.* **2021**, *113*, 106466. [CrossRef]

Article

In Vitro Evaluation of Curcumin Encapsulation in Gum Arabic Dispersions under Different Environments

Dwi Hudiyanti [1,*], Muhammad Fuad Al Khafiz [2], Khairul Anam [1], Parsaoran Siahaan [1] and Sherllyn Meida Christa [3]

[1] Department of Chemistry, Faculty of Science and Mathematics, Diponegoro University, Jl. Prof. Soedarto, Semarang 50275, Indonesia; k.anam@live.undip.ac.id (K.A.); siahaan.parsaoran@live.undip.ac.id (P.S.)
[2] Postgraduate Chemistry Program, Faculty of Science and Mathematics, Diponegoro University, Jl. Prof. Soedarto, Semarang 50275, Indonesia; queenfoe@gmail.com
[3] Chemistry Program, Faculty of Science and Mathematics, Diponegoro University, Jl. Prof. Soedarto, Semarang 50275, Indonesia; meidachrista@gmail.com
* Correspondence: dwi.hudiyanti@live.undip.ac.id; Tel.: +62-852-2506-4261

Citation: Hudiyanti, D.; Al Khafiz, M.F.; Anam, K.; Siahaan, P.; Christa, S.M. In Vitro Evaluation of Curcumin Encapsulation in Gum Arabic Dispersions under Different Environments. *Molecules* **2022**, *27*, 3855. https://doi.org/10.3390/molecules27123855

Academic Editor: Sylvain Caillol

Received: 15 May 2022
Accepted: 12 June 2022
Published: 16 June 2022

Publisher's Note: MDPI stays neutral with regard to jurisdictional claims in published maps and institutional affiliations.

Copyright: © 2022 by the authors. Licensee MDPI, Basel, Switzerland. This article is an open access article distributed under the terms and conditions of the Creative Commons Attribution (CC BY) license (https://creativecommons.org/licenses/by/4.0/).

Abstract: Biopolymers, especially polysaccharides (e.g., gum Arabic), are widely applied as drug carriers in drug delivery systems due to their advantages. Curcumin, with high antioxidant ability but limited solubility and bioavailability in the body, can be encapsulated in gum Arabic to improve its solubility and bioavailability. When curcumin is encapsulated in gum Arabic, it is essential to understand how it works in various conditions. As a result, in Simulated Intestinal Fluid and Simulated Gastric Fluid conditions, we investigated the potential of gum Arabic as the drug carrier of curcumin. This study was conducted by varying the gum Arabic concentrations, i.e., 5, 10, 15, 20, 30, and 40%, to encapsulate 0.1 mg/mL of curcumin. Under both conditions, the greater the gum Arabic concentration, the greater the encapsulation efficiency and antioxidant activity of curcumin, but the worse the gum Arabic loading capacity. To achieve excellent encapsulation efficiency, loading capacity, and antioxidant activity, the data advises that 10% is the best feasible gum Arabic concentration. Regarding the antioxidant activity of curcumin, the findings imply that a high concentration of gum Arabic was effective, and the Simulated Intestinal Fluid brought an excellent surrounding compared to the Simulated Gastric Fluid solution. Moreover, the gum Arabic releases curcumin faster in the Simulated Gastric Fluid condition.

Keywords: gum Arabic; curcumin; drug delivery system; Simulated Intestinal Fluid; Simulated Gastric Fluid; encapsulation efficiency; loading capacity; antioxidant activity; release rate

1. Introduction

Polymers are giant molecules with a high molecular weight (macromolecules) created by the covalent bonding of several smaller molecules or repeating units, called monomers. A natural polymer, found in plants, microorganisms, and animals, is one form of polymer based on its source of origin [1]. Natural polymers have several advantages over synthetic polymers, including homogeneous shapes and sizes, biodegradability, biocompatibility, non-toxicity, low cost, ease of modification, and accessibility [1–4]. Biopolymer is a natural polymer created directly by living organisms' cells. It comprises bio-based monomer units covalently bound together to form bigger bio-based polymer molecules [5]. Polynucleotides (made of nucleotide monomers, e.g., DNA and RNA), polypeptides (composed of amino acid monomers, e.g., collagen), and polysaccharides (containing carbohydrate structures, e.g., starch, cellulose, and gum Arabic) are the three types of biopolymers [6]. Due to their biocompatibility, processability, and other benefits, natural polymers and biopolymers, mainly polysaccharides, are commonly used as drug carriers in drug delivery systems (DDS) [7]. By stabilizing the drug, localizing the drug's action, and managing the release drug's rate, time, and location, DDS is claimed to provide better therapeutic effects of the encapsulated drug at specific disease sites with low toxicological effects [8,9].

Gum Arabic (GA) in Figure 1 [10], also known as acacia gum, is the hardened sap of the Leguminosae family of *Acacia senegal* and *Acacia seyal* trees. It is a complex mixture of glycoproteins and polysaccharides, branched heteropolysaccharides that are either neutral or slightly acidic, light-orange or pale white, and water-soluble [11,12]. The GA structure's mainframe comprises 1,3-linked β-D-galactopyranosyl units. At the same time, the side chains are made up of two to five 1,3-linked β-D-galactopyranosyl units that connect to the main chain via 1,6-linkages. Another study discovered that simple sugars such as D-galactose, L-arabinose, L-rhamnose, and D-glucuronic acid are also constituents of this heteropolysaccharide [13]. Because of biocompatibility, tastelessness, non-toxicity, and high-water solubility, GA is widely used as a drug carrier in DDS [14]. GA can also prevent aggregation as a drug carrier by forming a thick protective film around the encapsulated drug's core material and acting as an emulsifier. Several studies have shown that using GA to encapsulate drugs or active compounds with antioxidant properties can improve the drug's stability, encapsulation efficiency, and antioxidant capacity [15–19]. The ability of a mixture to scavenge free radicals by intervening in one of the three main steps of the oxidative process mediated by free radicals (i.e., initiation, propagation, and termination) is referred to as antioxidants [20,21].

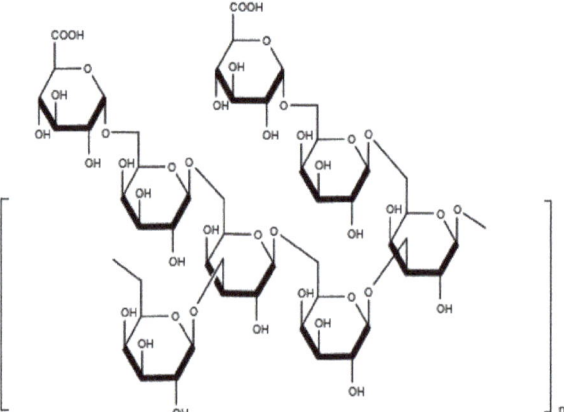

Figure 1. The structure of gum Arabic.

Curcumin is the curcuminoid active compound found in turmeric (*Curcuma longa* L.) It has numerous health benefits, one of which is an antioxidant [22,23]. Curcumin's unique reactive groups, which include two phenolic hydroxyl groups and an enol from a β-diketone moiety, are known to have potent free radical scavenging activity [24,25]. Due to their ability to directly react with free radicals and transform them into more stable or non-radical products, phenolic compounds with more than one hydroxyl group (–OH) are effective primary antioxidants [26,27]. Curcumin, poorly soluble in water (7.8 µg/mL), has low bioavailability in the body and a fast metabolism and excretion rate from the body's system [28–30]. Curcumin is rapidly degraded in alkaline conditions (pH > 7) but degrades slowly in acidic conditions, implying that its decomposition is pH-dependent [31]. As a result, finding a suitable DDS is critical to overcoming the problem of delivering curcumin into the body for therapeutic use.

Several studies have shown that liposomes as DDS can overcome curcumin's weaknesses, allowing curcumin to be well encapsulated and its effectiveness in the body to improve [32–36]. Other materials of DDS, as depicted in Figure 2, such as dendrimers, micelles, and microemulsions, emulsions and nanoemulsions, solid lipid nanoparticles (SLNs), nanoparticles (NPs) including polymeric nanoparticles, magnetic nanoparticles, biopolymer nanoparticles, microgels, and hydrogel beads, have also been used to increase the solubility and bioavailability of curcumin so that it can be delivered into the body [36,37].

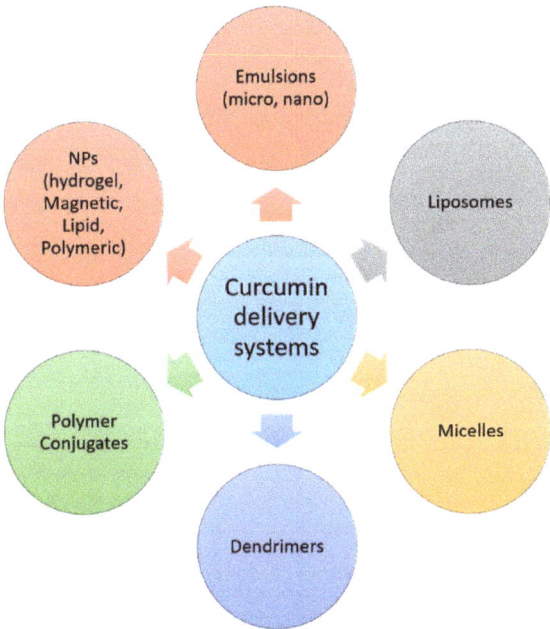

Figure 2. Representation of curcumin delivery systems.

As mentioned above, GA has been widely used as the primary drug carrier or additional stabilizing material to improve the ability of encapsulated drugs when delivered into the body [14,16]. Therefore, this study aimed to investigate GA's potential in encapsulating curcumin under two different oral drug delivery pathways, namely SIF and SGF solutions. Our new finding is that a 10% concentration of gum Arabic in both SIF and SGF solutions is the optimum concentration to achieve the optimal encapsulation efficiency of curcumin and the loading capacity of gum Arabic for curcumin.

2. Results and Discussion

Curcumin is a bioactive agent that is poorly soluble in water (7.8 µg/mL) [28], slightly improved under physiological pH conditions (0.0004 mg/mL) [36], easily soluble in organic solvents, including 96% ethanol (10 mg/mL) [38], and chemically unstable in gastric and intestinal environmental conditions [28]. Its decomposition depends on pH. Curcumin's half-life at pH 3–6.5 is ~100–200 min, while at pH 7.2–8.0 it decreases significantly to only 1–9 min [37]. Research has shown that encapsulation using polymeric micelles, liposomes, or surfactant micelles can increase curcumin solubility [29,39]. In this research we establish that DDS using GA matrices, which are biopolymer, provide promising results in SIF and SGF solutions.

2.1. Encapsulation Efficiency, Loading Capacity, Release Rate, and Antioxidant Activity in SIF and SGF Solutions

Encapsulation efficiency (*EE*) is an important parameter to consider when evaluating the success of a DDS. The percentage of an encapsulated material (e.g., active ingredients, drugs, etc.) successfully entrapped into drug carriers following an encapsulation process for protection, absorption, delivery in the body, and controlled release is defined as *EE* [40,41]. Therefore, we investigated the *EE* of curcumin encapsulated in GA and expressed it as a percentage. It represents the amount of the drug encapsulated. In this study, the *EE* of curcumin was calculated indirectly by measuring the amount of the unencapsulated curcumin (C_t) in the supernatant using UV-Vis spectrophotometer [42] and Equation (2).

The *EE* of curcumin increased as the concentration of GA (C_{GA}) increased in both SIF and SGF conditions, as shown in Figure 3. *EE* grew rapidly at low C_{GA} up to 20%, then relatively more slowly at higher C_{GA} up to 40%. The data in Figure 3 also revealed that the *EE* for each C_{GA} (5–40%) in SIF (range 32.3–72.8%) was more significant than that in SGF (range 10.47–49.97%). The higher the *EE* value obtained, the more curcumin was successfully encapsulated in GA. Thus, the C_{GA} to get the highest *EE* for better DDS was 40% for both SIF (*EE* = 72.8%) and SGF (*EE* = 49.97%) conditions.

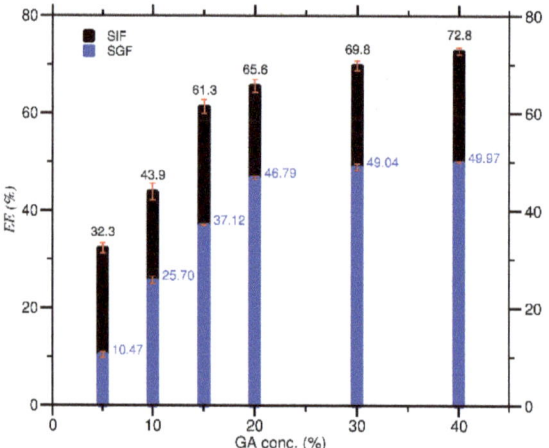

Figure 3. Encapsulation efficiency (*EE*) of curcumin at various C_{GA} in SIF and SGF solutions.

Loading capacity (*LC*) refers to a drug carrier's ability to encapsulate a specific encapsulated material. The percentage of drugs incorporated within the drug carrier relative to the total mass of the drug carrier is referred to as LC. The drug carrier's structural, physical, and chemical properties determine LC [41,43]. As shown in Equation (3), *LC* in this study can be calculated by dividing the total concentration of successfully encapsulated curcumin (C_0–C_t) by the total concentration of GA (C_{GA}). The higher the *LC* value of GA, the more curcumin was successfully encapsulated. This indicates that the best potential of GA as a drug carrier in DDS (composed of the drug carrier and the encapsulated material) can be obtained at this C_{GA} because curcumin can be maximally encapsulated [44,45].

As shown in Figure 4, the *LC* decreased as the C_{GA} increased in SIF and SGF conditions except for GA in SGF with a 5% to 10% concentration. The *LC* between these concentrations increased by 4.86%, from 21.35% to 26.21%. The data in Figure 4 also revealed that the *LC* for each C_{GA} (5–40%) in SGF (range 26.21–12.74%) was greater than that in SIF (range 6.58–1.86%). The lowering of the loading capacity of GA is assumed because the carboxylic group in GA has been wholly ionized to COO^-. The formation of this charge creates a repulsion force between the acid groups of GA, resulting in destabilization of the GA structure and a decrease in *LC* [16]. The C_{GA} for obtaining the highest *LC* value for DDS was 5% for SIF (6.58%) and 10% for SGF (26.21%) conditions.

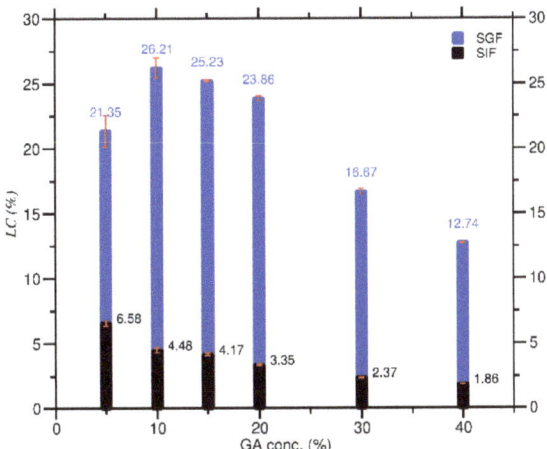

Figure 4. Loading capacity (LC) of GA at various C_{GA} in SIF and SGF solutions.

The rate of drug release (RR) from a DDS to the desired target tissues is a critical property associated with a drug's therapeutic activity in the body [46]. Hence, we investigated the release rate of curcumin encapsulated in GA. A controlled rate of release of a DDS is a delivery form in which the drug is released at a predetermined rate based on the desired therapeutic concentration and the drug's pharmacokinetic properties [47]. Because the release rate has been determined, the medication delivered can have a long lifetime ranging from days to months, with minimal side effects on the body.

The RR of curcumin encapsulated in GA varied as the C_{GA} increased in SIF and SGF conditions, as shown in Figure 5. The RR of curcumin encapsulated in GA dispersed in SIF was higher than that of SGF at 5% and 15% of C_{GA}, respectively. However, at other C_{GA} (10%, 20%, 30%, and 40%), the RR in SGF was higher than that in SIF. The observation of the RR for 12 days revealed that under SIF conditions, curcumin encapsulated in GA lasted the longest at 40% of C_{GA}, while, under SGF conditions, curcumin encapsulated in GA lasted the longest at 10% of C_{GA}. In general, curcumin would be released faster in the SGF compared to the SIF condition.

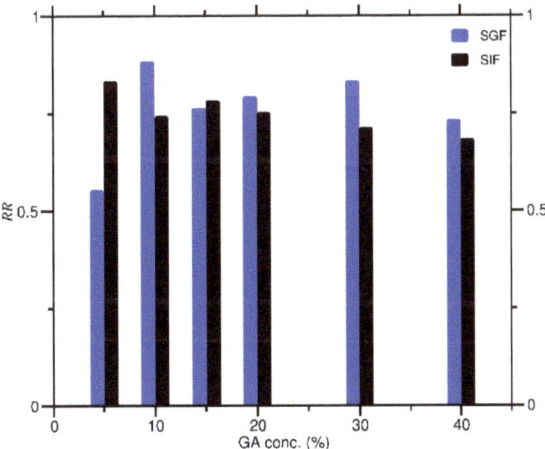

Figure 5. Release rate (RR) of curcumin encapsulated in GA at various C_{GA} in SIF and SGF solutions.

We estimate the mechanism of curcumin encapsulation in GA occurs due to non-covalent interactions between the –COOH group of GA and the –OH group of curcumin to form hydrogen bonds. The formation of hydrogen bonds affects both the encapsulation of curcumin in GA and the release of curcumin from GA. This hydrogen bond formation also explains why the *EE* and *LC* of GA to encapsulate curcumin are higher than that of tocopherol [16]. Curcumin has additional OH groups compared to tocopherol, so more hydrogen bonds are formed during the encapsulation process. Further studies are still needed to confirm the hydrogen bonding formation such as using X-ray Diffraction (XRD), Differential Scanning Calorimetry (DSC), or computational studies.

Antioxidant activity (*IR*) is defined as limiting or inhibiting the oxidation of nutrients (particularly lipids and proteins) by preventing oxidative chain reactions from occurring. We used the DPPH scavenging activity assay to assess the *IR* of curcumin encapsulated in GA, as shown in Equation (4). If the scavenging activity of DPPH is high, the value of *IR* will be increased. If the value of DPPH is higher, it means that the amount of antioxidant compounds in the related drug (e.g., curcumin) is smaller [21,32,48]. The lower the number of antioxidant compounds required to obtain a high value of IR, the better the compound's ability to defend against free radicals in its role as an antioxidant [49–51].

When the odd electron from the nitrogen atom in the radical form of DPPH accepts a hydrogen atom from the antioxidant, it undergoes reduction. It forms the corresponding hydrazine or non-radical form of DPPH [48,52]. Overall, the DPPH molecule is classified as a stable free radical due to the delocalization of the spare electron across the molecule, which prevents the molecule from dimerizing like most other free radicals. The presence of electron delocalization results in a deep violet color, with absorption in ethanol solution at around 515–517 nm. When the DPPH solution is mixed with an antioxidant compound that donates a hydrogen atom, such as curcumin, it loses its deep violet color (becomes colorless or pale yellow in color), as shown in Figure 6 [21,53,54].

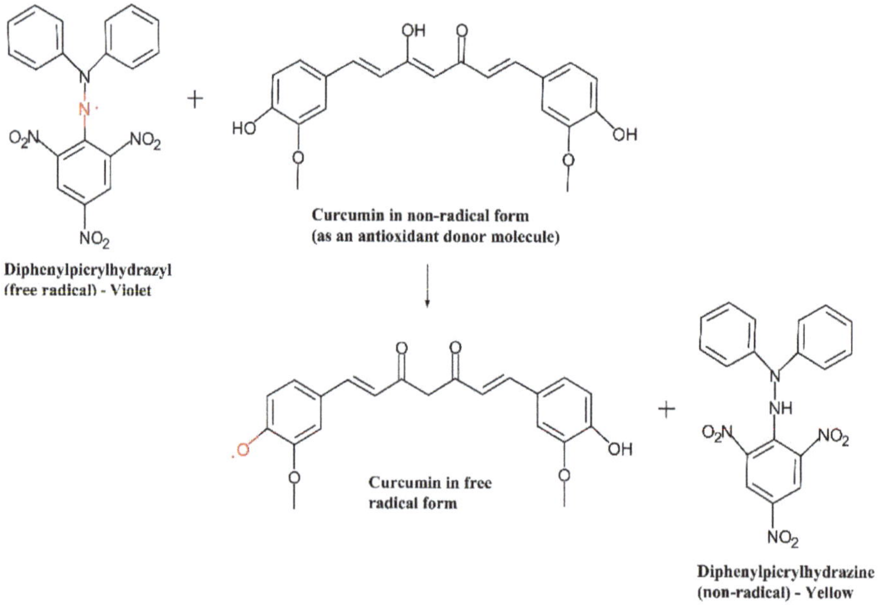

Figure 6. Antioxidant reaction mechanism of radical form DPPH with curcumin.

Equation (1) depicts the primary reaction in which the DPPH radical is $Z\bullet$, whose activity will be suppressed by AH as an antioxidant donor molecule, ZH is the reduced form of DPPH (non-radical), and $A\bullet$ is the antioxidant donor molecule's free radical form [21,55].

$$Z\bullet + AH = A\bullet + ZH \quad (1)$$

In terms of the number of electrons taken up, the decolorization in the DPPH molecule that reacts with antioxidants is stoichiometric. DPPH can respond with the entire sample, even if the antioxidants are weak. Therefore, the DPPH free radical scavenging assay method is widely used to assess a compound's ability to act as a free radical scavenger or hydrogen donor and its antioxidant activity (IR) [55,56].

In both SIF and SGF conditions, as shown in Figure 7, the IR increased as the C_{GA} increased. The IR in SIF (range 33.21–60.39%) was higher than in SGF (range 9.08–40.84%), indicating that curcumin's antioxidant activity was better in SIF than in SGF conditions. This is due to the nature of curcumin, which degrades quickly in alkaline but slowly in acidic conditions [31], resulting in a decrease in the amount of undegraded curcumin as an antioxidant compound that will react with DPPH. This decrease in the curcumin results in a high value of DPPH scavenging activity, which directly impacts the IR value in SIF rather than SGF. Furthermore, curcumin in SIF appears in the enolate form of the heptadienone chain (an electron donor), whereas curcumin in SGF appears in a protonated form (a hydrogen donor). Because only hydrogen donors can react with DPPH, SIF has a higher IR value than SGF [32,57]. The higher the IR value at a high C_{GA}, the better curcumin's antioxidant performance against free radicals. It is also aided by GA, which has antioxidant activity [17,19]. Therefore, the C_{GA} for obtaining the highest antioxidant activity of curcumin was 40% for both dispersions in SIF (60.39%) and SGF (40.84%) conditions.

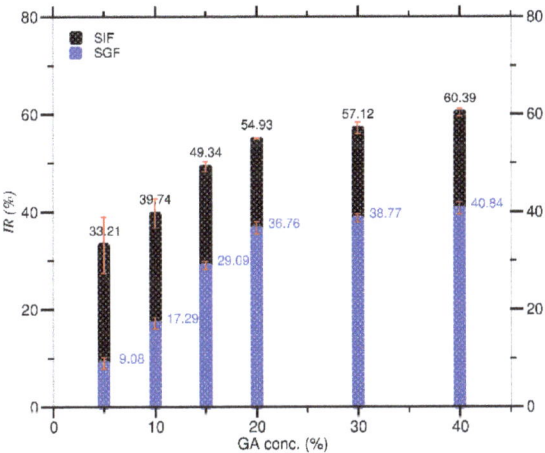

Figure 7. Antioxidant activity (IR) at various C_{GA} in SIF and SGF solutions.

2.2. Optimum Encapsulation Efficiency and Loading Capacity in SIF and SGF Solutions

The EE of the encapsulated material and the LC of the drug carrier are both parameters that are closely related to the ability of a DDS to encapsulate a drug for delivery to specific sites in the body. This study will compare the EE of curcumin, and the LC of GA dispersed in SIF and SGF to determine the optimum value between these two parameters.

EE increased while LC decreased as C_{GA} increased in SIF and SGF solutions as shown in Figure 8. Furthermore, Equation (2) demonstrates that the EE value was directly proportional to both the encapsulated curcumin concentration and the C_{GA}, whereas Equation (3) demonstrates that the LC value was directly proportional to the encapsulated curcumin concentration and inversely proportional to the C_{GA}. Therefore, EE is inversely proportional to *LC*, consistent with the results. The higher the C_{GA} used, the easier it was for curcumin to be encapsulated (higher value of *EE*), but it further reduced the space of GA to encapsulate curcumin again (lower value of *LC*).

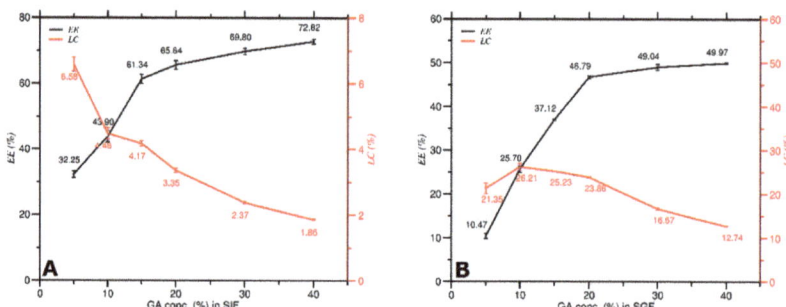

Figure 8. Encapsulation efficiency of curcumin vs. loading capacity of GA at various C_{GA} in, (**A**) SIF solution; (**B**) SGF solution.

The *EE* ranged from 32.25–72.82% for C_{GA} dispersed in SIF, and the *LC* ranged from 6.58–1.86%. The *EE* of curcumin increased significantly at 5–15% of C_{GA}, whereas the *LC* of GA decreased significantly at 5% of C_{GA} as shown in Figure 8A. The *EE* ranged from 10.47% to 49.97% for C_{GA} dispersed in SGF, and the *LC* ranged from 21.35 to 12.74%. The *EE* of curcumin increased significantly at 5–20% of C_{GA}, whereas the *LC* of GA decreased substantially at 20% of C_{GA} as shown in Figure 8B. The data recommend that 10% is the optimum C_{GA} to obtain the optimum *EE* of curcumin and *LC* of GA of the encapsulation process in SIF and SGF conditions.

2.3. Relationship of Encapsulation Efficiency and Antioxidant Activity of Curcumin in SIF and SGF Solutions

These two parameters relate to the amount of curcumin encapsulated in GA (*EE*) and the ability of curcumin to act as an antioxidant compound (*IR*) in different pH environments.

Both *EE* and *IR* increased as C_{GA} increased in SIF and SGF conditions as shown in Figure 9. According to Equation (2), the *EE* value was directly proportional to both the encapsulated curcumin concentration and the C_{GA}, whereas Equation (4) shows that the *IR* value was directly proportional only to the antioxidant activity of DPPH radical that reacts with GA-Curcumin to form non-radical DPPH and inversely proportional to the antioxidant activity of only DPPH radicals. Therefore, the results in which *EE* is directly proportional to *IR* align with the theoretical suggestion. The higher the C_{GA}, the more curcumin was successfully encapsulated in GA (higher value of *EE*) and the higher curcumin's ability as an antioxidant to ward off free radicals (higher value of *IR*).

The *EE* and *IR* progression values in both SIF (Figure 9A) and SGF (Figure 9B) showed that SGF gave a higher increment than SIF. This suggests that regarding *EE* and *IR*, SGF gave better environmental conditions for encapsulation of curcumin in GA. Furthermore, the results indicate that the highest increase in *EE* and *IR* in SIF and SGF occurred between 5% and 20% of C_{GA}.

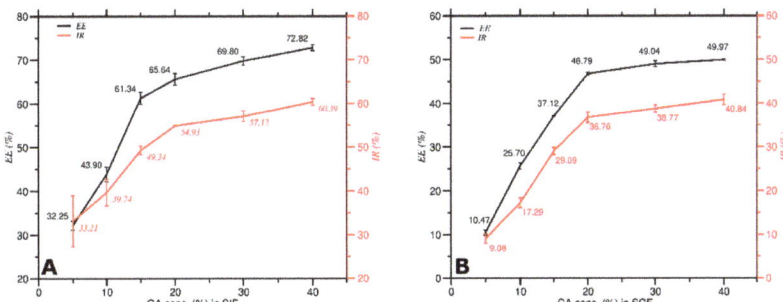

Figure 9. Encapsulation efficiency of curcumin vs. antioxidant activity at various C_{GA} in, (**A**) SIF solution; (**B**) SGF solution.

2.4. Relationship of Loading Capacity and Antioxidant Activity of Curcumin in SIF and SGF Solutions

These parameters relate to the ability of GA as a drug carrier for curcumin (*LC*) and the power of GA to stabilize further and enhance curcumin's ability as an antioxidant compound (*IR*) in two different pH surroundings.

As the C_{GA} increased in both SIF and SGF as shown in Figure 10, *LC* decreased while *IR* increased, implying that *LC* was inversely proportional to *IR*. The results suggest that a high amount of curcumin loading in GA was not adequate concerning the antioxidant activity of curcumin.

In SIF (Figure 10A), the *LC* ranged from 6.58 to 1.86%, and the *IR* ranged from 33.21 to 60.39%. The *LC* of GA decreased significantly at 5% of C_{GA}, whereas the *IR* increased significantly at 5–20% of C_{GA}. In SGF (Figure 10B), the *LC* ranged from 21.35 to 12.74%, and the *IR* ranged from 9.08 to 40.84%. The *LC* of GA decreased significantly at 20% of C_{GA}, while the *IR* increased significantly at 5–20% of C_{GA}. These outcomes propose that the optimum C_{GA} for obtaining the optimum *LC* of GA and *IR* of curcumin is around 10% to 20% in SIF and SGF conditions. The SGF provided a better atmosphere than the SIF solution regarding the antioxidant activity.

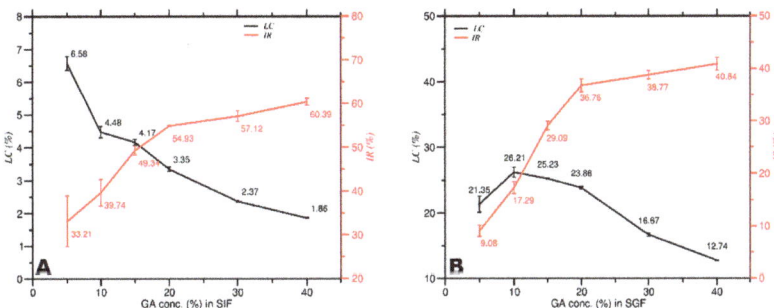

Figure 10. Loading capacity of GA vs. antioxidant activity at various C_{GA} in, (**A**) SIF solution; (**B**) SGF solution.

This study provides several parameters that feature the encapsulation of curcumin in GA, namely *EE*, *LC*, and release of curcumin encapsulation in GA. To understand more about the curcumin delivery system using GA, further study is necessary to analyze other physicochemical characteristics of the dispersion, including the particles' size, shape, and surface charge. These parameters are crucial for a successful delivery system.

3. Materials and Methods

3.1. Materials

The materials used were gum Arabic, curcumin, $Na_2HPO_4 \cdot 2H_2O$ (0.05 M), $NaH_2PO_4 \cdot 2H_2O$ (0.05 M), HCl (37%), NaCl, NaOH, ethanol, DPPH solution (40 µg/mL), and demineralized water.

3.2. Methods

3.2.1. Preparation of Simulated Intestinal Fluid (SIF)

A solution of 0.05 M was prepared from 7.5 g of $Na_2HPO_4 \cdot 2H_2O$ in 500 mL of demineralized water. Another solution of 0.05 M was also prepared from 3.9 g of $NaH_2PO_4 \cdot 2H_2O$ in 500 mL of demineralized water. A mixture of 9.5 mL of $NaH_2PO_4 \cdot 2H_2O$ (0.05 M) and 40.5 mL of $Na_2HPO_4 \cdot 2H_2O$ (0.05 M) was prepared and diluted into 100 mL. The pH was adjusted to 7.4.

3.2.2. Preparation of Simulated Gastric Fluid (SGF)

In 800 mL of demineralized water, about 2 g of NaCl was dissolved. Drop by drop, a total of 4.5 mL of 37% HCl solution was added into the NaCl solution, followed by demineralized water until the volume reached 1 L. The pH of the solution was tuned to 1.2.

3.2.3. Curcumin Encapsulation in Gum Arabic (GA)

A series of GA dispersions with concentrations (C_{GA}) of 5%, 10%, 15%, 20%, 30%, and 40% (w/v) were prepared in 100 mL of chloroform/methanol (9/1, v/v). Curcumin, at a concentration of up to 0.1 mg/mL per GA dispersion, was first dissolved in a small amount (a few drops) of ethanol before being added to each GA dispersion, then stirred for 10 min. The ethanol facilitates the mixing of the curcumin with the GA dispersion. In a test tube, 10 mL of GA-curcumin dispersion was streamed with nitrogen gas until a thin layer remained at the bottom. After that, 10 mL of the SIF solution was added to the test tube containing a thin layer, and the freeze-thawing process was continued in the test tube. The freeze-thawing cycle adapted from Hudiyanti's research [32,58–60] was carried out by cooling it at 4 °C and heating it at 45 °C repeatedly until the thin layer was completely dissolved. Then it was sonicated for 30 min at 27 °C. This procedure for repeated for each GA concentration in both SIF and SGF solutions. Next, 1 mL of sonicated GA-curcumin dispersion was dissolved in 5 mL of ethanol (w/v), then centrifuged at 3461× *g* for 40 min until two layers were formed in the test tube. The watery top layer (supernatant) containing unencapsulated curcumin was separated for analysis on *LC* and *EE*. The thick bottom layer (GA residue) was stored at −18 °C until it was reused for analysis of *IR* and *RR*. The sink condition was maintained at 30 °C for all compositions.

3.2.4. Curcumin Encapsulation Efficiency (*EE*) and GA Loading Capacity (*LC*)

The *EE* of curcumin and *LC* of GA were evaluated based on the concentration of unencapsulated curcumin in the supernatant. The unencapsulated curcumin in the supernatant, which had previously been dissolved in ethanol before the centrifugation process, was then analyzed using a UV-Vis spectrophotometer. The concentration of unencapsulated curcumin (C_t) was analyzed at a wavelength of 426 nm. The *EE* of curcumin was calculated with Equation (2), while the *LC* of GA was calculated with Equation (3) [16].

$$EE = \left[1 - \left(\frac{C_t}{C_0}\right)\right] \times 100\% \qquad (2)$$

$$LC = \left(\frac{C_0 - C_t}{C_{GA}}\right) \times 100\% \qquad (3)$$

3.2.5. Analysis on Curcumin Release Rate (*RR*)

The *RR* of curcumin was determined using the concentration of curcumin released from GA during storage. The GA residue obtained from the previous encapsulation procedure was dispersed in the buffer solution, i.e., SIF and SGF solutions (1/5, w/v), and stored in an incubator at 4 °C. Curcumin release from GA into the buffer solution was monitored for 12 days. Each dispersion was homogenized using an ultrasonic homogenizer for 5 min, followed by centrifugation at 4500 rpm for 15 min. The dispersion was then allowed to settle and form two layers. The absorbance of the supernatant separated from the GA residue was measured at a wavelength of 426 nm. This procedure was repeated for each GA dispersion.

3.2.6. Analysis of DPPH Free Radical Scavenging Assay (Antioxidant Activity, *IR*)

The Blois method [55] was used to perform the 1-Diphenyl-2-picrylhydrazyl (DPPH) free radical scavenging assay, in which 1 mL of each GA-Curcumin dispersion was mixed with 3 mL of DPPH (40 µg/mL) solution. The mixture was then incubated for 30 min at room temperature without exposure to light. The absorbance of the mixture was measured with a UV-Vis spectrophotometer at a maximum wavelength (λ_{max}) of 515 nm. The DPPH antioxidant activity (*IR*) was calculated using Equation (4) as follows:

$$IR = \left(\frac{A_0 - A_1}{A_0}\right) \times 100\% \tag{4}$$

where A_0 is the absorbance of the DPPH solution without the addition of GA-Curcumin dispersion and A_1 is the absorbance of the DPPH solution with the addition of GA-Curcumin dispersion after 30 min of incubation [32].

Statistical Analysis

All data presented in this article were acquired in triplicate. The data were presented as mean ± standard deviation (SD).

4. Conclusions

We have successfully encapsulated curcumin in gum Arabic dispersions in the SIF and SGF solutions. The results give a satisfactory outcome regarding the potency of gum Arabic for the encapsulation of curcumin in both environments. The higher the gum Arabic concentration, the higher the encapsulation efficiency and antioxidant activity of curcumin, but the lower the gum Arabic loading capacity. The data propose that 10% is the best possible gum Arabic concentration to achieve the optimal encapsulation efficiency of curcumin and the loading capacity of gum Arabic for curcumin. Regarding the antioxidant activity of curcumin, the results indicate that an excessive concentration of gum Arabic was effective, and the SIF delivered a superior milieu than the SGF solution. Moreover, the gum Arabic would release curcumin more quickly in the SGF setting.

Author Contributions: Conceptualization, D.H.; funding acquisition, D.H.; and P.S.; methodology, D.H., K.A. and M.F.A.K.; investigation, M.F.A.K. and K.A.; supervision, D.H., K.A. and P.S.; writing—original draft, D.H., M.F.A.K., K.A. and S.M.C.; writing—review and editing, D.H. and S.M.C.; resources, P.S.; project administration, S.M.C. All authors have read and agreed to the published version of the manuscript.

Funding: This research was supported by The Minister of Research and Technology and the Higher Education Republic Indonesia through PDUPT Research Scheme 2019. Grant No. 257-49/UN7.P4.3/PP/2019.

Institutional Review Board Statement: Not applicable.

Informed Consent Statement: Not applicable.

Data Availability Statement: All data generated or analysed during this study are included in this published article.

Acknowledgments: D.H. would like to express her gratitude to T. Widiarih for her advice in the statistical interpretation of the data.

Conflicts of Interest: The authors declare no conflict of interest.

Sample Availability: Samples of the compounds are available from the authors.

References

1. Kaushik, K.; Sharma, R.B.; Agarwal, S. Natural polymers and their applications. *Int. J. Pharm. Sci. Rev. Res.* **2016**, *37*, 30–36.
2. Deb, J.; Das, M.; Das, A. Excellency of natural polymer in drug delivery system: A Review. *IJPBA* **2017**, *5*, 17–22.
3. Gandhi, K.J.; Deshmane, S.V.; Biyani, K.R. Polymers in pharmaceutical drug delivery system: A review. *Int. J. Pharm. Sci. Rev. Res.* **2012**, *14*, 57–66.
4. Rajeswari, S.; Prasanthi, T.; Sudha, N.; Swain, R.P.; Panda, S.; Goka, V. Natural polymers: A recent review. *World J. Pharm. Pharm. Sci.* **2017**, *6*, 472–494. [CrossRef]
5. Thangavelu, K.; Subramani, K.B. Sustainable biopolymer fibers—Production, properties and applications. In *Sustainable Fibres for Fashion Industry*; Springer: Singapore, 2016; pp. 109–140. [CrossRef]
6. Karthik, T.; Rathinamoorthy, R. Sustainable biopolymers in textiles: An overview. In *Handb. Ecomater*; Springer: Cham, Switzerland, 2017; pp. 1–27. [CrossRef]
7. Tong, X.; Pan, W.; Su, T.; Zhang, M.; Dong, W.; Qi, X. Recent advances in natural polymer-based drug delivery systems. *React. Funct. Polym.* **2020**, *148*, 104501. [CrossRef]
8. Jacob, J.; Haponiuk, J.T.; Thomas, S.; Gopi, S. Biopolymer based nanomaterials in drug delivery systems: A review. *Mater. Today Chem.* **2018**, *9*, 43–55. [CrossRef]
9. Tibbitt, M.W.; Dahlman, J.E.; Langer, R. Emerging frontiers in drug delivery. *J. Am. Chem. Soc.* **2016**, *138*, 704–717. [CrossRef]
10. Ezera, E.J.; Nwufo, B.T.; Wapwera, J.A. Effects of graded quantities of xanthan gum on the physicochemical and flocculation properties of gum Arabic. *Int. J. Chem. Sci.* **2019**, *2*, 44–49.
11. Jahandideh, A.; Ashkani, M.; Moini, N. Biopolymers in textile industries. In *Biopolymers and Their Industrial Applications*; Elsevier: Amsterdam, The Netherlands, 2021; pp. 193–218. [CrossRef]
12. Patel, S.; Goyal, A. Applications of natural polymer gum arabic: A review. *Int. J. Food Prop.* **2015**, *18*, 986–998. [CrossRef]
13. Shirwaikar, A.; Shirwaikar, A.; Prabu, S.L.; Kumar, G.A. Herbal excipients in novel drug delivery systems. *Indian J. Pharm. Sci.* **2008**, *70*, 415. [CrossRef]
14. Dragostin, I.; Dragostin, O.; Pelin, A.-M.; Grigore, C.; Lăcrămioara Zamfir, C. The importance of polymers for encapsulation process and for enhanced cellular functions. *J. Macromol. Sci. Part A* **2017**, *54*, 489–493. [CrossRef]
15. Adsare, S.R.; Annapure, U.S. Microencapsulation of curcumin using coconut milk whey and Gum Arabic. *J. Food Eng.* **2021**, *298*, 110502. [CrossRef]
16. Al Khafiz, M.F.; Hikmahwati, Y.; Anam, K.; Hudiyanti, D. Key conditions of alpha-tocopherol encapsulation in gum Arabic dispersions. *ScopeIndex* **2019**, *10*, 2622–2627. [CrossRef]
17. Mirghani, M.E.S.; Elnour, A.A.M.; Kabbashi, N.A.; Alam, M.Z.; Musa, K.H.; Abdullah, A. Determination of antioxidant activity of gum arabic: An exudation from two different locations. *Sci. Asia* **2018**, *44*, 177–186. [CrossRef]
18. Mosquera, L.H.; Moraga, G.; Martínez-Navarrete, N. Critical water activity and critical water content of freeze-dried strawberry powder as affected by maltodextrin and arabic gum. *Food Res. Int.* **2012**, *47*, 201–206. [CrossRef]
19. Suhag, Y.; Nayik, G.A.; Nanda, V. Effect of gum arabic concentration and inlet temperature during spray drying on physical and antioxidant properties of honey powder. *J. Food Meas. Charact.* **2016**, *10*, 350–356. [CrossRef]
20. Cui, K.; Luo, X.; Xu, K.; Ven Murthy, M.R. Role of oxidative stress in neurodegeneration: Recent developments in assay methods for oxidative stress and nutraceutical antioxidants. *Prog. Neuro-Psychopharmacol. Biol. Psychiatry* **2004**, *28*, 771–799. [CrossRef]
21. Kedare, S.B.; Singh, R.P. Genesis and development of DPPH method of antioxidant assay. *J. Food Sci. Technol.* **2011**, *48*, 412–422. [CrossRef]
22. Carvalho, D.d.M.; Takeuchi, K.P.; Geraldine, R.M.; Moura, C.J.d.; Torres, M.C.L. Production, solubility and antioxidant activity of curcumin nanosuspension. *Food Sci. Technol.* **2015**, *35*, 115–119. [CrossRef]
23. Ghasemzadeh, A.; Jaafar, H.Z.E.; Juraimi, A.S.; Tayebi-Meigooni, A. Comparative evaluation of different extraction techniques and solvents for the assay of phytochemicals and antioxidant activity of hashemi rice bran. *Molecules* **2015**, *20*, 10822–10838. [CrossRef]
24. Chen, S.; Wu, J.; Tang, Q.; Xu, C.; Huang, Y.; Huang, D.; Luo, F.; Wu, Y.; Yan, F.; Weng, Z.; et al. Nano-micelles based on hydroxyethyl starch-curcumin conjugates for improved stability, antioxidant and anticancer activity of curcumin. *Carbohydr. Polym.* **2020**, *228*, 115398. [CrossRef] [PubMed]
25. O'Toole, M.G.; Soucy, P.A.; Chauhan, R.; Raju, M.V.R.; Patel, D.N.; Nunn, B.M.; Keynton, M.A.; Ehringer, W.D.; Nantz, M.H.; Keynton, R.S. Release-modulated antioxidant activity of a composite curcumin-chitosan polymer. *Biomacromolecules* **2016**, *17*, 1253–1260. [CrossRef]

26. Hermund, D.B. 10—Antioxidant Properties of Seaweed-Derived Substances. In *Bioactive Seaweeds for Food Applications*; Qin, Y., Ed.; Academic Press: Cambridge, MA, USA, 2018; pp. 201–221. [CrossRef]
27. Peter, K.V.; Shylaja, M.R. 1—Introduction to herbs and spices: Definitions, trade and applications. In *Handbook of Herbs and Spices*, 2nd ed.; Peter, K.V., Ed.; Woodhead Publishing: Sawston, UK, 2012; pp. 1–24. [CrossRef]
28. Suresh, K.; Nangia, A. Curcumin: Pharmaceutical solids as a platform to improve solubility and bioavailability. *CrystEngComm* **2018**, *20*, 3277–3296. [CrossRef]
29. Anand, P.; Kunnumakkara, A.B.; Newman, R.A.; Aggarwal, B.B. Bioavailability of curcumin: Problems and promises. *Mol. Pharm.* **2007**, *4*, 807–818. [CrossRef] [PubMed]
30. Siviero, A.; Gallo, E.; Maggini, V.; Gori, L.; Mugelli, A.; Firenzuoli, F.; Vannacci, A. Curcumin, a golden spice with a low bioavailability. *J. Hermed. Med.* **2015**, *5*, 57–70. [CrossRef]
31. Palanikumar, L.; Panneerselvam, N. Curcumin: A putative chemopreventive agent. *J. Life Sci.* **2009**, *3*, 47–53.
32. Hudiyanti, D.; Al Khafiz, M.F.; Anam, K.; Siahaan, P.; Suyati, L. Assessing encapsulation of curcumin in cocoliposome: In vitro study. *Open Chem.* **2021**, *19*, 358–366. [CrossRef]
33. Lamichhane, N.; Udayakumar, T.; D'Souza, W.; Simone, I. Liposomes: Clinical applications and potential for image-guided drug delivery. *Molecules* **2018**, *23*, 288. [CrossRef]
34. Lee, M.-K. Liposomes for enhanced bioavailability of water-insoluble drugs: In vivo evidence and recent approaches. *Pharmaceutics* **2020**, *12*, 264. [CrossRef]
35. Li, J.; Wang, X.; Zhang, T.; Wang, C.; Huang, Z.; Luo, X.; Deng, Y. A review on phospholipids and their main applications in drug delivery systems. *Asian J. Pharm. Sci.* **2015**, *10*, 81–98. [CrossRef]
36. Mohanty, C.; Das, M.; Sahoo, S.K. Emerging role of nanocarriers to increase the solubility and bioavailability of curcumin. *Expert Opin. Drug Deliv.* **2012**, *9*, 1347–1364. [CrossRef] [PubMed]
37. Sanidad, K.Z.; Sukamtoh, E.; Xiao, H.; McClements, D.J.; Zhang, G. Curcumin: Recent advances in the development of strategies to improve oral bioavailability. *Annu. Rev. Food Sci. Technol.* **2019**, *10*, 597–617. [CrossRef] [PubMed]
38. Farooqui, T.; Farooqui, A.A. Curcumin: Historical background, chemistry, pharmacological action, and potential therapeutic value. *Curcumin Neurol. Psychiatr. Disord.* **2019**, 23–44. [CrossRef]
39. Mondal, S.; Ghosh, S.; Moulik, S.P. Stability of curcumin in different solvent and solution media: UV–visible and steady-state fluorescence spectral study. *J. Photochem. Photobiol. B Biol.* **2016**, *158*, 212–218. [CrossRef] [PubMed]
40. Liu, R. *Water-Insoluble Drug Formulation*, 1st ed.; CRC Press: Boca Raton, FL, USA, 2000. [CrossRef]
41. Piacentini, E. Encapsulation Efficiency. In *Encyclopedia of Membranes*; Drioli, E., Giorno, L., Eds.; Springer: Berlin/Heidelberg, Germany, 2016; pp. 706–707. [CrossRef]
42. Kumari, A.; Singla, R.; Guliani, A.; Yadav, S.K. Nanoencapsulation for drug delivery. *EXCLI J.* **2014**, *13*, 265.
43. Shen, S.; Wu, Y.; Liu, Y.; Wu, D. High drug-loading nanomedicines: Progress, current status, and prospects. *Int. J. Nanomed.* **2017**, *12*, 4085–4109. [CrossRef]
44. Wilkosz, N.; Łazarski, G.; Kovacik, L.; Gargas, P.; Nowakowska, M.; Jamróz, D.; Kepczynski, M. Molecular insight into drug-loading capacity of PEG–PLGA nanoparticles for itraconazole. *J. Phys. Chem. B* **2018**, *122*, 7080–7090. [CrossRef]
45. Li, Y.; Yang, L. Driving forces for drug loading in drug carriers. *J. Microencapsul.* **2015**, *32*, 255–272. [CrossRef]
46. Gébleux, R.; Wulhfard, S.; Casi, G.; Neri, D. Antibody format and drug release rate determine the therapeutic activity of noninternalizing antibody—Drug conjugates. *Mol. Cancer Ther.* **2015**, *14*, 2606–2612. [CrossRef]
47. Bhowmik, D.; Gopinath, H.; Kumar, B.P.; Duraivel, S.; Kumar, K.P.S. Controlled release drug delivery systems. *Pharma Innov.* **2012**, *1*, 24–32.
48. Contreras-Guzmán, E.S.; Strong Iii, F.C. Determination of tocopherols (Vitamin E) by reduction of cupric ion. *J. Assoc. Off. Anal. Chem.* **1982**, *65*, 1215–1221. [CrossRef]
49. Hossain, S.; Rahaman, A.; Nahar, T.; Basunia, M.A.; Mowsumi, F.R.; Uddin, B.; Shahriar, M.; Mahmud, I. *Syzygium cumini* (L.) skeels seed extract ameliorates in vitro and in vivo oxidative potentials of the brain cerebral cortex of alcohol-treated rats. *Orient. Pharm. Exp. Med.* **2012**, *12*, 59–66. [CrossRef]
50. Peng, H.; Li, W.; Li, H.; Deng, Z.; Zhang, B. Extractable and non-extractable bound phenolic compositions and their antioxidant properties in seed coat and cotyledon of black soybean (*Glycinemax* (L.) merr). *J. Funct. Foods* **2017**, *32*, 296–312. [CrossRef]
51. Sukati, S.; Khobjai, W. Total Phenolic Content and DPPH Free Radical Scavenging Activity of Young Turmeric Grown in Southern Thailand. *Appl. Mech. Mater.* **2019**, *886*, 61–69. [CrossRef]
52. Correa, D.P.N.; Osorio, J.R.; Zúñiga, O.; Andica, R.A.S. Determinación del valor nutricional de la harina de cúrcuma y la actividad antioxidante de extractos del rizoma de *Curcuma longa* procedente de cultivos agroecológicos y convencionales del Valle del Cauca-Colombia/Determination of nutritional value of turmeric flour and the antioxidant activity of *Curcuma longa* rhizome extracts from agroecological and conventional crops of Valle del Cauca-Colombia. *Rev. Colomb. Química* **2020**, *49*, 26. [CrossRef]
53. Ak, T.; Gülçin, İ. Antioxidant and radical scavenging properties of curcumin. *Chem. Biol. Interact.* **2008**, *174*, 27–37. [CrossRef]
54. Warsi, W.; Sardjiman, S.; Riyanto, S. Synthesis and Antioxidant Activity of Curcumin Analogues. *J. Chem. Pharm. Res.* **2018**, *10*, 1–9.
55. Blois, M.S. Antioxidant determinations by the use of a stable free radical. *Nature* **1958**, *181*, 1199–1200. [CrossRef]
56. Prior, R.L.; Wu, X.; Schaich, K. Standardized Methods for the Determination of Antioxidant Capacity and Phenolics in Foods and Dietary Supplements. *J. Agric. Food Chem.* **2005**, *53*, 4290–4302. [CrossRef]

57. Lee, W.-H.; Loo, C.-Y.; Bebawy, M.; Luk, F.; Mason, R.S.; Rohanizadeh, R. Curcumin and its derivatives: Their application in neuropharmacology and neuroscience in the 21st century. *Curr. Neuropharmacol.* **2013**, *11*, 338–378. [CrossRef]
58. Hudiyanti, D.; Raharjo, T.J.; Narsito, N.; Noegrohati, S. Investigation on the morphology and properties of aggregate structures of natural phospholipids in aqueous system using cryo-tem. *Indones. J. Chem.* **2012**, *12*, 57–61. [CrossRef]
59. Hudiyanti, D.; Aminah, S.; Hikmahwati, Y.; Siahaan, P. Cholesterol implications on coconut liposomes encapsulation of beta-carotene and vitamin C. *IOP Conf. Ser. Mater. Sci. Eng.* **2019**, *509*, 012037. [CrossRef]
60. Hudiyanti, D.; Al-Khafiz, M.F.; Anam, K. Encapsulation of cinnamic acid and galangal extracts in coconut (*Cocos nucifera* L.) liposomes. *J. Phys. Conf. Ser.* **2020**, *1442*, 012056. [CrossRef]

Article

An Efficient Method of Birch Ethanol Lignin Sulfation with a Sulfaic Acid-Urea Mixture

Alexander V. Levdansky [1], Natalya Yu. Vasilyeva [1,2], Yuriy N. Malyar [1,2,*], Alexander A. Kondrasenko [1], Olga Yu. Fetisova [1], Aleksandr S. Kazachenko [1,2], Vladimir A. Levdansky [1] and Boris N. Kuznetsov [1,2,*]

[1] Institute of Chemistry and Chemical Technology, Krasnoyarsk Science Center, Siberian Branch, Russian Academy of Sciences, Akademgorodok 50/24, 660036 Krasnoyarsk, Russia
[2] School of Non-Ferrous Metals and Material Science, Siberian Federal University, Pr. Svobodny 79, 660041 Krasnoyarsk, Russia
* Correspondence: yumalyar@gmail.com (Y.N.M.); bnk@icct.ru (B.N.K.); Tel.: +7-908-2065-517 (Y.N.M.)

Abstract: For the first time, the process of birch ethanol lignin sulfation with a sulfamic acid-urea mixture in a 1,4-dioxane medium was optimized experimentally and numerically. The high yield of the sulfated ethanol lignin (more than 96%) and containing 7.1 and 7.9 wt % of sulfur was produced at process temperatures of 80 and 90 °C for 3 h. The sample with the highest sulfur content (8.1 wt %) was obtained at a temperature of 100 °C for 2 h. The structure and molecular weight distribution of the sulfated birch ethanol lignin was established by FTIR, 2D ^{1}H and ^{13}C NMR spectroscopy, and gel permeation chromatography. The introduction of sulfate groups into the lignin structure was confirmed by FTIR by the appearance of absorption bands characteristic of the vibrations of sulfate group bonds. According to 2D NMR spectroscopy data, both the alcohol and phenolic hydroxyl groups of the ethanol lignin were subjected to sulfation. The sulfated birch ethanol lignin with a weight average molecular weight of 7.6 kDa and a polydispersity index of 1.81 was obtained under the optimum process conditions. Differences in the structure of the phenylpropane units of birch ethanol lignin (syringyl-type predominates) and abies ethanol lignin (guaiacyl-type predominates) was manifested in the fact that the sulfation of the former proceeds more completely at moderate temperatures than the latter. In contrast to sulfated abies ethanol lignin, the sulfated birch ethanol lignin had a bimodal and wider molecular weight distribution, as well as less thermal stability. The introduction of sulfate groups into ethanol lignin reduced its thermal stability.

Keywords: birch ethanol lignin; sulfamic acid; urea; sulfation process optimization; sulfated product characterization; FTIR spectroscopy; 2D NMR spectroscopy; gel permeation chromatography; thermal analysis

Citation: Levdansky, A.V.; Vasilyeva, N.Y.; Malyar, Y.N.; Kondrasenko, A.A.; Fetisova, O.Y.; Kazachenko, A.S.; Levdansky, V.A.; Kuznetsov, B.N. An Efficient Method of Birch Ethanol Lignin Sulfation with a Sulfaic Acid-Urea Mixture. *Molecules* 2022, 27, 6356. https://doi.org/10.3390/molecules27196356

Academic Editor: Sylvain Caillol

Received: 30 August 2022
Accepted: 19 September 2022
Published: 26 September 2022

Publisher's Note: MDPI stays neutral with regard to jurisdictional claims in published maps and institutional affiliations.

Copyright: © 2022 by the authors. Licensee MDPI, Basel, Switzerland. This article is an open access article distributed under the terms and conditions of the Creative Commons Attribution (CC BY) license (https://creativecommons.org/licenses/by/4.0/).

1. Introduction

Lignin is the most widespread aromatic polymer on the earth. The lignin content in the dry mass of woody plants ranges within 15–40%, depending on the species [1,2]: 5–12% in herbaceous plants, 25–35% in coniferous wood, and 15–30% in deciduous wood. Lignin has an irregular 3D structure built from phenylpropane units (PPUs) with different numbers of methoxyl groups in the aromatic ring: p-hydroxyphenyl (H), guaiacyl (G), and syringyl (S). The PPUs are randomly cross-linked with simple ether C–O–C and C–C bonds [3]. In addition, the propane chains of the lignin PPUs can contain different functional groups: hydroxyl (–OH), carbonyl (C=O), carboxyl (–COOH), and double bonds (–CH=CH–).

The structure, chemical composition, and physicochemical properties of lignins vary within fairly wide limits, depending on the lignocellulosic raw material and lignin isolation method used [4]. The complex heterogeneous composition of lignins complicates the development of efficient techniques for utilization to obtain valuable products [5].

Recently, there has been an increased interest in organosolv methods for extracting cellulose from lignocellulosic biomass. The use of organic solvents makes the organosolv processes environmentally friendly and eliminates the contamination of lignin with sulfur [6]. Organosolv lignins have a higher content of hydroxyl and carbonyl groups [7] than conventional technical lignins and are soluble in organic solvents, which facilitates the chemical modification and processing of these lignins [8–10].

A promising direction in the valorization of lignin is its chemical modification to obtain bioactive derivatives [11–13]. In particular, sulfated lignins are known for their anticoagulant and antiplatelet activity and can be used in the treatment of thrombotic disorders [11].

The available methods for obtaining the sulfated lignin derivatives are based on the use of aggressive and environmentally hazardous sulfating reagents, e.g., sulfuric anhydride and its complexes with toxic amines [13,14]. The method for the enzymatic sulfation of organosolv and technical lignins was proposed in [15] and is based on the use of p-nitrophenyl sulfate (p-NPS) as a sulfate donor and aryl sulfotransferase (AST) as a catalyst. This method showed high selectivity for the phenolic hydroxyl groups, leaving the aliphatic hydroxyl groups in the lignin side chain intact. The main drawback of this method is the long sulfation time (96 h). We developed a new, simpler, environmentally safe method for producing water-soluble abies ethanol lignin sulfates, which uses low-toxic sulfamic acid mixed with urea as a sulfating agent [16]. The comparative 2D NMR spectroscopy analysis of the structure of the initial and sulfated abies ethanol lignins was used to establish the main structural units and moieties of lignin macromolecules.

It is known well that, in contrast to lignins of abies and other coniferous species, which consist mainly of the guaiacyl structural units, in the hardwood (e.g., birch) lignin structure, the syringyl units dominate [17,18]. Coniferous and hardwood lignins also contain different amounts of condensed structural units; hardwood lignins are less condensed [4]. Birch is widespread in Russia and other countries of the Northern Hemisphere, but its wood finds only limited application in the pulp, paper, and building industries. However, the high content of xylan in birch wood allows it to carry out the complex processing of its biomass with the production of xylose, levulinic acid, and ethanol lignin [19]. Birch ethanol lignin contains no sulfur and has a relatively low molecular weight and a fairly narrow molecular weight distribution, which makes it a convenient substrate for the synthesis of bioactive lignin sulfates.

The aim of this study was to experimentally and numerically optimize the process of the sulfation of ethanol lignin birch wood with a mixture of sulfamic acid and urea in a 1,4-dioxane medium and to characterize the structure and thermochemical properties of the sulfated ethanol lignin.

2. Results

2.1. Kinetic Study of the Process of Birch Ethanol Lignin Sulfation

The scheme of birch ethanol lignin sulfation with the low-toxic sulfamic acid–urea mixture in the 1,4-dioxane medium is shown in Figure 1. The isolation of sulfated lignin was isolated in the form of an ammonium salt.

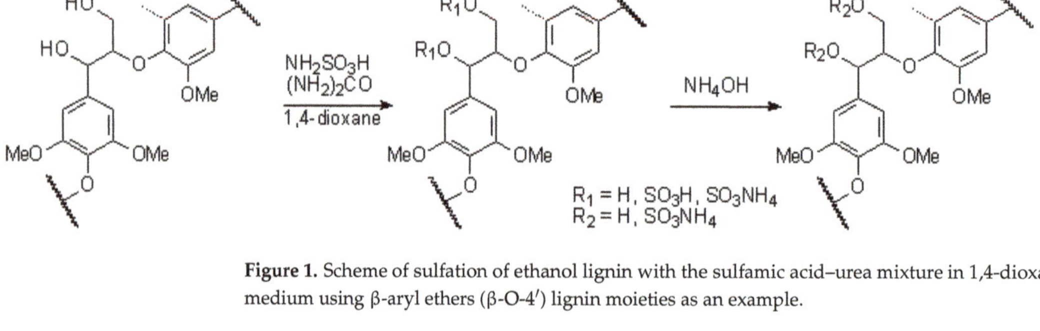

Figure 1. Scheme of sulfation of ethanol lignin with the sulfamic acid–urea mixture in 1,4-dioxane medium using β-aryl ethers (β-O-4′) lignin moieties as an example.

The use of the low-toxic and non-corrosive mixture of sulfamic acid and urea for the sulfation of abies ethanol lignin was previously proposed [16]. The yield of sulfated ethanol lignin of abies wood and sulfur content can be regulated by varying the sulfation process temperature, time, and the ratio of lignin to sulfating complex (sulfamic acid and urea mixture).

A high yield of sulfated lignin and sulfur content can be obtained at different combinations of the above-mentioned parameters of the sulfation process. Taking into account the data obtained when optimizing the process of abies ethanol lignin sulfation with a mixture of sulfamic acid and urea, the ratio of birch ethanol lignin to the sulfating complex was chosen in this study to be 1:3 mol/mol.

The sulfation process temperature ranged from 70 to 110 °C and the process time from 0.5 to 3.0 h. The data on the effect of the birch ethanol lignin sulfation conditions on the yield of sulfated lignin and sulfur content are given in Table 1.

Table 1. Effect of the conditions for ethanol lignin sulfation with the sulfamic acid–urea mixture in 1,4-dioxane on the yield of water-soluble sulfated lignin and sulfur content.

No.	L:SC, mol/mol	Temperature, °C	Time, min	Sulfur Content, wt %	Yield, wt %
1	1:3	70	30	2.31 ± 0.02	*
2	1:3	70	45	3.79 ± 0.02	*
3	1:3	70	60	5.22 ± 0.03	94.35 ± 4.04
4	1:3	70	90	5.79 ± 0.03	94.95 ± 3.96
5	1:3	70	120	6.03 ± 0.03	95.56 ± 3.93
6	1:3	70	180	6.31 ± 0.04	95.14 ± 3.88
7	1:3	80	30	2.62 ± 0.02	*
8	1:3	80	45	4.03 ± 0.03	*
9	1:3	80	60	6.08 ± 0.04	93.90 ± 3.91
10	1:3	80	90	6.73 ± 0.03	94.99 ± 3.83
11	1:3	80	120	6.92 ± 0.05	95.53 ± 3.80
12	1:3	80	180	7.09 ± 0.03	96.06 ± 3.77
13	1:3	90	30	3.42 ± 0.02	*
14	1:3	90	45	5.84 ± 0.04	95.61 ± 3.96
15	1:3	90	60	6.93 ± 0.05	96.17 ± 3.80
16	1:3	90	90	7.52 ± 0.05	96.43 ± 3.71
17	1:3	90	120	7.59 ± 0.05	94.82 ± 3.69
18	1:3	90	180	7.92 ± 0.05	94.31 ± 3.65
19	1:3	100	30	4.18 ± 0.03	*
20	1:3	100	45	6.43 ± 0.04	93.50 ± 3.87
21	1:3	100	60	7.44 ± 0.05	93.08 ± 3.72
22	1:3	100	90	7.93 ± 0.05	92.48 ± 3.65
23	1:3	100	120	8.15 ± 0.05	91.74 ± 3.62
24	1:3	100	180	8.14 ± 0.05	91.14 ± 4.07
25	1:3	110	30	5.02 ± 0.03	92.31 ± 3.93
26	1:3	110	45	6.91 ± 0.04	91.74 ± 3.80
27	1:3	110	60	7.90 ± 0.05	90.66 ± 3.65
28	1:3	110	90	8.11 ± 0.05	89.93 ± 3.62
29	1:3	110	120	8.10 ± 0.05	88.73 ± 3.62
30	1:3	110	180	8.13 ± 0.05	85.71 ± 3.62

*—Sulfated lignin with a sulfur content of ≤ 4.20 wt % is water-insoluble.

The high yield of sulfated birch ethanol lignin and the high sulfur content can be obtained at different combinations of the specified parameters of the sulfation process (temperature and time).

The sulfated ethanol lignin samples with a high yield (91.1–96.4%) and a sulfur content of 7.1–8.1 wt % were obtained at a process temperature of 90 °C and a time of 3 h or at a temperature of 100 °C and a time of 1.5–2.0 h. A further increase in the sulfation time and temperature did not significantly affect the sulfur content in the product but reduced the sulfated lignin yield. This may be due to the intensification of secondary condensation and destruction reactions under more severe conditions, which leads to the formation of products that are removed at the stage of sulfated lignin dialysis. It should be noted that the products of sulfation of birch ethanol lignin contain somewhat more sulfur than the products of sulfation of abies ethanol lignin under similar conditions [16]. This is possibly related to the lower reactivity of coniferous lignins that contain more condensed structural units than hardwood lignins [4].

The kinetics of the birch ethanol lignin sulfation with sulfamic acid–urea mixture in 1,4-dioxane medium was investigated in the temperature range of 70–100 °C (Figure 2). The apparent rates of the birch ethanol lignin sulfation were calculated from the change in the sulfur content in sulfated ethanol lignin. The calculation was made using the first-order equation. The activation energy of the sulfation process was determined from the temperature dependence of the rate constants in the Arrhenius coordinates (Figure 3).

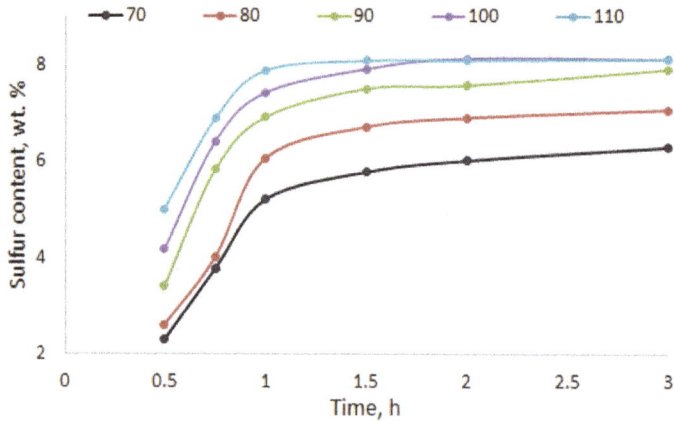

Figure 2. Dynamics of the sulfur content in the process of birch ethanol lignin sulfation with sulfamic acid–urea mixture at different temperatures.

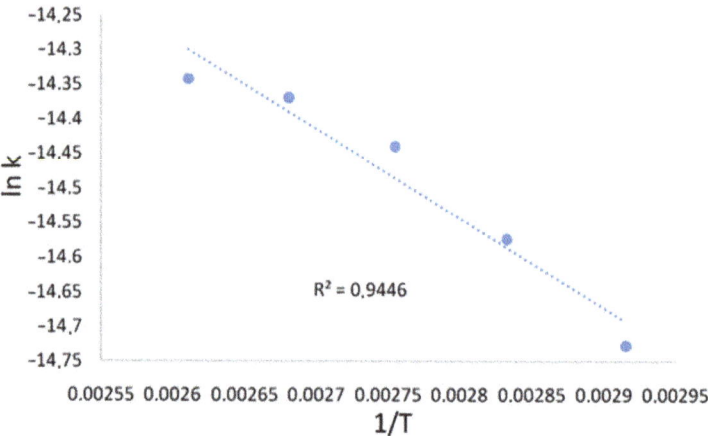

Figure 3. Temperature dependence of the rate constants of the birch ethanol lignin sulfation process.

The calculated apparent rate constants and activation energies of the birch ethanol lignin sulfation process are given in Table 2.

Table 2. Apparent rate constants and activation energies of the process of birch ethanol lignin sulfation with the sulfamic acid–urea mixture.

Temperature	Apparent Initial Rate Constant, $K \times 10^{-4}$ (s^{-1})	Activation Energy, kJ/mol
70	1.41	
80	1.80	10.7
90	2.05	
100	2.78	

The activation energies of the processes of sulfation of birch and abies ethanol lignins with sulfamic acid–urea mixture in 1,4-dioxane medium are similar: 10.7 kJ/mol for birch lignin (Table 2) and 8.4 kJ/mol for abies lignin [16]. It should be noted that, for the process of starch sulfation in a deep eutectic solvent (the sulfamic acid–urea mixture), the activation energy was 6.4 kJ/mol [20] and, for the sulfation of arabinogalactan with sulfamic acid in DMSO it was 13.1 kJ/mol [21]. It is known that the low activation energy of the process may indicate the presence of significant diffusion restrictions [22,23]. Taking this into account, we can conclude that, under the chosen conditions, the processes of biopolymer sulfation proceed under diffusion restrictions.

The solubility of the sulfated lignin in water increased with an increase in the content of sulfate groups. The maximum sulfur content in the sulfated ethanol lignin was estimated to be 10.6 wt %, taking into account the hypothetical structure of the Berkman spruce lignin [4], in which one phenylpropane unit has 0.9 mol of free OH groups capable of sulfating. In order to find the conditions that ensure the production of sulfated birch ethanol lignin with maximum yield and sulfur content, a numerical optimization of the sulfation process was carried out.

2.2. Numerical Optimization of the Process of Birch Ethanol Lignin Sulfation

As independent variables, we used two factors: process temperature X_1 (70, 80, 90, 100, 110 °C) and time X_2 (0.5, 0.75, 1.0, 1.5, 2.0, 3.0 h). The result of the sulfation process was characterized by two output parameters: sulfur content Y_1 (wt %) in the sulfated ethanol lignin and sulfated ethanol lignin yield Y_2 (wt %). The fixed parameter was the ratio L:SC = 1:3. A combined multilevel experiment plan (Users Design) was used in the calculations. The designations of the variables are listed in Table 3.

Table 3. Designations of independent variables (factors) and output parameters (experimental results).

Factors and Parameters	Designations in the Equations
Temperature, °C	X_1
Time, h	X_2
Sulfur content, %	Y_1
Product yield, %	Y_2

The experimental results given in Table 1 were used in the mathematical processing and optimization of the birch ethanol lignin sulfation process.

The dependences of the output parameters on the variable process factors were approximated by second-order regression equations. The results of the variance analysis are given in Table 4.

Table 4. Results of the variance analysis.

Variance Source	Output Parameters			
	Sulfur Content Y_1		Yield Y_2	
	Statistical Characteristics			
	Variance Relation F	Significance Level P	Variance Relation F	Significance Level P
X_1	78.74	0.0000	200.98	0.0000
X_2	24.87	0.0001	10.88	0.0042
X_1^2	9.84	0.0060	59.66	0.0000
$X_1 X_2$	0.00	0.9757	36.11	0.0000
X_2^2	21.54	0.0002	0.26	0.6184
R^2_{adj}	86.1		92.8	

The variance analysis showed that, within the limits of the experimental conditions used, the greatest contribution to the total variance of the output parameter was made by both factors: the temperature and time of the birch ethanol lignin sulfation process. This is indicated by the high variance ratios (F) for the main effects, which are called also the influence efficiencies. The data in the columns of Table 4 (P) are interpreted similarly. The influence of the variance source on the output parameter is considered to be statistically significant if its significance level is lower than a specified critical value (in our case, 0.05).

The dependence of the sulfur content Y_1 in the sulfated birch ethanol lignin on the process variables is approximated by the regression equation

$$Y_1 = -12.7119 + 0.3212 \times X_1 + 3.27557 \times X_2 - 0.00149576 \times X_1^2 + \\ +0.000225435 \times X_1 \times X_2 - 0.693718 \times X_2^2 \quad (1)$$

The predictive properties of Equation (1) are illustrated in Figure 4, in which the experimental values of the output parameter Y_1 are compared with its values calculated using Equation (1). The straight line corresponds to the calculated Y_1 values, and the dots correspond to the observed values. The proximity of the experimental points to the straight line confirms the good predictive properties of Equation (1).

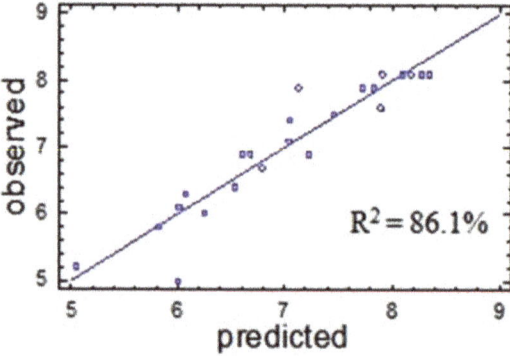

Figure 4. Output parameter Y_1 observed in the experiment (dots) and predicted by mathematical model (1) (solid line).

The approximation quality is characterized also by the determination coefficient R^2_{adj}. In the case under consideration, the value is $R^2_{adj} = 86.1\%$, which indicates acceptable approximation quality. This confirms the adequacy of Equation (1) for the experiment and makes it possible to use this equation as a mathematical model of the process under study.

Using the mathematical model, the dependence of the output parameter Y_1 on the variables X_1 and X_2 was plotted in the form of a response surface (Figure 5).

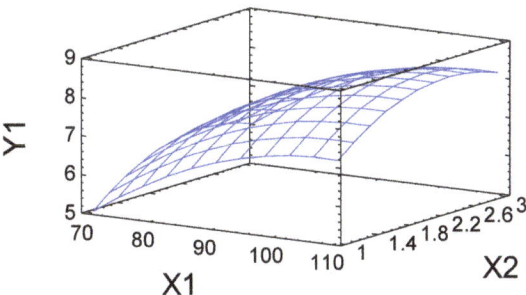

Figure 5. Response surface illustrating the dependence of sulfur content (wt %) in sulfated birch ethanol lignin on the process temperature (X_1) and time (X_2).

According to the calculation using mathematical model (1), the maximum predicted sulfur content (8.4 wt %) was obtained at the point corresponding to a process temperature of 107 °C and a process time of 2.3 h.

According to the results of the variance analysis within the limits of the chosen experimental conditions, the sulfation temperature contributes significantly to the total variance of the output parameter Y_2 (sulfated lignin yield, wt %). This is indicated by the high variance relation (F) corresponding to this factor and the small P criterion.

The dependence of Y_2 on the variable process factors is approximated by the regression equation

$$Y_2 = 36.8673 + 1.29494 \times X_1 + 7.57733 \times X_2 - 0.00715242 \times X_1^2 - \\ -0.0850839 \times X_1 \times X_2 - 0.14729 \times X_2^2 \qquad (2)$$

The determination coefficient is fairly high, R^2_{adj} = 92.8%, which evidences the good approximation quality. The latter is also confirmed by the good agreement between the output parameters calculated using Equation (2) and those obtained in the experimental measurements. This confirms the adequacy of Equation (2) for the experiment and its use as a mathematical model of the process under study (Figure 6).

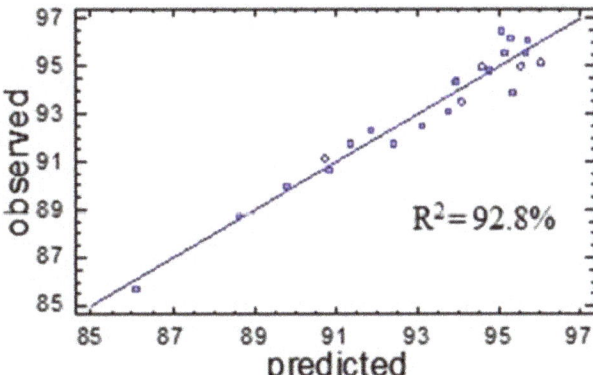

Figure 6. Comparison of the values of output parameter Y_2 observed in the experiment and those predicted by Equation (2).

Figure 7 shows the graphical representation of the dependence of the sulfated birch ethanol lignin yield on the variable factors X_1 and X_2.

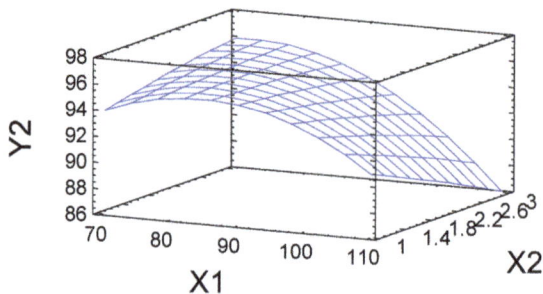

Figure 7. Response surface of the dependence of the sulfated birch ethanol lignin yield on the variable temperature (X_1) and time factors (X_2).

According to the above-described model, the optimum conditions for the sulfation of birch ethanol lignin that ensure the maximum yield of sulfated birch ethanol lignin (96.1 wt %) correspond to a process temperature of 78 °C and a time of 2.9 h.

2.3. Characterization of the Sulfated Birch Ethanol Lignin

The substitution of hydroxyl groups for sulfate groups during the sulfation of the birch ethanol lignin with the sulfamic acid–urea mixture was confirmed by FTIR and NMR spectroscopy.

The FTIR spectrum of birch ethanol lignin (Figure 8) contains absorption bands characteristic of hardwood lignins (GS) [24]. The band at 1123 cm^{-1} corresponds to planar bending vibrations of the C–H syringyl aromatic rings, and the C–O stretching vibrations in secondary alcohols are dominant in the spectrum. The pronounced band with a maximum at 1327 cm^{-1} belongs to the skeletal vibrations of the syringyl ring with the C–O stretching vibrations. In addition, the spectrum includes medium-intensity absorption bands around 1271 and 1034 cm^{-1}, characteristic of the vibrations of the guaiacyl units of lignin [4].

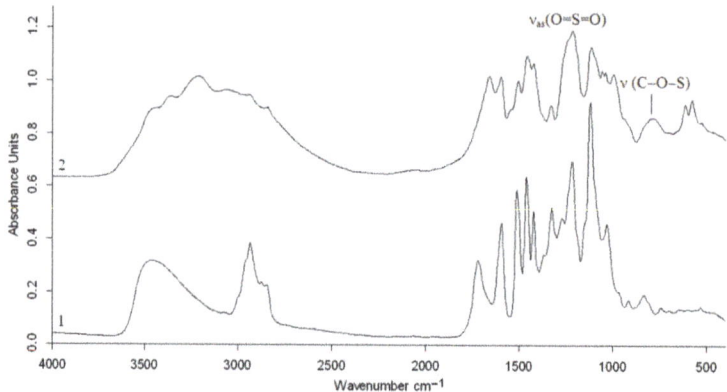

Figure 8. FTIR spectra of birch ethanol lignin (1) and sulfated ethanol lignin ammonium salt (2).

In contrast to the spectrum of the initial ethanol lignin, the FTIR spectrum of the ammonium salts of the ethanol lignin sulfates contained a new absorption band at 798 cm^{-1} (Figure 8), which corresponds to the stretching vibrations of the C–O–S bond of the sulfate group and a broad absorption band with the maximum at 1218 cm^{-1} corresponding to the asymmetric stretching vibrations v_{as}(O=S=O).

The FTIR spectra of birch and abies ethanol lignin sulfates [16] obtained in a similar way are almost identical, except for the presence in the spectrum of the sulfated birch ethanol lignin of the adsorption band at 1331 cm^{-1} characteristic of the syringyl structures.

The 2D HSQC NMR spectra of the initial and sulfated birch ethanol lignins are shown in Figures 9 and 10, respectively. The main ^{1}H–^{13}C peaks in the HSQC spectra identified using the literature data [25–27] are given in Table 5, together with the chemical shifts of some low-intensity peaks (not shown in Figures 9 and 10). The main structural units and fragments of the initial and sulfated birch ethanol lignins are presented in Figure 11.

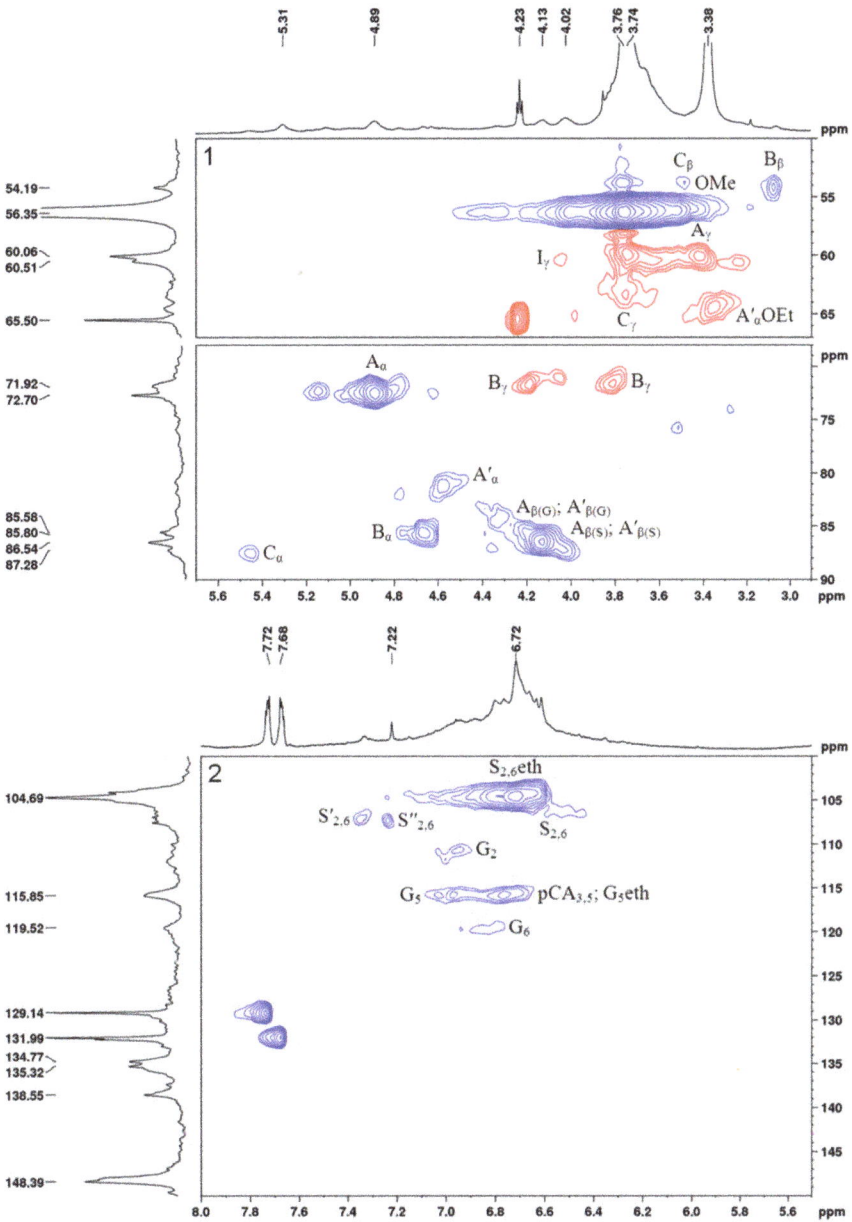

Figure 9. HSQCed spectrum of ethanol lignin: aliphatic oxygenated region 1 and aromatic region 2. The assignment of signals is given in Table 5, and the main identified structural units and fragments are shown in Figure 11.

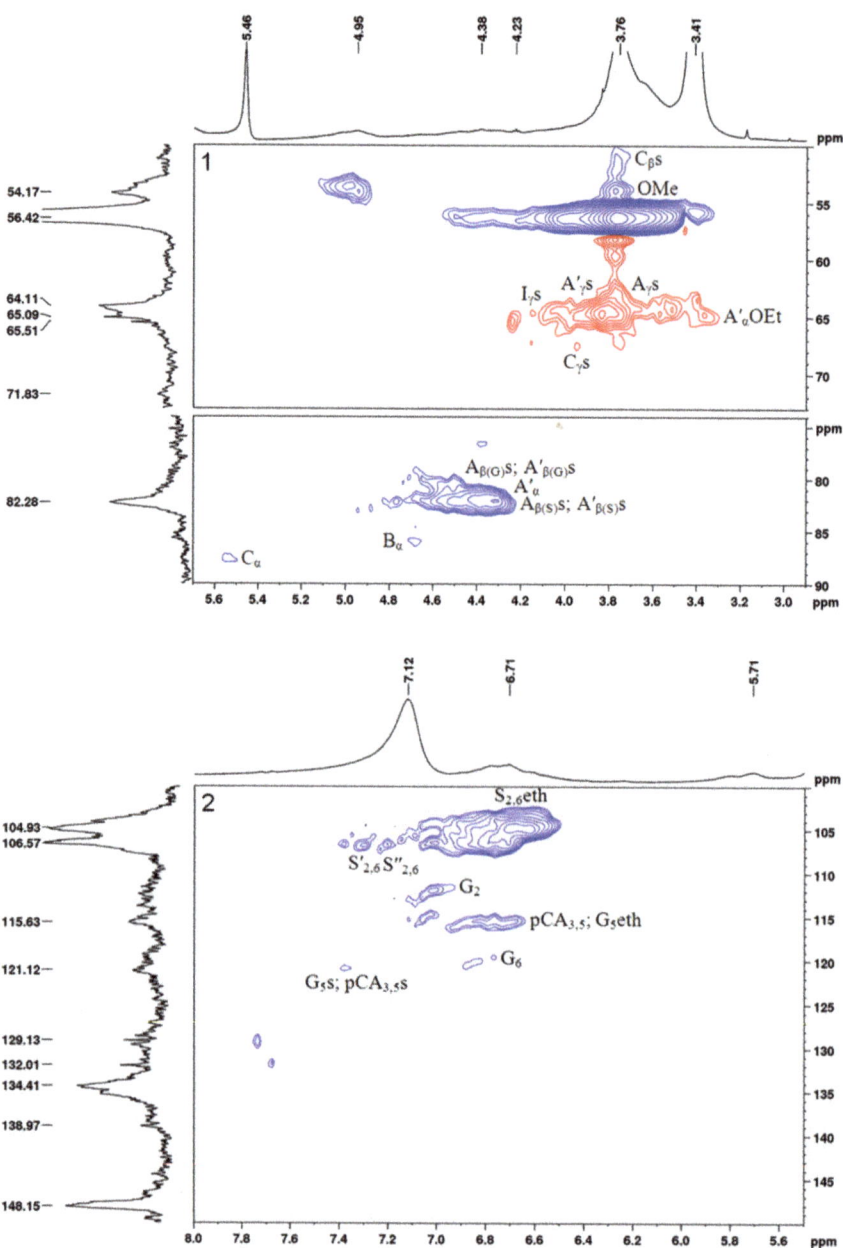

Figure 10. HSQCed spectrum of sulfated ethanol lignin: aliphatic oxygenated region 1 and aromatic region 2. The assignment of signals is given in Table 5, and the main identified structural units and fragments are shown in Figure 11.

Table 5. Assignment of the ^{1}H–^{13}C peaks in the HSQC spectra of the initial and sulfated birch ethanol lignins.

Designation	δ_C/δ_H, ppm (Initial Lignin)	δ_C/δ_H, ppm (Sulfated Lignin)	Assignment
OMe	56.3/3.74	56.3/3.76	C-H in the methoxy groups (OMe)
A$_\gamma$ and A'$_\gamma$	60.0/3.42–3.74	-	C$_\gamma$-H$_\gamma$ in the β-aryl ether (β-O-4') substructures (A) and α-ethoxylated (C$_\alpha$OEt) β-aryl ether (β-O-4') substructures (A')
A$_\gamma$s and A'$_\gamma$s	-	64.6/3.51 and 3.83	C$_\gamma$-H$_\gamma$ in the γ-sulfated (γ-OSO$_3$NH$_4$) β-aryl ether (β-O-4') substructures (As) and γ-sulfated (γ-OSO$_3$NH$_4$) α-ethoxylated (C$_\alpha$OEt) β-aryl ether (β-O-4') substructures (A's)
A$_{\beta(G)}$ and A'$_{\beta(G)}$	84.1/4.30 and 83.4/4.39	-	Cβ-H$_\beta$ in the β-aryl ether (β-O-4') substructures bonded to the G units (A) and α-ethoxylated (C$_\alpha$OEt) β-aryl ether (β-O-4') substructures bonded to the G units (A')
A$_{\beta(S)}$ and A'$_{\beta(S)}$	86.5/4.12 and 84.8/4.27	-	Cβ-H$_\beta$ in the β-aryl ether (β-O-4') substructures bonded to the S units (A) and α-ethoxylated (C$_\alpha$OEt) β-aryl ether (β-O-4') substructures boded to the S units (A')
A$_{\beta(G)}$s and A'$_{\beta(G)}$s	-	80.7/4.50 and 80.0/4.64	Cβ-H$_\beta$ in the γ-sulfated (γ-OSO$_3$NH$_4$) β-aryl ether (β-O-4') substructures bonded to the G units (As) and γ-sulfated (γ-OSO$_3$NH$_4$) α-ethoxylated (C$_\alpha$OEt) β-aryl ether (β-O-4') substructures bonded to the G units (A's)
A$_{\beta(S)}$s and A'$_{\beta(S)}$s	-	82.1/4.35 and 82.1/4.50	Cβ-H$_\beta$ in the γ-sulfated (γ-OSO$_3$NH$_4$) β-aryl ether (β-O-4') substructures boded to the S units (As) and γ-sulfated (γ-OSO$_3$NH$_4$) α-ethoxylated (C$_\alpha$OEt) β-aryl ether (β-O-4') substructures boded to the S units (A's)
A$_\alpha$	72.5/4.88	-	C$_\alpha$-H$_\alpha$ in the β-aryl ether (β-O-4') substructures (A)
A'$_\alpha$OEt	64.4/3.33	64.7/3.36	C-H of the methylene groups in the α-ethoxylated (C$_\alpha$OEt) β-aryl ether (β-O-4') substructures (A')
A'$_\alpha$	81.2/4.56	81.1/4.53	C$_\alpha$-H$_\alpha$ in the α-ethoxylated (C$_\alpha$OEt) β-aryl ether (β-O-4') substructures (A')
B$_\beta$	54.1/3.06	55.1/3.06	Cβ-H$_\beta$ in the pinoresinol (β–β') substructures (B)
B$_\gamma$	71.6/3.80 and 4.18	-	C$_\gamma$-H$_\gamma$ in the pinoresinol (β–β') substructures (B)
B$_\alpha$	85.6/4.68	85.9/4.68	C$_\alpha$-H$_\alpha$ in the pinoresinol (β–β') substructures (B)
C$_\beta$	53.7/3.48	-	Cβ-H$_\beta$ in the phenylcoumaran (β–5') moieties (C)
C$_\beta$s	-	51.1/3.65	Cβ-H$_\beta$ in the γ-sulfated (γ-OSO$_3$NH$_4$) phenyl coumaran (β–5') moieties (Cs)
C$_\gamma$	63.4/3.74	-	C$_\gamma$-H$_\gamma$ in the phenyl coumaran (β–5') substructures (C)
C$_\gamma$s	-	67.5/3.94	C$_\gamma$-H$_\gamma$ in the γ-sulfated (γ-OSO$_3$NH$_4$) phenylcoumaran (β–5') substructures (Cs)
C$_\alpha$	87.6/5.45	87.5/5.55	C$_\alpha$-H$_\alpha$ in the phenylcoumaran (β–5') substructures (C$_\alpha$)
I$_\gamma$	60.4/4.03	59.9/4.03	C$_\gamma$-H$_\gamma$ in the cinnamyl alcohol end groups (I)
I$_\gamma$s	-	64.6/4.15	C$_\gamma$-H$_\gamma$ in the γ-sulfated (γ-OSO$_3$NH$_4$) cinnamyl alcohol end groups (Is)
J$_{2,6(S)}$	107.0/7.08	106.4/7.00	C2,6-H$_{2,6}$ in the cinnamyl aldehyde end groups (J)
S$_{2,6}$eth	104.5/6.70	104.8/6.62	C2,6-H$_{2,6}$ in the 4-etherified syringyl units (Seth)
S$_{2,6}$	106.2/6.51	-	C2,6-H$_{2,6}$ in the 4-non-etherified syringyl units (S)
S'$_{2,6}$	107.1/7.35	106.7/7.30	C2,6-H$_{2,6}$ in the oxidized (C$_\alpha$=O) syringyl units (S')
S''$_{2,6}$	107.2/7.23	106.5/7.20	C2,6-H$_{2,6}$ in the oxidized (C$_\alpha$OOH) syringyl units (S'')
pCA$_{3,5}$	115.5/6.79	115.5/6.79	C3,5-H$_{3,5}$ in the p-coumarates (pCA)
pCA$_{3,5}$s	-	120.8/7.32	C3,5-H$_{3,5}$ in the 4-sulfated (4-OSO$_3$NH$_4$) p-coumarates (pCAs)
G$_2$	110.6/6.94	111.9/7.00	C2-H$_2$ in the 4-etherified guaiacyl units (Geth)
G$_5$	115.7/6.97	-	C5-H$_5$ in the 4-non-etherified guaiacyl units (G)
G$_5$s	-	120.8/7.38	C5-H$_5$ in the 4-sulfated (4-OSO$_3$NH$_4$) guaiacyl units (Gs)
G$_5$eth	115.7/6.76	115.4/6.76	C5-H$_5$ in the 4-etherified guaiacyl units (Geth)
G$_6$	119.4/6.79	119.5/6.76	C6-H$_6$ in the 4-etherified guaiacyl units (Geth)

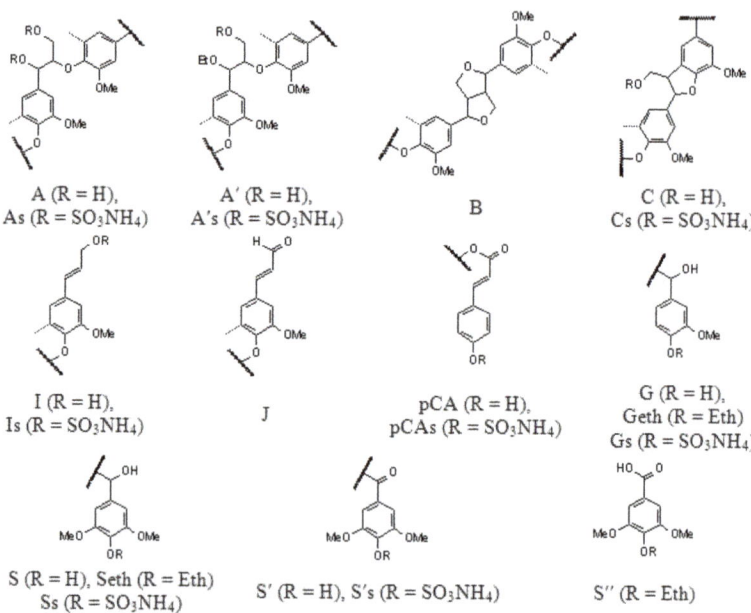

Figure 11. Main structural units and moieties of the initial and sulfated ethanol lignins: (**A**) β-aryl ethers, (**As**) α,γ-sulfated (α,γ-COSO₃NH₄) β-aryl ethers, (**A′**) α-ethoxylated (α-COEt) β-aryl ethers, (**A′s**) γ-sulfated (γ-COSO₃NH₄) α-ethoxylated (α-COEt) β-aryl ethers, (**B**) pinoresinols, (**C**) phenylcoumarans, (**Cs**) γ-sulfated (γ-COSO₃NH₄) phenylcoumarans, (**I**) cinnamyl alcohol end groups, (**Is**) γ-sulfated (γ-COSO₃NH₄) cinnamyl alcohol end groups, (**J**) cinnamyl aldehyde end groups, (**pCA**) p-coumarates, (**pCAs**) 4-sulfated (4-COSO₃NH₄) p-coumarates, (**S**) syringyl units, (**Seth**) 4-etherified (4-COEth) syringyl units, (**S′**) oxidized (α-C=O) syringyl units, (**S″**) oxidized (α-COOH) syringyl units, (**G**) guaiacyl units, (**Geth**) 4-etherified (4-COEth) guaiacyl units, and (**Gs**) 4-sulfated (4-COSO₃NH₄) guaiacyl units.

The HSQC spectra of the initial and sulfated ethanol lignin samples were compared in the regions of the chemical shifts of atoms from the lignin side chains (δ_C/δ_H 50–90/2.9–5.7 ppm) and aromatic rings (δ_C/δ_H 100–150/5.5–8.0 ppm).

Considering the region of the ^{1}H–^{13}C side chain signals in the HSQC spectrum of the birch ethanol lignin (Figure 9), we can see that it contains the intense correlation peaks of β-aryl ethers (A), pinoresinol (B), and phenylcoumaran fragments (C) (see Figure 11). A part of β-aryl ethers is ethoxylated in the α-position, judging by the fact that the spectra contain signals of the methylene group in the α-ethoxylated β-O-4′ bonds (δ_C/δ_H 64.4/3.33) and the α-position of the α-acylated β-O-4′ bonds (δ_C/δ_H 81.2/4.56). This assumption is confirmed by the presence of a correlation signal of the methyl group at δ_C/δ_H 14.3/1.00 ppm.

In the spectrum of the sulfated ethanol lignin (Figure 10), the group of peaks assigned to the phenylcoumaran fragments (C) is characterized by a change in the position of the correlation signals Cγ-Hγ and Cβ-Hβ (Figure 10), which is related to the effect of sulfation of OH groups in the γ position (Cγ-Hγ: a shift of $\Delta\delta_C$ ~ 4 ppm toward weak fields; Cβ-Hβ: a shift of $\Delta\delta_C$ ~ 3 ppm toward strong fields).

A similar change in the peak positions along the carbon atom axis in the sulfated ethanol lignin spectrum is observed for β-aryl ethers β-O-4′ (A) and α-ethoxylated β-aryl ethers (A′). The Cγ-Hγ correlation signals of the sulfated lignin are shifted relative to the initial lignin signals toward weak fields by ~5 ppm and the Cβ-Hβ signals toward strong fields by ~4 ppm. Such shifts of the signals in the spectra are most likely due to the sulfation of OH groups of β-aryl ethers of the lignin macromolecule in the γ position. In addition, free OH groups in the α position of β-aryl ethers (A) are probably subjected to the sulfation, since the Cα-Hα peak at δ_C/δ_H 72.5/4.88 ppm is missing in the sulfated lignin spectrum.

The shift of signals in the aliphatic region of the spectrum of the sulfated lignin sample as compared with the spectrum of the initial lignin was also found for the C_γ–H_γ correlations of the cinnamyl alcohol end groups (I). This is also indicative of the sulfation of OH groups bonded with carbon atoms C_γ of this lignin fragment.

Despite the presence of the fairly intense Cα-Hα, Cβ-Hβ, and Cγ-Hγ signals of pinoresinol (β–β′) fragments (B) in the spectrum of the initial ethanol lignin, the intensity of these signals in the spectrum of the sulfated sample drops dramatically. The peaks assigned to the Cγ-Hγ correlations (δ_C/δ_H 71.6/3.80 and 4.18 ppm) disappear almost completely.

It is important to note the appearance of a peak at δ_C/δ_H 53.6/5.00 ppm assigned to the CH_3 or CH groups in the aliphatic region of the sulfated ethanol lignin spectrum. We failed to establish an unambiguous correspondence of this peak to any structural fragment.

The ^1H–^{13}C aromatic region in the HSQC spectrum of the birch ethanol lignin (Figure 9) contains characteristic correlation peaks of the syringyl (S) and guaiacyl (G) units, p-coumarates (pCA), and cinnamyl aldehyde end groups (J). The syringyl units are of several types. In particular, using the assignments made in [28–30], we found S units with a substituent in the 4-position (δ_C/δ_H 104.5/6.70), S units with a free hydroxyl group in the 4-position (δ_C/δ_H 106.2/6.51), S units with a carbonyl group in the α-position (δ_C/δ_H 107.1/7.35), and S units with a carboxyl group in the α position (δ_C/δ_H 107.2/7.23).

In addition, there are high-intensity signals at δ_C/δ_H = 129.1/7.73 and 131.9/7.67 ppm corresponding to the CH or CH_3 groups. Some researchers believe that the signals located at these chemical shifts can be attributed to both $C_{\alpha,\beta}$–$H_{\alpha,\beta}$ in stilbenes [31] and $C2,6$-$H_{2,6}$ in p-benzoates [32].

In the aromatic region of the sulfated lignin spectrum containing the main part of the peaks of the syringyl ($S_{2,6}$eth, $S'_{2,6}$, $S''_{2,6}$) and guaiacyl units (G_2, G_5, G_6), p-coumarates ($pCA_{3,5}$), and cinnamyl aldehyde end groups ($J_{2,6}$) (see Table 6), the signal of the $C2,6$-$H_{2,6}$ syringyl units with a free hydroxyl group in the 4-position disappears almost completely. This is apparently due to the replacement of this hydroxyl group by the sulfate one. In addition, in this region of the spectrum, a new peak at δ_C/δ_H = 120.8/7.38 and 7.32 ppm appears, which is most likely a signal of the $C5$-H_5 guaiacyl (Gs) and $C3,5$-$H_{3,5}$ p-coumarate (pCAs) structural units with sulfate groups attached to the 4-position. These changes in the chemical shifts of adjacent positions 3 and 5 of the aromatic ring caused by esterification correspond to the expected change of the substituent in phenol [33]. The possible substitution of the sulfate group of the phenolic hydroxyls for the guaiacyl (G) units in the 4-position is evidenced also by the almost complete disappearance of the peak at δ_C/δ_H = 115.7/6.97 ppm characteristic of the $C5$-H_5 guaiacyl units (G_5) with unsubstituted hydroxyl in the 4-position [34].

Table 6. Average molecular weights M_n and M_w and polydispersity of the initial and sulfated birch ethanol lignin samples.

Sample	M_n (Da)	M_w (Da)	PD
Birch ethanol lignin	902	1828	2.02
Sulfated birch ethanol lignin	4199	7599	1.81

Based on the data obtained, it can be concluded that sulfation affects the acceptable aliphatic hydroxyl groups of lignin at the γ-positions of β-aryl ethers, α-ethoxylated β-aryl ethers, phenylcoumaran substructures, and cinnamyl alcohol end groups, as well as the unsubstituted hydroxyl groups in the α-position of β-aryl ethers. In addition, free phenolic hydroxyl groups in the 4-position of syringyl and guaiacyl units and p-coumarates can be subjected to sulfation.

The comparison of the 2D NMR spectroscopy data on birch and abies ethanol lignin sulfates [16] revealed higher structural diversity of the sulfated birch ethanol lignin. A significant difference between the HSQC spectra of birch ethanol lignin and abies ethanol lignin in the area of correlations of the aromatic fragments is the presence of several types of syringyl units in the former. However, both samples are built from the phenylpropane structural units linked by simple ether (β-O-4′) and C–C (β–β′, β–5′) bonds, while the sulfate groups are localized mainly in the γ and α positions of the side chains and, probably, in the 4-position of aromatic rings.

Data on the molecular weight distribution of the initial and sulfated birch ethanol lignin were obtained by the GPC method. The molecular weight distribution curves for the initial and sulfated birch ethanol lignin samples are shown in Figure 12. The average molecular weights and polydispersity of the initial and sulfated ethanol lignins are indicated in Table 6.

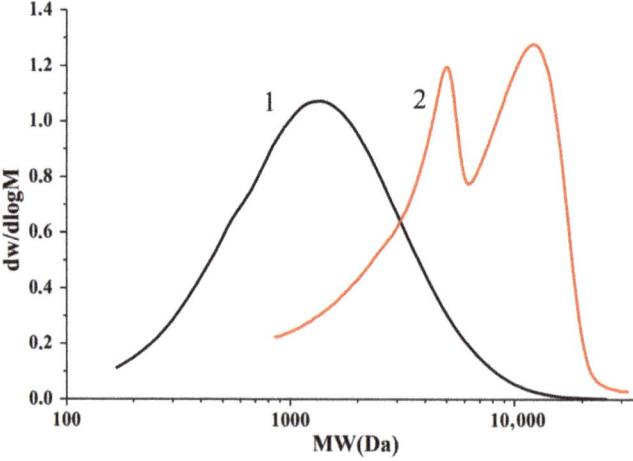

Figure 12. Molecular weight distribution curves for (1) the birch ethanol lignin and (2) sulfated ethanol lignin samples.

The birch ethanol lignin obtained in [35] had a low molecular weight (M_w = 1800 Da) and a monomodal distribution (PD = 2.02), which evidences higher homogeneity of the sample as compared to Alcell birch lignin (M_w = 3470 and PD = 4.1) [17].

As a result of birch ethanol lignin sulfation, the weight average molecular weight M_w of the samples increased from ~1800 to ~7600 Da. Such a significant growth is related to an increase in the weight of lignin macromolecules due to the introduction of sulfate groups and the removal of the low molecular weight fraction of the sulfated lignin along with inorganic impurities at the dialysis stage. A feature distinguishing sulfated birch ethanol lignin from abies ethanol lignin sulfated under similar conditions [16] is the bimodal molecular weight distribution (Figure 12). The molecular weight distribution curve of birch ethanol lignin has two pronounced peaks with molecular weights of ~5000 and ~12,000 Da. These peaks can be attributed to the heterogeneity of the initial ethanol lignin molecules, which enter into the sulfation reaction in different ways. The low molecular weight lignin fraction is possibly less sulfated than the high molecular weight fraction, which is reflected in the separation of the peaks in the molecular weight distribution curve. Sulfated birch ethanol lignin has a higher polydispersity and a higher average molecular weight than sulfated abies ethanol lignin (M_w~5300 Da, PD = 1.63) [16].

2.4. Thermochemical Properties of the Birch Ethanol Lignin

The thermochemical properties of the birch ethanol lignin were studied using the non-isothermal TG/DTG analysis in an argon medium in the temperature range of 30–900 °C.

The thermal decomposition of the ethanol lignin occurred over a wide temperature range, since its structure contains various functional groups with different thermal stabilities (Figure 13).

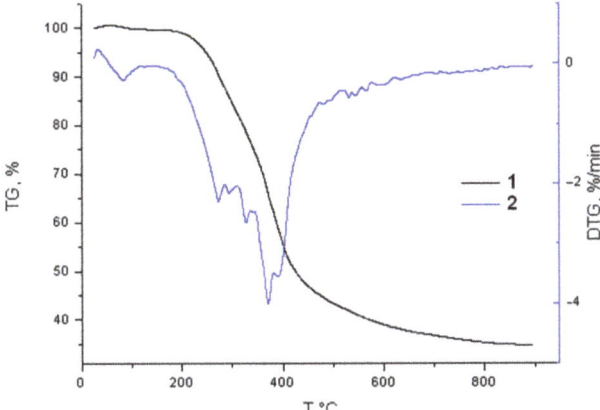

Figure 13. (1) TG and (2) DTG curves for the birch ethanol lignin sample.

The sample weight loss at temperatures of 30–180 °C was found to be less than 1%. This is explained by the loss of moisture and adsorbed gases. The main thermal decomposition of the ethanol lignin started after 200 °C and practically ended at 600 °C. The solid residue yield gradually decreased with an increase in temperature to 700 °C and then remained constant invariable. At a pyrolysis temperature of 900 °C, the solid residue yield was 34.6 wt %, which is somewhat less than in the case of pyrolysis of the abies ethanol lignin under similar conditions (36.2 wt %) [36]. As is known [37], coniferous lignins consist mainly of the guaiacyl structures, while in hardwood lignins, the syringyl structures dominate. The high yield of the carbon residue during the thermal decomposition of abies ethanol lignin is probably due to the tendency of the guaiacyl propane units to condensation reactions [38].

The DTG curve has a broad peak corresponding to the main thermal decomposition of ethanol lignin and an implicit peak. The maximum rate of thermal degradation of the birch ethanol lignin (4.1%/min) was reached at 372 °C. In the temperature range of 350–400 °C, the main lignin structural moieties (guaiacyl and syringyl) underwent cracking with the formation of phenol-type compounds of different molecular weights, the yield of which increased with temperature [39].

At this temperature range, the pyrolysis products represent a complex mixture of organic compounds containing the aromatic, hydroxyl, and alkyl groups and reflecting the composition and structural features of the initial lignin [40]. During the thermal decomposition of the lignin, the competing depolymerization reactions with the formation of lower molecular weight aromatic products and cross-linking reactions of aromatic compounds and their carbonization occurred [40]. In the temperature range of 450–600 °C, the birch ethanol lignin weight loss rate significantly decreased and the thermal decomposition was mainly completed at 600 °C. In this case, some of the aromatic rings in the lignin probably decomposed and condensed into carbon products [11].

The sulfation of the birch ethanol lignin noticeably changed the nature of its thermal transformation (Figure 14).

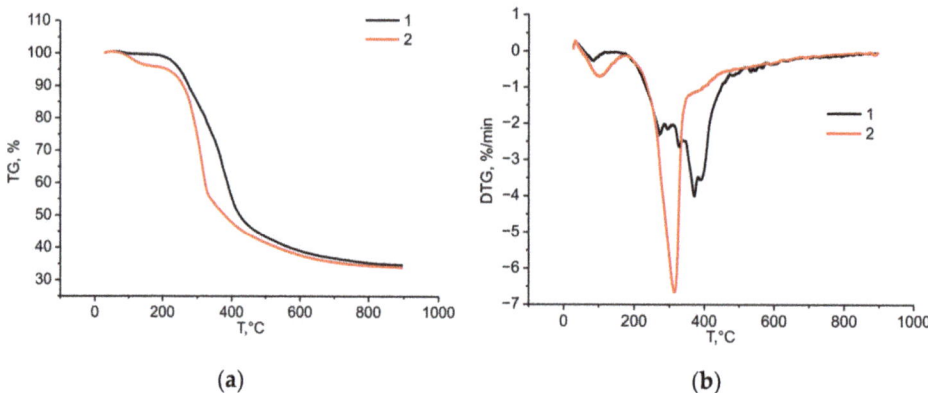

Figure 14. TG (**a**) and DTG (**b**) curves of the birch ethanol lignin (1) and sulfated birch ethanol lignin (2) samples.

According to the data presented in Figure 14a, the sulfation of the birch ethanol lignin reduced its thermal stability. At temperature 300 °C, the sulfated ethanol lignin lost 26.4% of its initial weight, while the initial lignin lost only 15.7%. This tendency continued until the completion of the pyrolysis process.

The sulfation of the birch ethanol lignin also changed its thermal transformation profile (Figure 14b). In the temperature range of 100–150 °C, the sulfated ethanol lignin weight loss rate was much higher than in the case of initial ethanol lignin. As was shown in [42], in this temperature range, the aliphatic hydroxyl groups, carbonyl groups, and C–C bonds in the lignin side chains are broken.

In the temperature range of 200–350 °C, an intense narrow peak appeared in the DTG curve of the sulfated birch ethanol lignin, with a maximum weight loss rate of 6.7%/min at 317 °C, which is attributed to the thermal decomposition of sulfate groups [43].

Thus, the TG/DTG study showed that the syringyl structure of hardwood (birch) ethanol lignin was thermally less stable than the guaiacyl structure dominating in softwood (abies) ethanol lignin. Additionally, the introduction of sulfate groups into the structure of birch ethanol lignin reduced its thermal stability.

3. Materials and Methods

3.1. Materials

The silver birch (*Betula pendula* Roth) wood harvested in the vicinity of Krasnoyarsk city was used as a feedstock for the isolation of ethanol lignin. The contents of the main birch wood components (% of the absolutely dry wood weight) were 47.3 cellulose, 28.5 hemicelluloses, 19.0 lignin, 4.9 extractives, and 0.3 ash.

3.2. Ethanol Lignin Isolation

The ethanol lignin was separated from birch wood by extraction with an ethanol–water (60:40) mixture in a Rexo Engineering autoclave reactor (Rexo Engineering Co., Ltd., Seoul, Korea) with a capacity of 3000 mL at a temperature of 185 °C under a working pressure of 0.75 MPa for 3 h and subsequent precipitation with cold water using the technique described in [35].

3.3. Ethanol Lignin Sulfation

The obtained ethanol lignin was sulfated with sulfamic acid in 1,4-dioxane in the presence of urea using the procedure described in [16,44].

The sulfation of ethanol lignin was carried out in a three-neck flask (150 mL) equipped with a reflux condenser, a thermometer, and a mechanical stirrer at temperatures 70–100 °C.

The ethanol lignin (1.25 g) was added to a mixture of sulfamic acid and urea (mol. ratio 1:1) in 15 mL of 1,4-dioxane. The mixture was intensively stirred for 30–180 min and cooled to room temperature. The solvent was decanted, and the remaining solid product was dissolved in a small amount of water and neutralized with aqueous ammonia to pH 8. To remove the excess reactants, the product was dialyzed against water in a plastic bag of MF-503-46 MFPI brand (USA) with a pore size of 3.5 kDa for 8–10 h. After dialysis, the aqueous solution of sulfated lignin was evaporated with the use of a rotary evaporator to obtain a solid residue—sulfated lignin in the form of an ammonium salt.

The sulfation process of the birch ethanol lignin with sulfamic acid was numerically optimized using the Statgraphics Centurion XVI software, DOE (Design of Experiment) block by the method described in [16].

The estimated sulfated ethanol lignin weight was calculated on the basis of the sulfur content [21] as

$$m_{calc} = \frac{32 \times m}{32 - 0.97 \times S} \quad (3)$$

The sulfated product Yield (%) was determined as

$$Yield\ (\%) = \frac{m_{actual}}{m_{calc}} \times 100\% \quad (4)$$

where m_{calc} is the calculated sulfated ethanol lignin weight (g), m is the weight of the initial ethanol lignin sample (g), m_{actual} is the sulfated ethanol lignin weight (g), and S is the sulfur content in the sulfated ethanol lignin (%).

3.4. Elemental Analysis

The elemental analysis of the sulfated ethanol lignin was carried out using a Thermo-Quest FlashEA-1112 analyzer (Milan, Italy).

3.5. FTIR Analysis

The FTIR spectra of the initial and sulfated ethanol lignin were recorded using a Shimadzu IRTracer-100 Fourier transform IR spectrophotometer (Tokyo, Japan) in the wavelength range of 400–4000 cm^{-1}. The spectral data were processed using the OPUS software (version 5.0). The solid samples for the analysis were tablets in a KBr matrix (2 mg of the sample/1000 mg of KBr).

3.6. NMR Analysis

The 2D NMR spectra were recorded at 25 °C in 5-mm ampoules using a Bruker Avance III 600 NMR spectrometer (Billerica, MA, USA) at working frequencies of 600 (^{1}H) and 150 MHz (^{13}C). Approximately 80 mg of lignin was dissolved in 0.6 mL of deuterated dimethyl sulfoxide, and then the spectra were recorded in the heteronuclear single quantum correlation (HSQC) experiments with editing (HSQCed) using the Bruker standard sequence library. The solvent signal was used as an internal standard (δ_C 40.1 and δ_H 2.5).

3.7. Gel Permeation Chromatography

The number average molecular weight M_n, weight average molecular weight M_w, and polydispersity index of the initial and sulfated ethanol lignin samples were determined by gel permeation chromatography (GPC) using an Agilent 1260 Infinity II Multi-Detector GPC/SEC System with triple detection: refractometer (RI), viscometer (VS), and light scattering (LS). The water-soluble samples were separated on two combined PL Aquagel-OH Mixed-M columns using 0.1 M NaNO$_3$ as the mobile phase. The tetrahydrofuran (THF)-soluble samples were separated on a PLgel 10 µm MIXED-E column with the THF mobile phase stabilized with 250 ppm of butylhydroxytoluene. The columns were calibrated using the polyethylene glycol and polystyrene polydisperse standards (Agilent, Santa Clara, CA, USA), respectively. The eluent flow rate was 1 mL/min, and the injected sample volume was 100 µL. Before the analysis, the ethanol lignin and sulfated ethanol lignin samples were dissolved in THF and water (5 mg/mL), respectively; after that, they were filtered through a 0.45-µm Millipore PTFE membrane filter. The data were collected and processed using the Agilent GPC/SEC MDS software.

3.8. Thermogravimetric Analysis

The thermogravimetry analysis was carried out using a Netzsch STA 449 F1 Jupiter instrument (Waldkraiburg, Germany). The thermal degradation of the lignin samples was studied in argon in the temperature range from 298 to 1173 K. The samples were heated in the dynamic mode at a heating rate of 10 °C/min in corundum crucibles. The measured data were processed using the Netzsch Proteus Thermal Analysis.5.1.0 software package supplied with the instrument.

3.9. Kinetic Calculations

The kinetics of the process of birch ethanol lignin sulfation was studied in the temperature range of 70–100 °C. The apparent initial rates and rate constants of the sulfation reaction were calculated from the change in the sulfur content in the sulfated ethanol lignin. The calculation was carried out according to the first-order equation:

$$V = k \times S = \frac{dS}{dt} \qquad (5)$$

where V is the rate of the sulfation reaction, wt %/s; k is the rate constant of the sulfation reaction, 1/s; dS is the change in the sulfur content in birch ethanol lignin sulfate, wt %; and dt is the change in time, s. The activation energy was found by the tangent of the slope of the dependence of $\ln k$ on $1/T$.

4. Conclusions

As a result of the accomplished study, the main regularities of the process of birch ethanol lignin sulfation with a sulfamic acid–urea mixture in a 1,4-dioxane medium at temperatures of 70–110 °C were established and the sulfating products were characterized using chemical and physical analysis methods.

It was found that similar to the sulfation of abies ethanol lignin under the same conditions [16], the process was complicated by diffusion restrictions due to the increased viscosity of the reaction medium. In the case of excess sulfating agent, the main factors affecting the yield of the sulfated product and sulfur content were the temperature and the duration of the process. Using experimental and computational methods, the optimal conditions for the process of birch ethanol lignin sulfation with a sulfamic acid–urea mixture to provide a high yield of sulfated product (more than 96 wt %) with a sulfur content of 8.1 wt % were established. As in the case of abies ethanol lignin, the sulfation increased the molecular weight of the birch ethanol lignin from 1800 Da to 7600 Da and decreased the polydispersity from 2.02 to 1.81. Moreover, aliphatic hydroxyl groups were more easily sulfated.

Some differences in the sulfation of ethanol lignins of birch and abies were established due to the presence of phenylpropane units of different compositions within these lignins. The sulfation of birch ethanol lignin in which syringyl structures predominate proceeds more completely at moderate temperatures than the abies ethanol lignin with guaiacyl structure. Additionally, in contrast to sulfated abies ethanol lignin, the sulfated birch ethanol lignin had a bimodal and wider molecular weight distribution, as well as less thermal stability.

The sulfated birch ethanol lignin has prospects for use in the production of new sorbents, biocomposites, and nanomaterials, as well as in the development of new anticoagulant and antiviral drugs [12,13,45].

Author Contributions: Conceptualization, A.V.L. and N.Y.V.; methodology, Y.N.M., A.A.K., O.Y.F., A.S.K. and V.A.L.; software, Y.N.M., A.A.K., O.Y.F. and A.S.K.; validation, A.V.L., N.Y.V. and B.N.K.; formal analysis, A.S.K.; investigation, Y.N.M., A.A.K., O.Y.F. and A.S.K.; resources, B.N.K.; data curation, A.V.L., N.Y.V., Y.N.M., A.A.K., O.Y.F. and B.N.K.; writing—original draft preparation, A.V.L., N.Y.V. and B.N.K.; writing—review and editing, A.V.L., N.Y.V. and B.N.K.; visualization, Y.N.M., A.A.K., O.Y.F. and A.S.K.; supervision, B.N.K.; project administration, B.N.K.; funding acquisition, B.N.K. All authors have read and agreed to the published version of the manuscript.

Funding: This study was supported by the Russian Science Foundation, project no. 21-13-00250, https://rscf.ru/project/21-13-00250/.

Institutional Review Board Statement: Not applicable.

Informed Consent Statement: Not applicable.

Data Availability Statement: All data generated during this study are included in the article.

Acknowledgments: This study was carried out using the equipment of the Krasnoyarsk Regional Center for Collective Use of the Krasnoyarsk Scientific Center, Siberian Branch of the Russian Academy of Sciences. The authors are grateful to Sergey V. Baryshnikov for his help in obtaining the lignin samples.

Conflicts of Interest: The authors declare no conflict of interest.

Sample Availability: Samples of the compounds are available from the authors.

References

1. Faruk, O.; Sain, M. *Lignin in Polymer Composites*; Elsevier: Oxford, UK, 2015. [CrossRef]
2. Poletto, M. *Lignin: Trends and Applications*; IntechOpen: Rijeka, Croatia, 2018. [CrossRef]
3. Calvo-Flores, F.G.; Dobado, J.A. Lignin as renewable raw material. *ChemSusChem* **2010**, *3*, 1227–1235. [CrossRef] [PubMed]
4. Heitner, C.; Dimmel, D.R.; Schmidt, J.A. *Lignin and Lignans: Advances in Chemistry*; CRC Press: Boca Raton, FL, USA, 2010. [CrossRef]
5. Feofilova, E.P.; Mysyakina, E.S. Lignin: Chemical structure, biodegradation, and practical application (a review). *Appl. Biochem. Microbiol.* **2016**, *52*, 573–581. [CrossRef]
6. Lobato-Peralta, D.R.; Duque-Brito, E.; Villafán-Vidales, H.I.; Longoria, A.; Sebastian, P.J.; Cuentas-Gallegos, A.K.; Arancibia-Bulnes, C.A.; Okoye, P.U. A review on trends in lignin extraction and valorization of lignocellulosic biomass for energy applications. *J. Clean. Prod.* **2021**, *293*, 126123. [CrossRef]
7. De la Torre, M.J.; Moral, A.; Hernández, M.D.; Cabeza, E.; Tijero, A. Organosolv lignin for biofuel. *Ind. Crops Prod.* **2013**, *45*, 58–63. [CrossRef]
8. Inamuddin. *Green Polymer Composites Technology. Properties and Applications*; CRC Press: Boca Raton, FL, USA, 2016. [CrossRef]
9. Kai, D.; Tan, M.J.; Chee, P.L.; Chua, Y.K.; Yap, Y.L.; Loh, X.J. Towards lignin-based functional materials in a sustainable world. *Green Chem.* **2016**, *18*, 1175–1200. [CrossRef]
10. Shrotri, A.; Kobayashi, H.; Fukuoka, A. Catalytic conversion of structural carbohydrates and lignin to chemicals. *Adv. Catal.* **2017**, *60*, 59–123. [CrossRef]
11. Mehta, A.Y.; Mohammed, B.M.; Martin, E.J.; Brophy, D.F.; Gailani, D.; Desai, U.R. Allosterism-based simultaneous, dual anticoagulant and antiplatelet action: Allosteric inhibitor targeting the glycoprotein Ibα-binding and heparin-binding site of thrombin. *J. Thromb. Haemost.* **2016**, *14*, 828–838. [CrossRef] [PubMed]
12. Raghuraman, A.; Tiwari, V.; Zhao, Q.; Shukla, D.; Debnath, A.K.; Desai, U.R. Viral inhibition studies on sulfated lignin, a chemically modified biopolymer and a potential mimic of heparan sulfate. *Biomacromolecules* **2007**, *8*, 1759–1763. [CrossRef]
13. Raghuraman, A.; Tiwari, V.; Thakkar, J.N.; Gunnarsson, G.T.; Shukla, D.; Hindle, M.; Desai, U.R. Structural characterization of a serendipitously discovered bioactive macromolecule, lignin sulfate. *Biomacromolecules* **2005**, *6*, 2822–2832. [CrossRef] [PubMed]
14. Malyar, Y.N.; Vasil'yeva, N.Y.; Kazachenko, A.S.; Skvortsova, G.P.; Korol'kova, I.V.; Kuznetsova, S.A. Sulfation of abies ethanol lignin by complexes of sulfur trioxide with 1,4-dioxane and pyridine. *Russ. J. Bioorg. Chem.* **2021**, *47*, 1368–1375. [CrossRef]
15. Prinsen, P.; Narani, A.; Hartog, A.F.; Wever, R.; Rothenberg, G. Dissolving lignin in water through enzymatic sulfation with aryl sulfotransferase. *ChemSusChem* **2017**, *10*, 2267–2273. [CrossRef]
16. Kuznetsov, B.N.; Vasilyeva, N.Y.; Kazachenko, A.S.; Levdansky, V.A.; Kondrasenko, A.A.; Malyar, Y.N.; Skvortsova, G.P.; Lutoshkin, M.A. Optimization of the process of abies ethanol lignin sulfation by sulfamic acid–urea mixture in 1,4-dioxane medium. *Wood Sci. Technol.* **2020**, *54*, 365–381. [CrossRef]
17. Gabov, K.; Gosselink, R.J.A.; Smeds, A.I.; Fardim, P. Characterization of lignin extracted from birch wood by a modified hydrotropic process. *J. Agric. Food Chem.* **2014**, *62*, 10759–10767. [CrossRef] [PubMed]
18. Thoresen, P.P.; Lange, H.; Crestini, C.; Rova, U.; Matsakas, L.; Christakopoulos, P. Characterization of organosolv birch lignins: Toward application-specific lignin production. *ACS Omega* **2021**, *6*, 4374–4385. [CrossRef]

19. Kuznetsov, B.N.; Sudakova, I.G.; Chudina, A.I.; Garyntseva, N.V.; Kazachenko, A.S.; Skripnikov, A.M.; Malyar, Y.N.; Ivanov, I.P. Fractionation of birch wood biomass into valuable chemicals by the extraction and catalytic processes. *Biomass Conv. Bioref.* **2022**, 1–15. [CrossRef]
20. Akman, F.; Kazachenko, A.S.; Vasilyeva, N.Y.; Malyar, Y.N. Synthesis and characterization of starch sulfates obtained by the sulfamic acid-urea complex. *J. Mol. Struct.* **2020**, *1208*, 127899. [CrossRef]
21. Levdansky, A.V.; Vasilyeva, N.Y.; Kondrasenko, A.A.; Levdansky, V.A.; Malyar, Y.N.; Kazachenko, A.S.; Kuznetsov, B.N. Sulfation of arabinogalactan with sulfamic acid under homogeneous conditions in dimethylsulfoxide medium. *Wood Sci. Technol.* **2021**, *55*, 1725–1744. [CrossRef] [PubMed]
22. Karger, J.; Grinberg, F.; Heitjans, P. *Diffusion Fundamentals*; Leipziger University: Leipzig, Germany, 2005.
23. Lente, G. *Deterministic Kinetics in Chemistry and Systems Biology*; Springer: New York, NY, USA, 2015. [CrossRef]
24. Faix, O. Classification of lignin from different botanical origins by FTIR spectroscopy. *Holzforschung* **1991**, *45*, 21–27. [CrossRef]
25. You, T.-T.; Mao, J.-Z.; Yuan, T.-Q.; Wen, J.-L.; Xu, F. Structural elucidation of the lignins from stems and foliage of Arundo donax Linn. *J. Agric. Food Chem.* **2013**, *61*, 5361–5370. [CrossRef] [PubMed]
26. Wen, J.-L.; Sun, S.-L.; Yuan, T.-Q.; Xu, F.; Sun, R.-C. Structural elucidation of lignin polymers of Eucalyptus chips during organosolv pretreatment and extended delignification. *J. Agric. Food Chem.* **2013**, *61*, 11067–11075. [CrossRef] [PubMed]
27. Bauer, S.; Sorek, H.; Mitchell, V.D.; Ibáñez, A.B.; Wemmer, D.E. Characterization of Miscanthus giganteus lignin isolated by ethanol organosolv process under reflux condition. *J. Agric. Food Chem.* **2012**, *60*, 8203–8212. [CrossRef] [PubMed]
28. Lagerquist, L.; Pranovich, A.; Smeds, A.; von Schoultz, S.; Vähäsalo, L.; Rahkila, J.; Kilpeläinen, I.; Tamminend, T.; Willför, S.; Eklund, P. Structural characterization of birch lignin isolated from a pressurized hot water extraction and mild alkali pulped biorefinery process. *Ind. Crops Prod.* **2018**, *111*, 306–316. [CrossRef]
29. Rencoret, J.; Marques, G.; Gutiérrez, A.; Ibarra, D.; Li, J.; Gellerstedt, G.; Santos, J.I.; Jiménez-Barbero, J.; Martínez, A.T.; del Río, J.C. Structural characterization of milled wood lifnins from different eucalypt species. *Holzforschung* **2008**, *62*, 514–526. [CrossRef]
30. del Río, J.C.; Rencoret, J.; Marques, G.; Li, J.; Gellerstedt, G.; Jiménez-Barbero, J.; Martínez, Á.T.; Gutiérrez, A. Structural characterization of the lignin from jute (*Corchorus capsularis*) fibers. *J. Agric. Food Chem.* **2009**, *57*, 10271–10281. [CrossRef]
31. Kangas, H.; Liitiä, T.; Rovio, S.; Ohra-aho, T.; Heikkinen, H.; Tamminen, T.; Poppius-Levlin, K. Characterization of dissolved lignins from acetic acid Lignofibre (LGF) organosolv pulping and discussion of its delignification mechanisms. *Holzforschung* **2014**, *69*, 247–256. [CrossRef]
32. Lu, F.; Karlen, S.D.; Regner, M.; Kim, H.; Ralph, S.A.; Sun, R.-C.; Kuroda, K.-i.; Augustin, M.A.; Mawson, R.; Sabarez, H.; et al. Naturally p-hydroxybenzoylated lignins in palms. *BioEnergy Res.* **2015**, *8*, 934–952. [CrossRef]
33. Ragan, M.A. Phenol sulfate esters: Ultraviolet, infrared, ^{1}H and ^{13}C nuclear magnetic resonance spectroscopic investigation. *Can. J. Chem.* **1978**, *56*, 2681–2685. [CrossRef]
34. Rencoret, J.; Marques, G.; Gutiérrez, A.; Nieto, L.; Santos, J.I.; Jiménez-Barbero, J.; Martínez, A.T.; del Río, J.C. HSQC-NMR analysis of lignin in woody (*Eucalyptus globulus* and *Picea abies*) and non-woody (*Agave sisalana*) ball-milled plant materials at the gel state. *Holzforschung* **2009**, *63*, 691–698. [CrossRef]
35. Kuznetsov, B.N.; Chesnokov, N.V.; Sudakova, I.G.; Garyntseva, N.V.; Kuznetsova, S.A.; Malyar, Y.N.; Yakovlev, V.A.; Djakovitch, L. Green catalytic processing of native and organosolv lignins. *Catal. Today* **2018**, *309*, 18–30. [CrossRef]
36. Fetisova, O.Y.; Mikova, N.M.; Chesnokov, N.V. A kinetic study of the thermal degradation of fir and aspen ethanol lignins. *Kinet. Catal.* **2019**, *60*, 273–280. [CrossRef]
37. Moustaqim, M.E.; Kaihal, A.E.; Marouani, M.E.; Men-La-Yakhaf, S.; Taibi, M.; Sebbahi, S.; Hajjaji, S.E.; Kifani-Sahban, F. Thermal and thermomechanical analyses of lignin. *Sustain. Chem. Pharm.* **2018**, *9*, 63–68. [CrossRef]
38. Poletto, M. Assessment of the thermal behavior of lignins from softwood and hardwood species. *Maderas Cienc. Tecnol.* **2017**, *19*, 63–74. [CrossRef]
39. Liu, Q.; Wang, S.; Zheng, Y.; Luo, Z.; Cen, K. Mechanism study of wood lignin pyrolysis by using TG–FTIR analysis. *J. Anal. Appl. Pyrol.* **2008**, *82*, 170–177. [CrossRef]
40. Brebu, M.; Vasile, C. Thermal degradation of lignin—A review. *Cellul. Chem. Technol.* **2010**, *44*, 353–363.
41. Nakamura, T.; Kawamoto, H.; Saka, S. Condensation reactions of some lignin related compounds at relatively low pyrolysis temperature. *J. Wood Chem. Technol.* **2007**, *27*, 121–133. [CrossRef]
42. Ji, X.; Guo, M.; Zhu, L.; Du, W.; Wang, H. Synthesis mechanism of an environment-friendly sodium lignosulfonate/chitosan medium-density fiberboard adhesive and response of bonding performance to synthesis mechanism. *Materials* **2020**, *13*, 5697. [CrossRef] [PubMed]
43. Roman, M.; Winter, W.T. Effect of sulfate groups from sulfuric acid hydrolysis on the thermal degradation behavior of bacterial cellulose. *Biomacromolecules* **2004**, *5*, 1671–1677. [CrossRef] [PubMed]
44. Malyar, Y.N.; Kazachenko, A.S.; Vasilyeva, N.Y.; Fetisova, O.Y.; Borovkova, V.S.; Miroshnikova, A.V.; Levdansky, A.V.; Skripnikov, A.M. Sulfation of wheat straw soda lignin: Role of solvents and catalysts. *Catal. Today* **2022**, *397–399*, 397–406. [CrossRef]
45. Henry, B.L.; Desai, U.R. Sulfated low molecular weight lignins, allosteric inhibitors of coagulation proteinases via the heparin binding site, significantly alter the active site of thrombin and factor xa compared to heparin. *Thromb. Res.* **2014**, *134*, 1123–1129. [CrossRef] [PubMed]

Article

Biosourced Poly(lactic acid)/polyamide-11 Blends: Effect of an Elastomer on the Morphology and Mechanical Properties

Ali Fazli and Denis Rodrigue *

Department of Chemical Engineering, Laval University, Quebec, QC G1V 0A6, Canada
* Correspondence: denis.rodrigue@gch.ulaval.ca; Tel.: +1-418-656-2903

Abstract: Fully biobased polylactide (PLA)/polyamide-11 (PA11) blends were prepared by melt mixing with an elastomer intermediate phase to address the low elasticity and brittleness of PLA blends. The incorporation of a biobased elastomer made of poly(butylene adipate-co-terephthalate) (PBAT) and polyethylene oxide (PEO) copolymers was found to change the rigid interface between PLA and PA11 into a much more elastic/deformable one as well as promote interfacial compatibility. The interfacial tension of the polymer pairs and spreading coefficients revealed a high tendency of PEO to spread at the PLA/PA11 interface, resulting in a complete wetting regime (interfacial tension of 0.56 mN/m). A fully percolated rubbery phase (PEO) layer at the PLA/PA11 interface with enhanced interfacial interactions and PLA chain mobility contributed to a better distribution of the stress around the dispersed phase, leading to shear yielding of the matrix. The results also show that both the morphological modification and improved compatibility upon PEO addition (up to 20 wt %) contributed to the improved elongation at break (up to 104%) and impact strength (up to 292%) of the ternary PLA/PA11/PEO blends to obtain a super-tough multiphase system.

Keywords: biobased blends; poly(lactic acid); polyamide-11; morphology; interfacial tension; impact strength

1. Introduction

Poly(lactic acid) (PLA), as an aliphatic biobased polyester made from fermented plant starch, has attracted significant attention in the industry and academia as an alternative material to petroleum-based plastics to develop sustainable polymeric materials [1,2]. PLA is one of the most produced and consumed biobased, biodegradable and biocompatible plastics because of its high tensile strength (59 MPa) and tensile modulus (1280 MPa) combined with good biocompatibility, transparency, compostability and processability [3,4]. Despite these performances and the wide availability of different grades, PLA has low heat distortion temperature (HDT) (55 °C), inherent brittleness (less than 10% elongation at break) and poor impact strength (Notch Izod impact strength of 26 J/m), limiting its practical applications in biomedical, packaging and automotive industries [3,5]. To overcome these limitations, several studies reported on different options to improve the heat resistance and impact strength of PLA. The most common methods are based on copolymerization [6], plasticization [7] and physical blending [8]. For the latter, PLA has been melt blended with other bioplastics, such as polyhydroxyalkanoate (PHA) [9], poly(3-hydroxybutyrate-co-3-hydroxyvalerate) (PHBV) [10], poly(butylene adipate-co-terephthalate) (PBAT) [11] and poly(butylene succinate) (PBS) [12], which are seen as practical and economical approaches to overcome the PLA shortcomings.

On the other hand, polyamide-11 (PA11), as a bioplastic produced from castor oil, has good HDT, impact strength and elongation-at-break, chemical resistance, and excellent dimensional stability, exhibiting a similar glass transition temperature ($T_g \approx 45$ °C) and melting point ($T_m \approx 190$ °C) to PLA [13,14]. However, the superior mechanical performance of PA11 is attributed to intermolecular hydrogen bonding in the crystalline and amorphous state [7].

Few studies have been carried out on PLA/PA11 blends as fully biobased materials for applications such as medical implants and devices, food packaging, agricultural films, etc. Feng and Ye [15] observed a partial miscibility between PLA and PA11 due to hydrogen bonding between the amino (NH) groups of PA11 and the carbonyl (CO) groups of PLA. Despite interfacial interactions between PLA and PA11, Heshmati and Favis [7] reported very low elongation at break (6%) and impact strength (11 J/m) of a binary PLA/PA11 (50/50) blend in line with similar studies highlighting the necessity to perform compatibilization for PLA-based blends [16,17].

In general, compatibilization studies focused on the development of binary polymer systems (PLA = matrix), while the presence of an intermediate phase in ternary blends is more effective to provide good interfacial interactions between the phases to generate super-tough PLA-based systems [18]. For example, Mehrabi et al. [19] developed highly tough PLA-based ternary systems by the inclusion of a core–shell impact modifier based on polybutadiene-g-poly(styrene-co-acrylonitrile) (PB-g-SAN) and poly(methyl methacrylate) (PMMA). The improved dispersion state of PB-g-SAN terpolymer and its interfacial interactions with PLA as a result of the partial miscibility of PMMA with PLA and terpolymer (PMMA miscibility with SAN) was shown to increase the impact strength and elongation at break of PLA/PB-g-SAN/PMMA (45/30/25) from 25.6 J/m and 6% to 500 J/m and 80%, respectively, upon the addition of 25% PMMA. Following this concept, Zhang et al. [20] optimized the blend composition of multiphase systems based on PLA/ethylenemethyl acrylate-glycidyl methacrylate (EMA-GMA)/poly(ether-b-amide) (PEBA) (70/20/10) where PEBA encapsulated the EMA-GMA, exhibiting a core–shell structure leading to an impact strength of 410 J/m with an elongation at break of 72.7%. However, the majority of studies on ternary immiscible polymer blends deal with core–shell morphologies, while multiphase systems with co-continuous and tri-continuous phase morphology also have the potential to modify the mechanical properties, i.e., the impact and tensile properties of PLA blends [21]. Controlling the morphologies and interfacial properties (complex interaction between all the components) can lead to appropriate stress transfer between the components by the creation of intermediate phases to generate superior properties in multicomponent systems [19,22]. Li and Shimizu [23] reported on PLA impact modification by blending with acrylonitrile–butadiene–styrene copolymer (ABS) with the help of 5% styrene/acrylonitrile/glycidyl methacrylate copolymer (SAN-GMA) as a reactive compatibilizer. The epoxide groups of SAN-GMA can react either with –COOH or –OH on the PLA end groups to form an intermediate layer, decreasing the ABS domain size with a narrow size distribution improving the mixing state (uniformity). Thus, PLA/ABS (50/50) compatibilized with 5 wt % SAN-GMA could undergo stretching up to 23.5% before breaking and showed higher impact strength (162.8 kJ/m^2) compared with the neat PLA/ABS (70/30) blend (48.3 kJ/m^2). Generally, based on the equilibrium of interfacial forces in multicomponent immiscible systems and the values of the spreading coefficient, complete wetting or partial wetting may occur, leading to the generation of core–shell, tri-continuous structures (complete wetting) or multiple stacked morphologies where none of the phases completely spread at the interface of the other two (partial wetting) [24]. According to Zolali and Favis [22], the development of a tri-continuous morphology contributed to a noticeable improvement in the impact strength of ternary PLA-based blends. Possible interactions between the polyether blocks of PEBA with PLA and its amide affinity toward PA11 decreased the interfacial tension of ternary PLA/PA11/PEBA blends, and a positive spreading coefficient ($\lambda_{PLA/PEBA/PA}$ = +0.3) suggested the complete assembling of the PEBA elastomer at the PLA/PA11 interface. The substitution of a rigid PLA/PA11 interface with a thick (350 nm) deformable interface of PEBA led to the development of a tri-continuous structure of interconnected phases, increasing the impact strength by about 8 times (from 17.3 to 142.4 J/m) in the ternary PLA/PA11/PEBA (45/45/10) system [22]. In another work, Ravati et al. [25] observed a tri-continuous morphology for a PLA/PBAT/PBS (33/33/33) blend with a positive spreading coefficient ($\lambda_{PLA/PBAT/PBS}$ = 0.3 ± 0.2 mN/m), indicating that the PBAT phase completely wet the interface between PLA and PBS by locating between both phases, leading to high impact strength (271 J/m) and elongation at break (567.9%) due to enhanced shear yielding and plastic deformation of the PLA matrix.

Despite a number of studies addressing the issues of low deformation and the poor impact resistance of PLA blends without sacrificing stiffness and strength, it is still not clear how the phase structure and phase interaction in multicomponent blends contribute to the impact modification of ternary blends. The objective of this article is to report on the effect of adding an intermediate elastomer phase and the blend composition on the morphology development of fully biobased PLA/PA11 blends prepared by melt blending. In particular, polyethylene oxide (PEO) and PBAT are used to modify the blends. For this purpose, a theoretical prediction of interfacial adhesion between the components of ternary systems and spreading coefficients are evaluated to assess their potential to locate at the interface and generate a partial or complete wetting morphology, forming a core–shell or tri-continuous structure. Based on the blend structures produced, their effect on the tensile and impact properties is reported.

2. Results and Discussion

2.1. Theoretical Prediction of Interfacial Tension

The morphology of multiphase systems thermodynamically depends on the interfacial tension between polymer pairs [26]. Torza and Mason [27] were the first to use the concept of a spreading coefficient to predict the phase morphology and wetting phenomenon of ternary blends through the calculation of spreading coefficients. The spreading coefficient model is a commonly used theoretical model to evaluate the tendency of a component to segregate the other two components and to predict the morphology of immiscible ternary blends [27]. This method is based on the interfacial tensions (γ_{ij}) between each pair of components, which can be calculated via contact angle measurement because of the simplicity and facility of measuring the parameters. The spreading theory first was developed by Harkins in the 1920s to predict the spreading coefficients and wetting regime of ternary polymer systems by the following equation [27,28]:

$$\lambda_{ijk} = \gamma_{ik} - \gamma_{ij} - \gamma_{jk} \tag{1}$$

where λ is the spreading coefficient and γ represents the interfacial tensions between the polymer pairs indicated by sub-indices. Here, the spreading coefficient (λ_{ijk}) predicts the tendency of component (j) to segregate the other phases and be located at the interface between components (i) and (k) [28]. For a ternary blend, three spreading coefficients are required to predict the morphology. A positive value of λ_{ijk} implies a tendency of one phase (j) to separate two other phases (i) and (k) and form a core–shell or tri-continuous structure (k encapsulated by j) corresponding to a complete wetting morphology (two-phase contact only). On the other hand, negative values of spreading coefficients imply a three-phase contact, as none of the phases are located fully between the other two, and only one-phase droplets are located at the interface of the other two polymers, implying partial wetting [25,29]. A schematic representation of both wetting scenarios is presented in Figure 1 to better understand the difference between complete wetting and partial wetting regimes in a ternary blend with the corresponding coefficients [24].

The well-known harmonic mean equation can be used to determine the interfacial tension (γ) between different components (i, j and k) as follows [26]:

$$\gamma_{ij} = \gamma_i + \gamma_j - \frac{4\gamma_i^d \gamma_j^d}{\gamma_i^d + \gamma_j^d} - \frac{4\gamma_i^p \gamma_j^p}{\gamma_i^p + \gamma_j^p} \tag{2}$$

where γ_i and γ_j represent the surface tensions of component (i) and (j), respectively, while γ^p and γ^d are, respectively, the polar and dispersive components of surface tension ($\gamma_l = \gamma_l^d + \gamma_l^p$) calculated from the contact angle (Equations (1) and (2)) [26]. Based on data obtained from the sessile drop method, the surface tension of PLA, PA11, PEO and

PBAT, as well as the interfacial tension values and the spreading coefficients, are reported in Table 1.

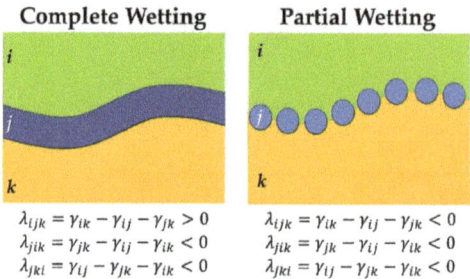

Figure 1. Schematic representation of the blend structure associated with complete and partial wetting morphologies in ternary polymer blends.

Table 1. Surface tension and interfacial tension values at 25 °C calculated from contact angle and spreading coefficients (all values are in mN/m).

Polymer	γ	γ^d	γ^p	Interfacial Tension	Spreading Coefficient (λ)	
PLA	39.7 ± 0.3	26.3 ± 0.2	13.4 ± 0.1	$\gamma_{PLA/PA11}$ = 3.65 ± 0.7	λ_{ijk} = −2.71	λ_{jih} = −3.16
PA11	42.0 ± 0.4	35.4 ± 0.2	6.6 ± 0.2	$\gamma_{PLA/PBAT}$ = 2.87 ± 0.6	λ_{jik} = −4.59	λ_{ijh} = −4.14
PEO	33.7 ± 0.2	25.1 ± 0.1	8.6 ± 0.1	$\gamma_{PA11/PBAT}$ = 2.38 ± 0.7	λ_{ikj} = 0.56	λ_{ihj} = −1.6
PBAT	53.1 ± 0.3	40.1 ± 0.2	13.0 ± 0.1	$\gamma_{PLA/PEO}$ = 1.17 ± 0.5	Complete wetting	Partial wetting
				$\gamma_{PA11/PEO}$ = 2.10 ± 0.6	PLA (i), PA11 (j), PEO (k) and PBAT (h)	

The interfacial tension results for the ternary PLA/PA11/PEO blend show an almost equal affinity of PEO toward PLA to PA11 because of very low and similar interfacial tensions between PEO and the other two polymers as of $\gamma_{PLA/PEO}$ = 1.17 ± 0.5 mN/m and $\gamma_{PA11/PEO}$ = 2.10 ± 0.6 mN/m with respect to $\gamma_{PLA/PA11}$ = 3.65 ± 0.7 mN/m, while PBAT showed slightly higher interfacial values with PLA and PA11 as $\gamma_{PLA/PBAT}$ = 2.87 ± 0.6 mN/m and $\gamma_{PA11/PBAT}$ = 2.38 ± 0.7 mN/m. The incorporation of PEO (plasticizer) is expected to decrease the interfacial tension between PLA and PA11 in line with literature results for which the interfacial tension of PLA/PA11, determined by the in situ Neuman triangle method, decreased from 3.2 to 2.1 ± 0.3 mN/m upon PEO addition (10%) in PLA/PEO/PA11 45/10/45 [13]. The miscibility of PEO with PLA and its increased chain mobility due to the presence of PEO can further enhance the generation of PEO droplets at the interface between PLA and PA11, which underlines the higher thermodynamic driving force of PEO for interfacial wetting than that of PBAT in these ternary systems. In other words, the positive spreading coefficient of $\lambda_{PA11/PEO/PLA}$ (0.56 mN/m) indicates the possibility of a high concentration of intermediate phase assembling at the interface and small sizes of the dispersed phases preventing droplet coalescence, which correlates with a higher tendency of PEO to be located between PLA and PA11, thus stabilizing the morphology. This is in agreement with the complete wetting morphology of PEO predicted by the positive spreading coefficients of PLA/PEO/PA11 leading to a continuous layer of elastomer between PLA and PA11 compared to scattered PBAT droplets as a consequence of all three spreading coefficients in PLA/PBAT/PA11 being negative (see Table 1). From Table 1, a negative spreading coefficient ($\lambda_{PLA/PBAT/PA11}$ = −1.6 mN/m) predicts a partial wetting morphology for PLA/PBAT/PA11 ternary blend in agreement with Fu et al. [24] for the prediction of a partial wetting morphology (all spreading coefficients being negative), as the PBAT phase has similar affinity with both PLA and PA11. In agreement with our observation, Zolali and Favis [30] also reported negative spreading coefficients for

PLA/PBAT/PA11 blends; thus, all the phases met each other at a three-phase line of contact (partial wetting regime) as a result of the high interfacial tension between each polymer pairs.

2.2. Morphological Observation

The morphological behavior of multiphase systems is of great importance since the overall macroscopic properties highly depend on the morphology development and interface quality. SEM micrographs (Figure 2) present the morphological structure of the binary PL70/PA30 and PL50/PA50 blends depending on the PA11 content. As shown in Figure 2A,B, PL70/PA30 shows a two-phase morphology with a nodular structure (sea-island) of PA11 as the minor phase with low affinity (gaps at the interface) dispersed in the PLA matrix. A clear interfacial region between PLA and PA11 droplets indicates poor interfacial bonding and incompatibility for this system, resulting in dispersion problems and a heterogenous structure in agreement with similar reports [7,31]. Increasing the PA11 content from 30 to 50% contributed to an increase in the size of the dispersed PA11 phase changing from a spherical-like domain in PL70/PA30 (Figure 2A,B) to a more elongated structure in PL50/PA50 (Figure 2C,D). Comparing the phase morphology for different blend compositions, large voids around the spherical PA11 nodules (Figure 2B) and the pull-out of weekly embedded PA11 particles from the PLA matrix during fracture leaving empty cavities (voids) indicate the immiscibility of the binary blend and high interfacial tension between the polymers (high surface energy of the polymer pairs; $\gamma_{PLA/PA11} = 3.65 \pm 0.7$ mN/m), which is a confirmation of their poor compatibility and is in agreement with similar observations [7,31,32].

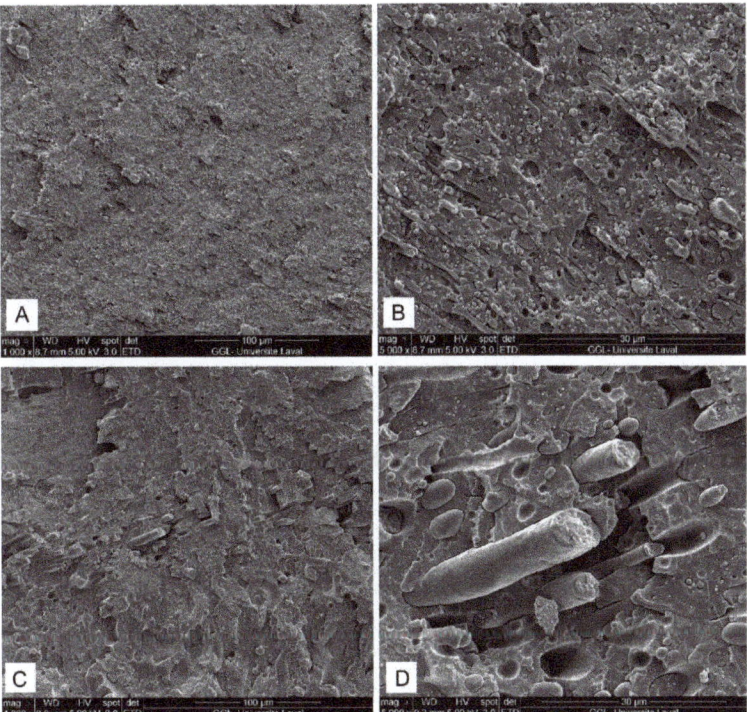

Figure 2. SEM micrographs of PL70/PA30 (**A**,**B**) and PL50/PA50 (**C**,**D**) blends (see Section 3.2 for definition).

The incorporation of premade block copolymer elastomers, such as SAN-GMA, EMA-GMA or PEBA, is commonly used to compatibilize PLA-based blends, to adjust interfacial adhesion and stabilize the morphology of dispersed droplets against coalescence during polymer blending as well as improve the macroscopic properties of the final blends, such as impact strength and elasticity [19,20,22]. Here, SEM images (Figure 3) show that the morphologies match the spreading coefficient predictions for PLA/PBAT/PA11 and PLA/PEO/PA11 ternary blends. Figure 3 clearly shows differences in the morphology of these multiphase blends depending on which elastomer (PEO or PBAT) was added (10 or 20 wt %). It is well known that the breakup and coalescence of dispersed droplets are common phenomena in polymer blends processing, which can be minimized by decreasing the diameter of the dispersed phase and/or lowering the interfacial tension upon the addition of a third component (compatibilizer) if such copolymers can be located at the interface and stabilize the morphology of multiphase systems. Figures 3 and 4 show a more homogenous structure for ternary systems compared to binary ones (Figure 2). This observation supports the thermodynamic predictions, as the positive value of $\lambda_{PA11/PEO/PLA}$ (0.56 mN/m) suggested that PEO should wet the interfacial region between PLA and PA11 (complete wetting), resulting in improved interfacial adhesion and the elimination of voids at the PLA/PA11 interface via morphology stabilization, thus preventing PA11 droplet coalescence. The adhesion between PLA and PA11 was improved by creating a thick/soft interphase of PEO particles around PA11 droplets, promoting a more uniform dispersion of spherical particles of the minor phase (Figure 3C,D). In agreement with this observation, Heshmati et al. [13] observed higher PLA chain mobility and better interfacial interaction with PA11, thus decreasing the dispersed phase size and limiting coalescence upon the addition of PEO (5%) in a PLA/PA11 (70/30) blend. They claimed that a plasticization/miscibility effect of PEO with PLA occurred by increasing the diffusion of PLA chains toward the interface due to higher free volume improving the interfacial interactions in these polymer blends [13].

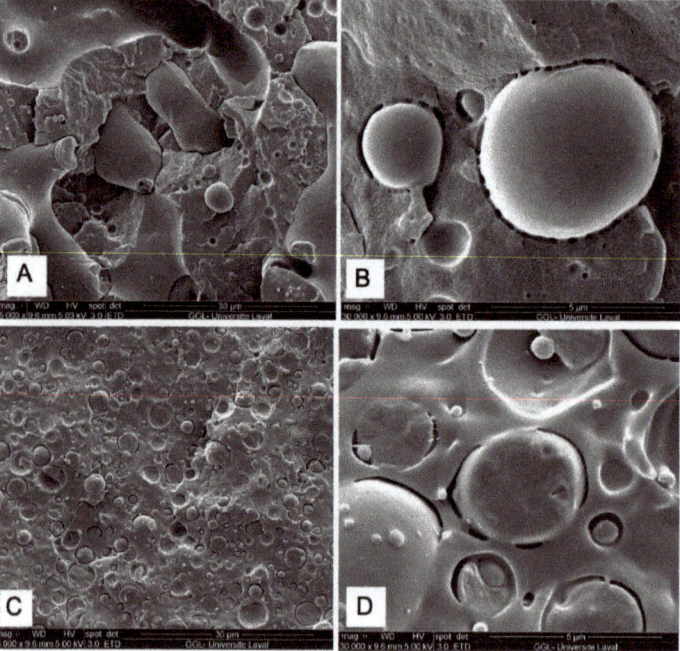

Figure 3. SEM micrographs of PA20/(PBAT20) (**A,B**) and PA20/(PEO20) (**C,D**) ternary blends (see Section 3.2 for definition).

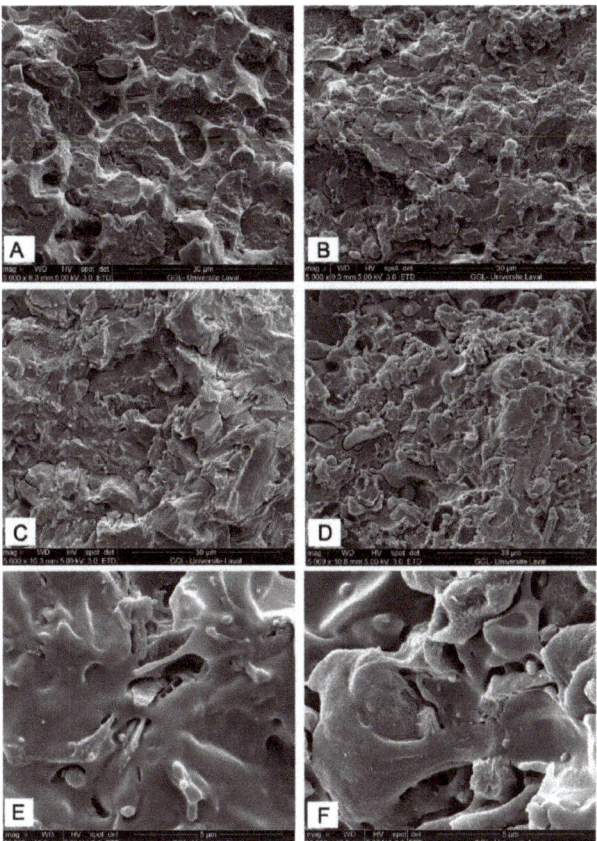

Figure 4. SEM micrographs of PA45/(PBAT10*) (**A**,**C**,**E**) and PA45(PEO10*) (**B**,**D**,**F**) ternary blends (see Section 3.2 for definition).

Figure 4B,D show that increasing the PEO content from 10 to 20% enhanced the interfacial interactions and wetting of PEO with PLA and PA11. Compared to binary PL50/PA50 (Figure 2C,D), it is more difficult to detect the homogeneously distributed PA11 phase, while agglomerated PA11 domains are collapsed, and much more homogeneous dispersion is observed, as the PA11 phase is strongly embedded into the PLA matrix (Figure 4D,F). On the other hand, Figure 3A,B and Figure 4A,C present a clean and featureless fractured surface with a slightly more pronounced pull-out of PBAT and PA11 particles as well as a large number of cavities due to the stress concentration around the agglomerated phase as a consequence of weak interfacial adhesion due to the poor wetting ability of PBAT, especially for the PL50/PA50 system (Figure 4A,C). In a similar work, Zolali and Favis [30] observed the negligible plastic deformation of the PLA phase upon PBAT addition due to its low interfacial interactions with PLA and PA11. They also observed that PBAT poorly wet the PLA/PA11 interface compared to EMA-GMA, which generated a finer distribution of dispersed droplets with smaller sizes due to reactions between the epoxy groups of EMA-GMA with either PLA or PA11 (reactive compatibilization). Similar findings support the higher coalescence rate of PLA/PE/PBAT blends than that of PLA/PE/PHBV by about 5% [33], as PBAT addition was not able to prevent droplet coalescence and surround the dispersed phase/located between the other two polymers due to its weak partial wetting behavior in line with the interfacial energy and spreading coefficient measurements in Table 1. It can be concluded that similar to the effect of solid particles in Pickering emulsions,

PEO as soft copolymers can effectively promote the compatibility between PLA and PA11. As the interface becomes better bonded with less interfacial gaps, the blends can better transfer the applied stresses as described later in terms of mechanical properties [30].

2.3. Mechanical Properties

2.3.1. Tensile Properties

The mechanical properties (tensile strength, elongation at break and tensile modulus) of the neat polymers (Table 2) as well as the binary PLA/PA11 and ternary PLA/PA11/elastomer systems (Figures 5–7) are presented. As expected, Table 2 shows that PLA has a brittle behavior with high rigidity from its high modulus (1590 MPa) and low elongation at break (6.1%), while PA11 has lower rigidity (tensile modulus of 1024 MPa) and higher elongation (194.7%). The low tensile strength of the binary PLA/PA11 systems for both mixing ratios of 70/30 (56.4 MPa) and 50/50 (49.7 MPa) implies incompatibility and the presence of a poor and rigid interface failing to effectively transfer the stresses between the components, which is the weakest point leading to fracture [31]. The very low elongation (6.8%) of PL70/PA30 indicates that the brittle behavior of PLA plays a dominant role in the tensile properties of the blends (PLA is the predominant phase = matrix), while increasing the PA11 content up to 50% slightly increased the elasticity of PL50/PA50 to 7.9%.

Table 2. Tensile properties of the neat polymers.

Sample Code	Tensile Strength (MPa)	Tensile Modulus (MPa)	Elongation at Break (%)
PLA	64.5 (0.4)	1190 (25)	6.1 (2.5)
PA11	48.3 (0.3)	1024 (11)	194.7 (7.3)
PBAT	19.7 (0.5)	105 (7)	486.1 (11.5)
PEO	24.6 (0.3)	74 (5)	516.4 (9.1)

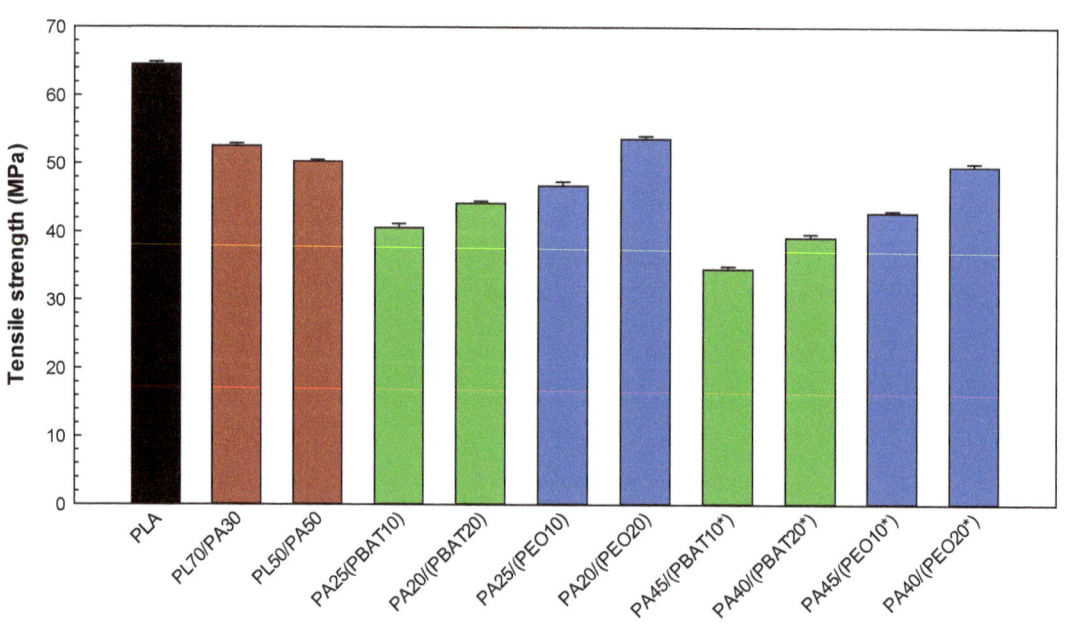

Figure 5. Tensile strength of the blends (see Section 3.2 for definition).

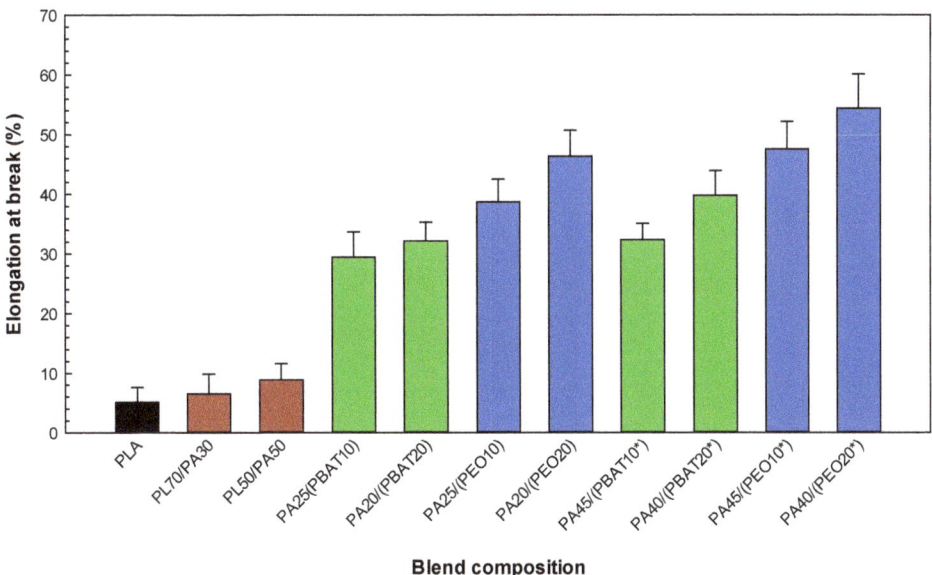

Figure 6. Elongation at break of the blends (see Section 3.2 for definition).

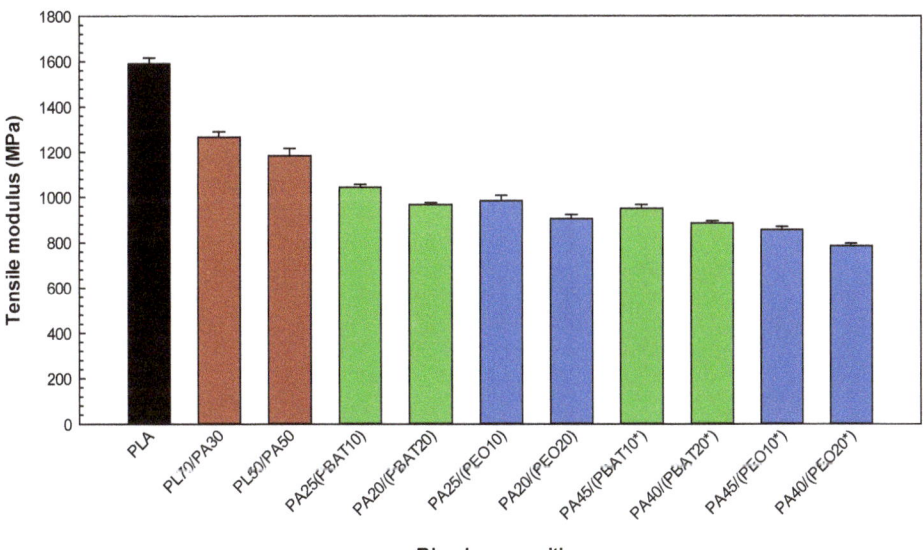

Figure 7. Tensile modulus of the blends (see Section 3.2 for definition).

According to the literature and in agreement with our tensile results for the ternary systems, it is often observed that the localization of an elastomer copolymer at the interface can convert glassy polymer blends into a much more deformable/ductile material having a distinct phase structure and mechanical properties [22,34]. Based on the results of the interfacial analysis (Table 1) and phase morphology (Figures 2–4), ternary blends based on PEO have a finer morphology, leading to superior tensile properties compared to the blends containing PBAT (Figure 5). For the same blend composition, the higher tensile strength of the PA25/(PEO10) blend (49.6 MPa) compared to that of PA25/(PBAT10) with 46.8 MPa may be related to the possibility of dipole–dipole interactions between the ester groups of PLA and the ether groups of PEO as well as hydrogen bonding between the amide groups of PA11 and the ether oxygens of PEO, inducing better compatibility between PLA and PA11 and leading to the higher tensile strength of the corresponding sample [35,36]. According to the spreading coefficient predictions and SEM images, PEO is able to spread at the interface in PL50/PA50, providing a soft and deformable interfacial area or surrounding PA11 droplets in PL70/PA30, forming a core–shell structure (PA11 encapsulated by PEO) in the PLA matrix. This action promotes uniform particles distribution and consequently a better mixing state, leading to improved stress transfer between PLA and PA11 under tension (delayed fracture process). The ternary blends containing PEO yield higher deformation than PBAT due to the better interactions between PLA and PA11 and increased PLA chain mobility, as PEO addition resulted in a significant increase (104%) in elongation at break of PA40/(PEO20*) compared to that of PA40/(PBAT20*) with an elongation at break of 42.8% and PL50/PA50 with 7.9% (Figure 6). It has been reported that increasing the PLA chain mobility through the plasticization effect of PEO may lead to better interfacial adhesion and a more homogenous phase structure generating smooth stress transfer between PLA and PA11 as well as a homogenous structure with less particles coalescence and smaller size in PLA/PA11 blends. Heshmati and Favis [7] also observed substantial elongation at break increases for a PL50/PA50 blend from 6% to 100% after the addition of 10% PEO copolymer, which was attributed to its plasticization effect and the presence of a more elastic content (elongation at break of PEO = 516.4%, see Table 2) inducing higher deformation/elasticity, which was previously reported in similar works [37]. The decreasing trend of tensile moduli can be ascribed to the weaker intermolecular interactions of PLA upon PEO addition and improved rearrangement between polymer chains under external force to induce flexibility to PLA [38]. Figure 7 shows a decrease in the tensile modulus of the ternary blends upon addition of the elastomer phase with low inherent modulus (PEO = 73.6 MPa and PBAT = 105 MPa, see Table 2) and their very low glass transition temperatures (PEO = −67 °C [39] and PBAT = −30 °C [30]), which is typically reported for rubber-toughened plastic blends. For example, increasing the PEO content from 10 to 20 wt % showed a decreasing trend of tensile modulus for PL70/PA30 from 1167 to 1017 MPa and 946 MPa for the ternary systems attributed to the substitution of the rigid components (PLA and PA11) with a soft rubbery component (PEO) of low rigidity. In similar reports, Nofar et al. [40] observed important drops in the tensile modulus of PLA from 1800 to 1200–1250 MPa upon the addition of 25% PBAT or poly(butylene succinate-co-butylene adipate) (PBSA) (minor dispersed phases) due to the very low modulus of these elastomers, which is in line with the findings of Wu and Zhang [41] reporting significant decreases in PLA tensile modulus from 1.8 to 1.1 GPa upon the addition of 10% acrylonitrile–butadiene–styrene (ABS).

2.3.2. Impact Properties

The impact strength as a function of blend composition is presented in Figure 8. The concentration of elastomer (PEO and PBAT) was changed from 0 to 20 wt % in the ternary blends to examine its effect on interfacial interaction and impact strength. As expected, PLA, as a very brittle polymer, shows low impact strength (15.4 J/m). A partial miscibility of PLA with polyamide is reported in the literature due to hydrogen bonding between the ester groups of PLA and amine groups of polyamides. However, such interactions are not able to induce enough compatibility and good deformation to improve the impact strength of neat PLA/PA11 blends [15]. The blending of PLA and PA11, as very rigid polymers, results in the formation of a rigid interface that cannot smoothly transfer interfacial stresses in PLA/PA11 blends, resulting in easy crack initiation and propagation along these interfaces and leading to the low impact strength of the 70/30 (22.5 J/m) and 50/50 (37.4 J/m) blends. As expected, the notched Charpy impact strength follows the same trend as the elongation at break upon elastomer (PEO or PBAT) addition. The presence of an elastomer phase changes the rigid interface into a much more deformable area if the copolymer is localized at the PLA/PA11 interface, thereby increasing failure resistance through effective load transfer, especially when PEO is used [21]. The notched Charpy impact strength of PL70/PA30 and PL50/PA50 blends increases by 2.7 fold (from 22.5 to 84.5 J/m) and about 3 fold (from 37.4 to 147 J/m), respectively, with the addition of PEO (20 wt %). When PBAT is added to the PLA/PA11 blend, the ternary systems show slightly lower impact strength improvement with values of 71.4 J/m and 119.3 J/m upon the addition of PBAT (20 wt %) into PLA/PA11 systems at mixing ratios of 70/30 and 50/50, respectively. The theoretical values of interfacial tension measurements (Table 1) and morphological observations (Figures 2–4) support phase debonding and the low increase in impact strength of the ternary blends modified with PBAT. In similar works, Zolali and Favis [30] reported that the impact strength of PLA/PA11 (50/50) increased from 15 to about 50 J/m upon the addition of 10% PBAT, triggering plastic deformation of the PLA and PA11 matrices, while Kanzawa and Tokumitsu [42] claimed that adding 18% PBAT as an impact modifier only increased the impact strength of PLA/polycarbonate (PC) (60/40) from 1.9 to 2.1 kJ/m^2. Our results underline the relations between the phase structure and impact properties for the blends, which are directly influenced by the introduction of an elastomer. Here, the PA40/(PEO20*) system shows an impact strength (147 J/m) over 10 times higher than that of the virgin PLA (15.4 J/m), thus being considered as a super-tough material [43]. This behavior is in line with similar findings reporting a high level of interfacial interactions for PEO with PLA and PA11 leading to a homogenous phase morphology with fibrillated structures and percolation of the stress field around the cavitated PEO contributing to interfacial shear yielding of the matrix and the increased Izod impact strength of PLA/PA11 (50/50) from 17.3 to 58.3 J/m upon the addition of 20 wt % PEO [22].The presence of PEO as a plasticizer promotes PLA chain mobility and plastic deformation of the PLA matrix, thus increasing the Izod impact strength and elongation at break of binary blends by about 263% (from 11 ± 2 to 40 ± 8 J/m) and 15 fold (from 6 ± 1 to 100 ± 20 %), respectively [7].

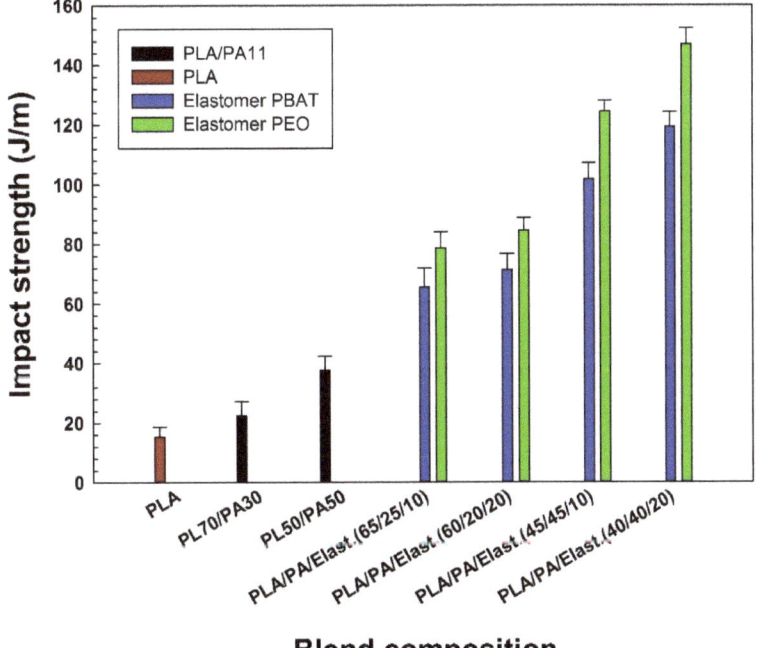

Figure 8. Impact strength of the blends (see Section 3.2 for definition).

More important increases in the impact strength of PLA/PA11/PEO systems can be related to the presence of partially wet PEO droplets spreading at the interface between PLA and PA11 phases modifying the stress state at the interface of the other two phases to locally release the triaxial stresses and contributing to plastic deformation [22]. It is of high importance to create internal rubber cavitation ahead of a crack and involve the matrix in the plastic deformation to delocalize the triaxial stress, leading to more impact energy dissipation, and to obtain rubber toughened plastic blends [18]. The PEO assembled at the interface between PLA and PA11 contributes to internal rubber cavitation; hence, a significant increase in the impact strength of the ternary blends is observed, while less cavitation is expected to occur upon PBAT addition [22,30]. A suitable interfacial adhesion, better wetting of the particles, finer droplets morphology and lower resistance to the cavitation (low moduli of 50 MPa) of PEO localized at the PLA/PA11 interface provides interfacial interactions to improve the shear yielding and postpone phase debonding or interfacial void formation, leading to significant impact strength improvement.

3. Materials and Methods

3.1. Materials

Table 3 summarizes the grades, melting point (T_m), density at 25 °C, melt flow rate (MFI) at 210 °C, and chemical structure of all the materials used in this study. In all cases, the resins were used as received. Residual moisture was removed from the samples prior to the experiments using oven drying at 70 °C for at least 8 h except for PEO, which was dried at 40 °C.

Table 3. Main characteristics of the materials used.

Polymer	Supplier	Grade	T_m (°C)	Density (g/cm^3)	MFI (g/10 min)	Chemical Structure
PLA	Nature Works	2003D	175	1.24	6	
PA11	Arkema	Rilsan BMNO	178	1.03	11	
PBAT	TUHNE	TH801T	116	1.21	4.5	
PEO	Dow Chemicals	Polyox WSR-N10	65	1.13	2.5	

3.2. Processing

Binary PLA/PA11 blends and ternary blends of PLA/PA11/elastomer (PEO or PBAT) were prepared via melt mixing in a co-rotating twin-screw extruder Leistritz ZSE-27 with an L/D ratio of 40 and 10 heating zones coupled to a circular die (2.7 mm in diameter). All the samples were prepared with a screw speed of 100 rpm to give a total flow rate of 3 kg/h under a temperature profile of 175/175/180/180/190/190/200/200/200/200 °C from the feed hopper to the die. The extrudates were quenched in a cold-water bath and pelletized using a model 304 pelletizer (Conair, Stanford, CT, USA) and dried (60 °C for 12 h) to eliminate any residual water before molding. Injection molding was performed using a PN60 (Nissei, Japan) injection molding machine with a temperature profile of 200–200–190–180 °C (nozzle, front, middle and rear). The mold had four cavities to directly produce standard geometries (Type IV of ASTM D638) and impact test bars (dimensions

12.7 × 63.5 × 3.2 mm³) for mechanical characterization. The injection pressure was adjusted (45 to 65 MPa) depending on the compound viscosity, while the mold temperature was fixed at 30 °C. Table 4 summarizes the samples prepared in this study and the ratio of each component for binary and ternary blends.

Table 4. Coding and formulation of the samples produced.

Sample Code	PLA (wt %)	PA11 (wt %)	PBAT (wt %)	PEO (wt %)
PLA	100	-	-	-
PL70/PA30	70	30	-	-
PL50/PA50	50	50	-	-
PA25(PBAT10)	65	25	10	-
PA20/(PBAT20)	60	20	20	-
PA25/(PEO10)	65	25	-	10
PA20/(PEO20)	60	20	-	20
PA45/(PBAT10*)	45	45	10	-
PA40/(PBAT20*)	40	40	20	-
PA45/(PEO10*)	45	45	-	10
PA40/(PEO20*)	40	40	-	20

3.3. Characterization

3.3.1. Contact Angle Measurements

An optical contact angle analyzer (OCA 15 Plus, Future Digital Scientific Corp., Westbury NC, USA) was used at room temperature to measure the contact angle of the materials based on the sessile drop method. Water and ethylene glycol were used as liquids to obtain the average values of five replicates for each sample. The surface tension values were calculated using the following equations [26]:

$$(1 + cos\theta_x)\gamma_x = 4\left(\frac{\gamma_x^d \gamma^d}{\gamma_x^d + \gamma^d} + \frac{\gamma_x^p \gamma^p}{\gamma_x^p + \gamma^p}\right) \quad (3)$$

$$(1 + cos\theta_y)\gamma_y = 4\left(\frac{\gamma_y^d \gamma^d}{\gamma_y^d + \gamma^d} + \frac{\gamma_y^p \gamma^p}{\gamma_y^p + \gamma^p}\right) \quad (4)$$

where γ^d and γ^p are the dispersion and polar components of surface tension, respectively ($\gamma = \gamma^d + \gamma^p$). In addition, θ_x and θ_y are the contact angles of the polymer with water and ethylene glycol, respectively.

3.3.2. Morphological Observation

An Inspect F50 scanning electron microscope (SEM) (FEI, Hillsboro, OR, USA) was used at 15 kV to take micrographs and observe the interfacial adhesion quality between all the phases. The cryogenically fractured specimens in liquid nitrogen were coated with gold/palladium to be observed at different magnifications.

3.3.3. Mechanical Testing

An Instron (Norwood, MA, USA) universal mechanical tester model 5565 was used to perform tensile tests according to ASTM D638 using a 5 kN load cell at a rate of 10 mm/min and room temperature. The average values of the tensile strength (σ_Y), Young's modulus (E) and elongation at break (ε_b) were reported for five dog bone specimens (type IV) with 3 mm thickness for each formulation.

The notched Charpy impact strength was measured on a Tinius Olsen (Horsham, PA, USA) model 104 at room temperature according to ASTM D256 using 10 samples (60×12.7 mm^2) for each composition. Before testing, all the samples were automatically V-notched on a Dynisco (Franklin, MA, USA) model ASN 120 m sample notcher 24 h before testing.

4. Conclusions

This work underlined the importance of the presence of an intermediate elastomeric phase to modify the interfacial area and interactions in multiphase systems and improve their properties. In this particular study, the mechanical properties (elasticity and toughness) of brittle PLA-PA11 systems were investigated by the addition of PBAT and PEO.

The results clearly indicated the importance of interfacial interaction and morphology development to produce super-tough PLA-based materials. The morphological results for the ternary PLA/PA11/elastomer blend strongly correlate with the theoretical prediction of interfacial values and spreading coefficients. A positive spreading coefficient of $\lambda_{PA11/PEO/PLA} = 0.56$ mN/m and a lower interfacial tension of polymer pairs in PLA/PA11/PEO compared to PLA/PA11/PBAT suggested a complete wetting of the interface by PEO compared to the partial wetting of PBAT droplets. PEO tended to completely wet the interface of PLA and PA11, thus generating a soft and deformable interfacial area. It was found that PEO addition generated a more homogenous structure with few voids/defects at the PLA/PA11 interface, while the addition of PBAT led to poorly distributed droplets which did not prevent pull-out of the dispersed particles and the presence of a large number of interfacial voids due to the weak interfacial adhesion and poor wetting ability of PBAT.

A brittle-to-ductile transition was achieved for fully biobased PLA/PA11 blend upon the addition of PEO (20 wt %) between PLA and PA11 resulting in a smooth stress transfer at the interface, thus improving shear yielding and more deformation/elasticity as well as leading to improved energy absorption/dissipation before complete parts failure. The addition of PEO (up to 20 wt %) to the PLA/PA11 (70/30) and (50/50) blends increased the elongation at break of up to 45.6% and 104%, while the impact strength was improved by 2.7 fold (from 22.5 to 84.5 J/m) and 3 fold (from 37.4 to 147 J/m), respectively.

Finally, PEO was more effective than PBAT in modifying the interfacial interactions, phase morphology and mechanical properties of the immiscible PLA/PA11 blend because of its partial miscibility with PLA and better affinity toward PA11 as well as increased PLA chain mobility, which further enhanced the interfacial interactions with PA11.

Author Contributions: Conceptualization, A.F. and D.R.; methodology, A.F.; formal analysis, A.F.; resources, D.R.; writing—original draft preparation, A.F.; writing—review and editing, D.R.; supervision, D.R.; project administration, D.R.; funding acquisition, D.R. All authors have read and agreed to the published version of the manuscript.

Funding: This research was funded by the Natural Sciences and Engineering Research Council of Canada (NSERC) project RGPIN-2016-05958.

Institutional Review Board Statement: Not applicable.

Informed Consent Statement: Not applicable.

Data Availability Statement: Not applicable.

Acknowledgments: The authors acknowledge the technical support of the Research Center on Advanced Materials (CERMA).

Conflicts of Interest: The authors declare no conflict of interest.

Sample Availability: Not available.

References

1. Coudane, J.; Van Den Berghe, H.; Mouton, J.; Garric, X.; Nottelet, B. Poly (Lactic Acid)-Based Graft Copolymers: Syntheses Strategies and Improvement of Properties for Biomedical and Environmentally Friendly Applications: A Review. *Molecules* **2022**, *27*, 4135. [CrossRef] [PubMed]
2. Nofar, M.; Sacligil, D.; Carreau, P.J.; Kamal, M.R.; Heuzey, M.-C. Poly (lactic acid) blends: Processing, properties and applications. *Int. J. Biol. Macromol.* **2019**, *125*, 307–360. [CrossRef]
3. Farah, S.; Anderson, D.G.; Langer, R. Physical and mechanical properties of PLA, and their functions in widespread applications—A comprehensive review. *Adv. Drug Del. Rev.* **2016**, *107*, 367–392. [CrossRef]
4. Cusson, E.; Mougeot, J.-C.; Maho, M.; Lacroix, F.; Fazli, A.; Rodrigue, D. Poly (Lactic Acid)(PLA)/Recycled Styrene Butadiene Rubber (rSBR) Composites. *Adv. Environ. Eng. Res.* **2022**, *3*, 1. [CrossRef]
5. Laaziz, S.A.; Raji, M.; Hilali, E.; Essabir, H.; Rodrigue, D.; Bouhfid, R.; Qaiss, A.E.K. Bio-composites based on polylactic acid and argan nut shell: Production and properties. *Int. J. Biol. Macromol.* **2017**, *104*, 30–42. [CrossRef]
6. Zhang, N.; Zhao, M.; Liu, G.; Wang, J.; Chen, Y.; Zhang, Z. Alkylated lignin with graft copolymerization for enhancing toughness of PLA. *J. Mater. Sci.* **2022**, *57*, 8687–8700. [CrossRef]
7. Heshmati, V.; Favis, B.D. High performance poly (lactic acid)/bio-polyamide11 through controlled chain mobility. *Polymer* **2017**, *123*, 184–193. [CrossRef]
8. Tejada-Oliveros, R.; Balart, R.; Ivorra-Martinez, J.; Gomez-Caturla, J.; Montanes, N.; Quiles-Carrillo, L. Improvement of impact strength of polylactide blends with a thermoplastic elastomer compatibilized with biobased maleinized linseed oil for applications in rigid packaging. *Molecules* **2021**, *26*, 240. [CrossRef]
9. Burzic, I.; Pretschuh, C.; Kaineder, D.; Eder, G.; Smilek, J.; Másilko, J.; Kateryna, W. Impact modification of PLA using biobased biodegradable PHA biopolymers. *Eur. Polym. J.* **2019**, *114*, 32–38. [CrossRef]
10. Ma, P.; Spoelstra, A.; Schmit, P.; Lemstra, P. Toughening of poly (lactic acid) by poly (β-hydroxybutyrate-co-β-hydroxyvalerate) with high β-hydroxyvalerate content. *Eur. Polym. J.* **2013**, *49*, 1523–1531. [CrossRef]
11. Fu, Y.; Wu, G.; Bian, X.; Zeng, J.; Weng, Y. Biodegradation behavior of poly (butylene adipate-co-terephthalate)(PBAT), poly (lactic acid)(PLA), and their blend in freshwater with sediment. *Molecules* **2020**, *25*, 3946. [CrossRef]
12. Xue, B.; He, H.-Z.; Huang, Z.-X.; Zhu, Z.; Xue, F.; Liu, S.; Liu, B. Fabrication of super-tough ternary blends by melt compounding of poly (lactic acid) with poly (butylene succinate) and ethylene-methyl acrylate-glycidyl methacrylate. *Compos. B Eng.* **2019**, *172*, 743–749. [CrossRef]
13. Heshmati, V.; Zolali, A.M.; Favis, B.D. Morphology development in poly (lactic acid)/polyamide11 biobased blends: Chain mobility and interfacial interactions. *Polymer* **2017**, *120*, 197–208. [CrossRef]
14. Rashmi, B.J.; Prashantha, K.; Lacrampe, M.-F.; Krawczak, P. Toughening of poly (lactic acid) without sacrificing stiffness and strength by melt-blending with polyamide 11 and selective localization of halloysite nanotubes. In Proceedings of the 31st International Conference of the Polymer Processing Society, Jeju Island, Korea, 7–11 June 2015; AIP Publishing LLC: Huntington, NY, USA, 2016.
15. Feng, F.; Ye, L. Structure and property of polylactide/polyamide blends. *J. Macromol. Sci. Part B Phys.* **2010**, *49*, 1117–1127. [CrossRef]
16. Stoclet, G.; Seguela, R.; Lefebvre, J.-M. Morphology, thermal behavior and mechanical properties of binary blends of compatible biosourced polymers: Polylactide/polyamide11. *Polymer* **2011**, *52*, 1417–1425. [CrossRef]
17. Patel, R.; Ruehle, D.A.; Dorgan, J.R.; Halley, P.; Martin, D. Biorenewable blends of polyamide-11 and polylactide. *Polym. Eng. Sci.* **2014**, *54*, 1523–1532. [CrossRef]
18. Zhao, X.; Hu, H.; Wang, X.; Yu, X.; Zhou, W.; Peng, S. Super tough poly (lactic acid) blends: A comprehensive review. *RSC Adv.* **2020**, *10*, 13316–13368. [CrossRef]
19. Mazidi, M.M.; Edalat, A.; Berahman, R.; Hosseini, F.S. Highly-toughened polylactide-(PLA-) based ternary blends with significantly enhanced glass transition and melt strength: Tailoring the interfacial interactions, phase morphology, and performance. *Macromolecules* **2018**, *51*, 4298–4314. [CrossRef]
20. Zhang, K.; Nagarajan, V.; Misra, M.; Mohanty, A.K. Supertoughened renewable PLA reactive multiphase blends system: Phase morphology and performance. *ACS Appl. Mater. Interfaces* **2014**, *6*, 12436–12448. [CrossRef]
21. Zolali, A.M.; Favis, B.D. Toughening of cocontinuous polylactide/polyethylene blends via an interfacially percolated intermediate phase. *Macromolecules* **2018**, *51*, 3572–3581. [CrossRef]
22. Zolali, A.M.; Heshmati, V.; Favis, B.D. Ultratough co-continuous PLA/PA11 by interfacially percolated poly (ether-b-amide). *Macromolecules* **2017**, *50*, 264–274. [CrossRef]
23. Li, Y.; Shimizu, H. Improvement in toughness of poly (l-lactide)(PLLA) through reactive blending with acrylonitrile–butadiene–styrene copolymer (ABS): Morphology and properties. *Eur. Polym. J.* **2009**, *45*, 738–746. [CrossRef]
24. Fu, Y.; Fodorean, G.; Navard, P.; Peuvrel-Disdier, E. Study of the partial wetting morphology in polylactide/poly [(butylene adipate)-co-terephthalate]/polyamide ternary blends: Case of composite droplets. *Polym. Int.* **2018**, *67*, 1378–1385. [CrossRef]
25. Ravati, S.; Beaulieu, C.; Zolali, A.M.; Favis, B.D. High performance materials based on a self-assembled multiple-percolated ternary blend. *AICHE J.* **2014**, *60*, 3005–3012. [CrossRef]
26. Wu, S. *Polymer Interface and Adhesion*; CRC Press—Routledge: Boca Raton, FL, USA, 1982.
27. Torza, S.; Mason, S. Three-phase interactions in shear and electrical fields. *J. Colloid Interface Sci.* **1970**, *33*, 67–83. [CrossRef]

28. Harkins, W.D.; Feldman, A. Films. The spreading of liquids and the spreading coefficient. *J. Am. Chem. Soc.* **1922**, *44*, 2665–2685. [CrossRef]
29. Kolahchi, A.R.; Ajji, A.; Carreau, P.J. Surface morphology and properties of ternary polymer blends: Effect of the migration of minor components. *J. Phys. Chem. B* **2014**, *118*, 6316–6323. [CrossRef]
30. Zolali, A.; Favis, B.D. Compatibilization and toughening of co-continuous ternary blends via partially wet droplets at the interface. *Polymer* **2017**, *114*, 277–288. [CrossRef]
31. Yu, X.; Wang, X.; Zhang, Z.; Peng, S.; Chen, H.; Zhao, X. High-performance fully bio-based poly (lactic acid)/polyamide11 (PLA/PA11) blends by reactive blending with multi-functionalized epoxy. *Polym. Test.* **2019**, *78*, 105980. [CrossRef]
32. Walha, F.; Lamnawar, K.; Maazouz, A.; Jaziri, M. Rheological, morphological and mechanical studies of sustainably sourced polymer blends based on poly (lactic acid) and polyamide 11. *Polymers* **2016**, *8*, 61. [CrossRef]
33. Zolali, A.M.; Favis, B.D. Partial to complete wetting transitions in immiscible ternary blends with PLA: The influence of interfacial confinement. *Soft Matter* **2017**, *13*, 2844–2856. [CrossRef] [PubMed]
34. Fazli, A.; Rodrigue, D. Thermoplastic Elastomer based on Recycled HDPE/Ground Tire Rubber Interfacially Modified with an Elastomer: Effect of Mixing Sequence and Elastomer Type/Content. *Polym. Plast. Technol. Eng.* **2022**, *61*, 1021–1038. [CrossRef]
35. Oliveira, J.E.; Moraes, E.A.; Marconcini, J.M.; Mattoso, L.H.C.; Glenn, G.M.; Medeiros, E.S. Properties of poly (lactic acid) and poly (ethylene oxide) solvent polymer mixtures and nanofibers made by solution blow spinning. *J. Appl. Polym. Sci.* **2013**, *129*, 3672–3681. [CrossRef]
36. Halldén, Å.; Ohlsson, B.; Wesslén, B. Poly (ethylene-graft-ethylene oxide)(PE-PEO) and poly (ethylene-co-acrylic acid)(PEAA) as compatibilizers in blends of LDPE and polyamide-6. *J. Appl. Polym. Sci.* **2000**, *78*, 2416–2424. [CrossRef]
37. Buddhiranon, S.; Kim, N.; Kyu, T. Morphology development in relation to the ternary phase diagram of biodegradable PDLLA/PCL/PEO blends. *Macromol. Chem. Phys.* **2011**, *212*, 1379–1391. [CrossRef]
38. Khanteesa, R.; Threepopnatkul, P.; Sittattrakul, A. Effect of poly (ethylene oxide) on the properties of poly (lactic acid)-based blends. In *IOP Conference Series: Materials Science and Engineering*; IOP Publishing: Bristol, UK, 2020.
39. Henry, S.; De Vadder, L.; Decorte, M.; Francia, S.; Van Steenkiste, M.; Saevels, J.; Vanhoorne, V.; Vervaet, C. Development of a 3D-printed dosing platform to aid in Zolpidem withdrawal therapy. *Pharmaceutics* **2021**, *13*, 1684. [CrossRef]
40. Nofar, M.; Salehiyan, R.; Ciftci, U.; Jalali, A.; Durmus, A. Ductility improvements of PLA-based binary and ternary blends with controlled morphology using PBAT, PBSA, and nanoclay. *Compos. B Eng.* **2020**, *182*, 107661. [CrossRef]
41. Wu, N.; Zhang, H. Toughening of poly (l-lactide) modified by a small amount of acrylonitrile-butadiene-styrene core-shell copolymer. *J. Appl. Polym. Sci.* **2015**, *132*, 42554. [CrossRef]
42. Kanzawa, T.; Tokumitsu, K. Mechanical properties and morphological changes of poly (lactic acid)/polycarbonate/poly (butylene adipate-co-terephthalate) blend through reactive processing. *J. Appl. Polym. Sci.* **2011**, *121*, 2908–2918. [CrossRef]
43. Wu, F.; Misra, M.; Mohanty, A.K. Super toughened poly (lactic acid)-based ternary blends via enhancing interfacial compatibility. *ACS Omega* **2019**, *4*, 1955–1968. [CrossRef]

Article

Assessment of the Properties of Giant Reed Particleboards Agglomerated with Gypsum Plaster and Starch

Maria Teresa Ferrandez-Garcia, Antonio Ferrandez-Garcia, Teresa Garcia-Ortuño and Manuel Ferrandez-Villena *

Department of Engineering, Universidad Miguel Hernández, 03300 Orihuela, Spain
* Correspondence: m.ferrandez@umh.es; Tel.: +34-966-749-716

Abstract: This paper analyzes the properties of composite particleboards made from a mix of giant reed with gypsum plaster and starch as binders. Experimental boards were manufactured with a 10:2 weight ratio of giant reed/gypsum plaster particles and different amounts of starch. Giant reed particles used were ≤ 0.25 mm. The mix was pressed at a temperature of 110 °C with a pressure of 2.6 MPa for 1, 2, and 3 h. The results showed that the boards manufactured with longer times in the press and with 10 wt.% starch achieved the best physical and mechanical properties, obtaining a modulus of rupture (MOR) of 17.5 N/mm^2, a modulus of elasticity (MOE) of 3196 N/mm^2, and an internal bounding strength (IB) of 0.62 N/mm^2. Thickness swelling (TS) at 24 h of the panels was reduced from 36.16% to 28.37% when 10 wt.% starch was added. These results showed that giant reed–gypsum–starch particleboards can be manufactured with physical and mechanical properties that comply with European standards for use in building construction.

Keywords: *Arundo donax* L.; valorization; composite; panel; plant residues; waste

1. Introduction

The world is facing a shortage of wood as a row material. In the last 30 years, forest area decreased from 32.5% to 30.8% of the total land, representing a net loss of 178 million hectares [1], and it is forecasted that, by 2030, there will be a deficit of 300 million m^3. The use of wood is only expected to increase by 30% in the energy sector within this decade [2]. The scarcity of wood has a negative influence on the performance and the quality of the commercial boards. In this present situation, the construction and furniture industries must find a reliable substitute to traditional wood.

The most common wood board on the market is particleboard, which is obtained by binding wood particles with an adhesive, usually urea–formaldehyde (UF). In the construction sector, several types of particleboards include gypsum plaster as an inorganic glue (gypsum plasterboard—GPB). These panels have the advantages of no formaldehyde emission, combustion blocking off, and less loss of heat exchange [3]. However, they also have low mechanical properties, resistance to withdrawal of screws, hardness, and dimensional stability in response to changes in humidity.

In order to reduce the use of wood and replace it with other lignocellulosic materials in GPB, several investigations have been focused on using agricultural residues: banana fiber and rice straw [4], paper cellulose and abacá fiber [5,6], bagasse [7], jute and cabuya [8], sisal [9], recycled cellulose [10], and bagasse with natural rubber [11]. However, in general, these panels have low mechanical behavior and need to be reinforced with cement [12–16].

Currently, gypsum boards have a high market share since they are commonly used in the finishing of interior walls and ceilings. GPB made from agricultural residues is more environmentally friendly and could replace traditional gypsum boards. It is, however, necessary to improve their quality by making them more resistant and durable.

Giant reed (*Arundo donax* L.) is one of the largest types of grass growing in the Mediterranean region; it is a wild plant that grows annually, reaching average heights of 4 m and a mean thickness of 4 cm. It was used in building construction in many Mediterranean countries; however, it is today in disuse and has been replaced by different types of wood. Currently, reeds grow unchecked and at a great rate. When reeds colonize riverbanks, the riverside vegetation is impoverished, and infrastructures that cross watercourses are blocked, making drainage difficult. When the water level rises more than usual, this leads to flooding that causes environmental, economic, and material damage. These problems force the competent authorities to make significant economic investments in cleaning and clearing this plant waste [17]. Some investigations focused on valorizing giant reed with different adhesives such as cement [18], UF [19], citric acid [20], and starch [21].

Starch is a macroconstituent of many foods, being one of the most important plant products [22]. With a production of approximately 60 million tons, 60% is used in food and 40% is used in nonfood industries [23] as an additive in cement, paper, gypsum, adhesives, bioplastics, composites, etc. [24]. Due to their nature, starches have great potential as substitutes for synthetic polymeric materials for environmental purposes. In particleboards, they are used as a substitute for binders such as UF, phenol–formaldehyde (PF), and other petroleum derivatives.

The aim of the present research was to study the manufacturability of particleboards made from giant reed (*Arundo donax* L.) with gypsum plaster and starch following a method based on the wood industry dry process but with variations (no pretreatments, different parameters at the hot press, and adhesives) so that it can be produced in the particleboard industry and, therefore, counteract the high dependence on wood imports using an easily renewable resource such as giant reed. By controlling this reed, a more sustainable and formaldehyde-free particleboard can be obtained, which would reduce pressure on forest resources and create new job opportunities.

2. Materials and Methods

2.1. Materials

The materials used in the manufacturing of the particleboards were giant reed (*Arundo donax* L.), calcium sulfate hemihydrate (gypsum plaster), and different amounts of potato starch (*Solanum Tuberosum* L.).

Giant reed biomass was provided by the authorities in charge of clearing the banks of the Segura river, in southeast Spain. Reeds were cleaned from impurities and left outside for 12 months for air-drying (Figure 1). They were then cut and shredded in a blade mill. In order to obtain the particles, a vibrating sieve was used, and only those which passed through the 0.25 mm sieve were selected. Reed particles had a moisture content approximately of 8 wt.% and an apparent density of 537 ± 38 kg/m^3. Particles were kept in ambient laboratory conditions with an average temperature of 20 °C and an approximate ambient humidity of 65%.

Commercial gypsum plaster ($CaSO_4 \cdot \frac{1}{2}H_2O$) without any additives with an apparent density of 2330 ± 20 kg/m^3 was mixed with the giant reed. Potato starch from the food industry with a purity of 90% and a bulk density of 320 ± 15 kg/m^3 was added to the gypsum and giant reed. This starch is a mixture of amylose and amylopectin (both polysaccharides), and it typically contains large oval granules and gels at a temperature of 58–65 °C.

Figure 1. Giant reed left outside for 12 months for air drying.

2.2. Methods

2.2.1. Board Manufacture

The method followed was based on the wood industry dry process but with variations (no pretreatments and different parameters at the hot press).

The experimental boards were manufactured with a 100:20 proportion by weight of giant reed/gypsum, adding different amounts of starch (0, 5, and 10 wt.%). Figure 2 shows the raw materials of a panel type 20:10.

Figure 2. Giant reed (left), gypsum (center), and starch (right) particles used for manufacturing the experimental board type 20:10.

First, particles of giant reed, starch, and gypsum were dry-mixed. Then, 10% water was sprayed in relation to the weight of the giant reed particles in a laboratory glue blender mixer with rotating blades and a volumetric capacity of 100 dm^3 (model LGB100, IMAL S.R.L, Modena, Italy).

Afterward, the mat was formed manually in a mold with dimensions of 400 × 600 mm (Figure 3). Then, it was hot-pressed with the following processing parameters: 110 °C pressing temperature, 2.6 MPa pressing pressure, and three different pressing times (1, 2, and 3 h).

Figure 3. Mat formed manually to manufacture the particleboards.

Particleboards were single-layered with an approximately thickness of 7 mm. Characteristics of the panels are shown in Table 1.

Table 1. Characteristics of the experimental panels.

Panel Type	Ratio Giant Reed/Gypsum Plaster/H$_2$O in Weight	Ratio Giant Reed/Starch in Weight	Particle Size (mm)	Time (h)	Temperature (°C)
20:0		100:0			
20:5	100:20:10	100:5	<0.25	1, 2, and 3	110
20:10		100:10			

For each panel type and time, four replicates were made. Therefore, 36 particleboards were manufactured in total.

2.2.2. Experimental Tests

The method followed was experimental. The tests were conducted in the Materials Laboratory of the Higher Technical College of Orihuela at the Miguel Hernández University of Elche.

In order to characterize the mechanical, physical, and thermal properties of each of the boards being studied, samples (Figure 4) were cut to the appropriate dimensions to carry out the tests [25] and to determine their values according to the European standards established for wood particleboards [26]. Tests performed were as follows: density [27], thickness swelling (TS) and water absorption (WA) after 2 and 24 h immersed in water [28], internal bonding strength (IB) [29], modulus of elasticity (MOE) and modulus of rupture (MOR) [30], and thermal conductivity [31].

Figure 4. Samples of the experimental particleboards.

The moisture content of the material was measured with a laboratory moisture meter (model UM2000, Imal S.R.L, Modena, Italy), the water immersion test was carried out in a heated tank, and the mechanical tests and density were performed with a universal testing machine (model IB700, Imal, S.R.L., Modena, Italy).

For the statistical analysis, SPSS v. 28.0 software (IBM, Chicago, IL, USA) was used. Analysis of variance (ANOVA) was performed, and, for the mean values of the tests, the standard deviation was obtained.

3. Results and Discussion

3.1. Physical Properties

Table 2 shows the mean results and the standard deviation of thickness, density, TS, and WA after 2 and 24 h, as well as the thermal conductivity of each type of particleboard manufactured.

Table 2. Physical properties of the experimental panels.

Panel Type	Time (h)	Thickness (mm)	Density (kg/m^3)	TS2h (%)	TS24h (%)	WA2h (%)	WA24h (%)	T. Cond. (W/mK)
20:0	1	7.70 (0.57)	925 (18)	29.16 (0.63)	47.60 (0.73)	75.75 (0.19)	84.19 (4.88)	0.072 (0.003)
	2	7.72 (0.34)	984 (28)	30.64 (1.91)	44.05 (3.87)	66.79 (2.84)	76.71 (0.12)	0.068 (0.005)
	3	7.11 (0.16)	994 (18)	31.32 (1.45)	39.66 (2.83)	63.66 (8.53)	75.11 (2.41)	0.064 (0.003)
20:5	1	7.36 (0.11)	929 (12)	37.06 (4.81)	44.47 (0.89)	74.14 (2.88)	85.53 (1.01)	0.063 (0.002)
	2	7.18 (0.14)	974. (59)	29.40 (3.12)	40.99 (1.45)	57.50 (7.58)	71.21 (4.23)	0.061 (0.002)
	3	6.89 (0.35)	1052 (11)	29.34 (0.00)	35.72 (3.93)	66.55 (0.35)	71.07 (5.28)	0.061 (0.001)
20:10	1	7.50 (0.55)	1024 (42)	25.72 (5.91)	36.16 (6.84)	48.96 (3.77)	58.85 (6.83)	0.066 (0.003)
	2	7.27 (0.41)	1053 (60)	26.34 (8.17)	32.19 (8.04)	43.84 (9.07)	57.67 (9.31)	0.063 (0.002)
	3	6.89 (0.14)	1095 (46)	23.19 (6.17)	28.37 (6.09)	41.47 (8.93)	52.25 (7.23)	0.061 (0.002)

Values in parentheses are the standard deviation.

Mean density of the boards ranged between 925 and 1095 kg/m^3; therefore, they can be considered as high-density particleboards. The panels with the highest density were the 20:10 type, which included 10% starch in their manufacture. Longer times in the hot press resulted in more density.

Meanwhile, the 20:10 type had lower values of TS and WA than the remaining experimental panels. Increasing time in the hot press also improved the TS and WA results. This was probably related to the increase in density, since there were fewer gaps between the particles that could be filled with water.

Thermal conductivity seemingly decreased with the addition of starch and longer pressing times, but this could not be concluded. Panels manufactured with more than 5% starch for 2 h had the lowest conductivity values. Given the standard deviation of the results, it is possible that pressing the panels for more than 2 h would not have improved the results when starch was added.

3.2. Mechanical Properties

MOR values of the experimental panels can be seen in Figure 5. Increasing time in the hot press improved the results in types 20:0 and 20:5. In type 20:10, the standard deviation overlapped between 2 and 3 h of pressing time; therefore, this statement cannot be confirmed. The addition of starch resulted in better MOR results in all types, reaching a maximum in type 20:10 3 h with a mean value of 17.5 N/mm^2.

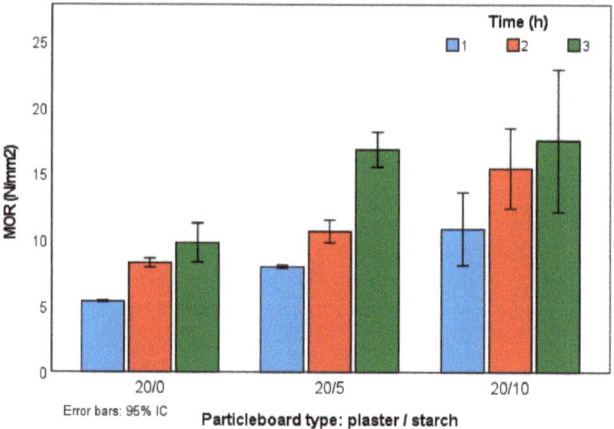

Figure 5. MOR values of the experimental panels.

The MOE results showed in Figure 6 followed a similar trend to MOR values but more pronounced. Increasing the pressing time and the starch proportion resulted in higher MOE values. Type 20:10 3 h reached 3196 N/mm^2 of MOE.

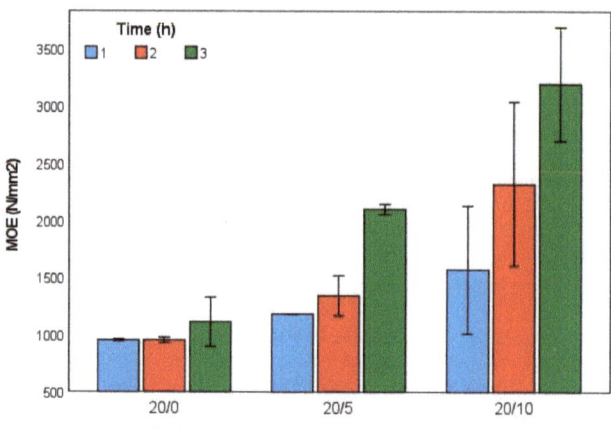

Figure 6. MOE values of the experimental panels.

From the results of IB in Figure 7, it can be concluded that increasing the proportion of starch resulted in better mechanical behavior in the three categories: MOR, MOE, and IB. Type 20:10 3 h had the highest IB value with 0.65 N/mm^2. The standard deviation in types 20:0 and 20:5 of the results prevent concluding that increasing the pressing time between 2 and 3 h improved the results.

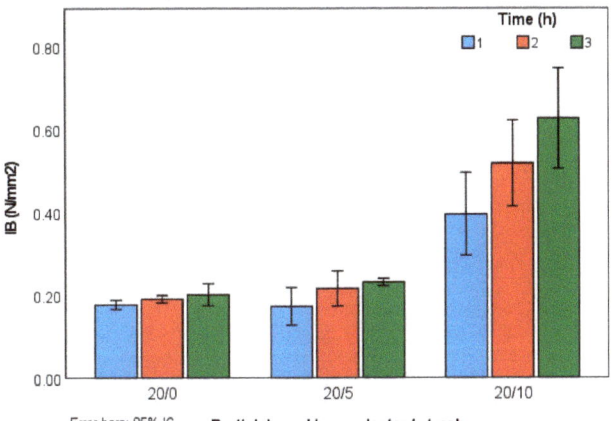

Figure 7. IB values of the experimental panels.

3.3. Statistical Analysis

Statistical analysis of Table 3 indicates that all the properties had dependency relationships (p-value < 0.05) with the panel type and the pressing time except for IB, which was only influenced by the panel type.

Table 3. ANOVA of the results of the tests.

Factor	Properties	Sum of Squares	d.f.	Half Quadratic	F	p-Value
Panel Type	Density (kg/m^3)	84,534.65	2	33,660.139	12.248	0.000
	TS 24 h (%)	1551.53	2	524.107	17.039	0.000
	WA 24 h (%)	1213.56	2	2254.850	44.949	0.000
	T. Cond. (W/mK)	0.001	2	0.001	11.57	0.000
	MOR (N/mm^2)	15.08	2	175.442	14.294	0.000
	MOE (N/mm^2)	754,422.53	2	6,978,605.131	23.840	0.000
	IB (N/mm^2)	0.03	2	0.491	82.366	0.000
Pressing Time	Density (kg/m^3)	115,661.14	2	29,069.446	9.781	0.000
	TS 24 h (%)	7.93	2	249.800	5.647	0.007
	WA 24 h (%)	38.76	2	417.170	2.984	0.042
	T. Cond. (W/mK)	0.001	2	0.001	5.580	0.007
	MOR (N/mm^2)	13.36	2	166.638	13.118	0.000
	MOE (N/mm^2)	130,082.85	2	3,131,162.189	6.518	0.000
	IB (N/mm^2)	0.083	2	0.042	1.490	0.237

d.f.: degrees of freedom. F: Fisher–Snedecor distribution.

3.4. Discussion of the Results

The European standards [32] establish a minimum requirement for particleboards according to their thickness. Panels from 6 to 13 mm with an MOR value of 10.5 N/mm^2 and an IB value of 0.28 N/mm^2 can be considered as particleboards for general use in a dry environment (grade P1). Particleboards that can be used in the manufacture of furniture (grade P2) have to reach an MOR of 11.0 N/mm^2, an MOE of 1800 N/mm^2, and an IB of 0.40 N/mm^2. For nonstructural boards in a humid environment, the minimum requirements are 15.0 N/mm² for MOR, 2050 N/mm² for MOE, 0.45 N/mm² for IB, and 17% for TS at 24 h.

In the present study, boards manufactured with less than 10% starch did not reach the minimum mechanical values to be commercialized. Table 4 shows a comparison of type 20:10 results regarding European standards. Increasing the pressing time improved the performance of the particleboards. Type 20:10 with 1 h of pressing time could be classified as P1, whereas board types 20:10 with 2 and 3 h of pressing time reached the P2 grade. None of them could be classified as P3 since they did not meet the required TS value at 24 h. Further research is needed to take into account the density of the panels, the peak time, the amount of added starch, and the addition of hydrophobic substances in the manufacturing process.

Table 4. Comparison of the properties of the experimental panels with 10% starch and the European standards [32].

Panel Type	MOR (N/mm²)	MOE (N/mm²)	IB (N/mm²)	WA 24 h (%)
20:10—1 h	10.82	1573	0.39	58.85
20:10—2 h	15.40	2324	0.52	57.67
20:10—3 h	17.49	3196	0.62	52.25
Grade P1 [32]	10.50	-	0.28	-
Grade P2 [32]	11.00	1800	0.40	-
Grade P3 [32]	15.00	2050	0.45	17.00

The selection of a particle size of less than 0.25 mm for the manufacturing of the boards was based on our previous experience and the existing literature, since smaller particle sizes resulted in better properties [33].

The process followed in the research was slightly different since GPBs are not usually pressed or heated. Particleboards that follow a dry process usually require a pretreatment that consists of drying the material until it reaches a maximum of 4% moisture content. In the present study, there was no need for pretreatments, and the material could be used with a 8% moisture content.

Kim et al. [34] manufactured GPBs with *Eucalyptus* sp. and *Pinus massoniata* with citric acid (CA) with a targeted density of 1200 kg/m³ and dimensions of 300 × 300 × 10 mm. They reduced the moisture content of the GPB mats to about 2–3% in a dryer at 45 °C before pressing at room temperature the particleboards. Kim [35] used the same process for manufacturing particleboards with rice husk particles of 2 mm in different ratios (0%, 10%, 20%, 30%, and 40%). MOR and MOE values of the particleboards improved with the addition of rice husk with a peak content of 30% and then declined. IB reached its maximum with 20% and had a lower value with 40%. These two studies obtained similar results and are in line with commercial GPB, as shown in Table 5. The experimental boards of the present study achieved better mechanical results with the difference of applying heat in the pressing (110 °C) and only adding 20% gypsum.

Table 5. Comparison with other particleboards made from gypsum and other lignocellulosic materials.

Material	Ratio Material/ Gypsum/H₂O in Weight	Other Additions	Press. (MPa)	Time (h)	MOR (N/mm²)	MOE (N/mm²)	IB (N/mm²)	WA 24 h (%)
This study (type 20:10—2 h)	100:20:10	10% Starch	2.6	2	10.82	1573	0.39	58.85
Eucalyptus sp. [34]	30:100:40	0.05% CA	3	2	6.80	2700	0.24	29.50
Pinus massoniata [34]	30:100:40	0.05% CA	3	2	5.50	2100	0.43	28.00
Rice husk [35]	40:100:40	0.05% CA	3	2	6.70-	3800	0.34	17.00
Commercial GPB [34]	-	-	-	-	9.20	4500	0.30	28.00

This lower gypsum content is not enough for the binding of the particles. If the method used was cold pressing, the particles would have disaggregated. Therefore, since type 20:0 resulted in stable particleboards, some self-binding could have happened in the hot press within the giant reed in the absence of starch. The elucidation of this process is very complex with multiple theories such as the glass transition of cellulose, hemicellulose, and lignin [36], the starches, gluten, sorbitol, and sugars contained in the lignocellulosic material [37], and the presence of furfural [38,39].

In comparison with binderless particleboards, some authors indicated that a high temperature of at least 180 °C and high pressures of 3.5 MPa are needed in order to manufacture the boards [37]. Others concluded that, by extending the time in the hot press, better results can be achieved [40]. In the present investigation, the temperatures of the press (110 °C) and the pressure (2.1 MPa) were lower; therefore, this method consumes less energy and is more environment friendly. Good MOR and MOE values were obtained with a pressing time of 2 and 3 h. It may be possible for the mechanical behavior to be improved with longer pressing times; however, the energy required in the manufacture of the particleboards would increase, which is counterproductive, and the panels already reached the P2 grade values within 2 h.

4. Conclusions

It is feasible to manufacture particleboards made from giant reed, gypsum plaster, and starch that have good mechanical properties according to the European standards.

Through the addition of starch and increasing the pressing time of the experimental panels, better mechanical and physical properties were obtained. With 10% starch and a pressing time of 2 or 3 h, the panels reached the P2 classification for furniture production.

Using giant reed waste in manufacturing particleboards would be beneficial in terms of the environment and energy efficiency since less energy is required than the conventional process.

Author Contributions: Conceptualization and methodology, M.F.-V. and T.G.-O.; experiments, T.G.-O. and M.T.F.-G.; resources, M.F.-V.; statistics, T.G.-O.; project administration, A.F.-G.; supervision, M.T.F.-G.; writing, A.F.-G. and M.F.-V.; review, M.T.F.-G. All authors have read and agreed to the published version of the manuscript.

Funding: This research was funded through Agreement No. 4/20 between the company Aitana, Actividades de Construcciones y Servicios, S.L. and Universidad Miguel Hernandez, Elche.

Institutional Review Board Statement: Not applicable.

Informed Consent Statement: Not applicable.

Data Availability Statement: The data presented in this study are available within the article.

Acknowledgments: The authors would like to thank the company Aitana, Actividades de Construcciones y Servicios, S.L. for its support by signing Agreement No. 4/20 with Universidad Miguel Hernández, Elche on 20 December 2019.

Conflicts of Interest: The authors declare no conflict of interest.

References

1. FAO; UNEP. *The State of the World's Forests 2020. Forests, Biodiversity and People*; FAO & UNEP: Rome, Italy, 2020; ISBN 978-92-5-132419-6. [CrossRef]
2. Mantau, U.; Saal, U.; Prins, K.; Steierer, F.; Lindner, M.; Verkerk, H.; Eggers, J.; Leek, N.; Oldenburguer, J.; Asıkaınen, A.; et al. *Real Potential for Changes in Growth and Use of EU Forests*; EUwood: Hamburg, Germany, 2010.
3. Feng, Q.; Deng, Y.; Kim, H.; Lei, W.; Sun, Z.; Jia, Y.; Lin, X.; Kim, S. Observation and analysis of gypsum particleboard using SEM. *J. Wuhan Univ. Technol.-Mater. Sci. Ed.* **2007**, *22*, 44–47. [CrossRef]
4. Alfonzo Salinas, J.M.; Alarcón Ramírez, M.M. Paneles de Yeso con Fibra de Banano y Cáscara de Arroz para Cielo Raso de Edificaciones. Bachelor's Thesis, ULVR, Guayaquil, Ecuador, 2020.
5. Sinchire Cartuche, D.C. Ecomateriales: Biocompuesto de Aglomerantes de Cemento, Yeso con Partículas de Celulosa de Papel y Fibra de Abacá. Bachelor's Thesis, UTPL, Loja, Ecuador, 2017.

6. Iñiguez Rojas, C.M. Ecomateriales Material Compuesto de Matriz de Aglomerantes, Celulosa de Cartón, y Refuerzo de Fibra Vegetal de Abacá. Bachelor's Thesis, Universidad Técnica Particular de Loja, Loja, Ecuador, 2019.
7. Medina Alvarado, R.; Burneo Valdivieso, X.; Hernández-Olivares, F.; Zúñiga Suárez, A. Reuse of organic waste type in the development of ecoefficient and sustainable composites. In *Congreso Internacional de Construcción Sostenible y Soluciones Ecoeficientes*; Universidad de Sevilla, Departamento de Construcciones Arquitectónicas I.: Sevilla, Spain, 2015.
8. Oteiza San José, I. *Estudio del Comportamiento de la Escayola Reforzada con Fibras de Sisal, para Componentes en Viviendas de bajo Coste*; CSIC-Instituto de Ciencias de la Construcción Eduardo Torroja (IETCC): Madrid, Spain, 1993.
9. Guzmán Castillo, W. Comportamiento Mecánico de la Matriz cal-yeso con dos Fibras Vegetales, yute (*Corchorus capsularis Linn-Corchorus olitorius Linn*) y Cabuya (*Furcroya andina trelease*) (No. N10 G8-T). Bachelor's Thesis, Facultad de Ingeniería Agrícola, Universidad Nacional Agraria La Molina, La Molina, Lima, Peru, 1992.
10. Muñoz Muñoz, D.R.; Narváez Pupiales, J.I. Construcción Sostenible a Partir de Paneles Prefabricados Utilizando yeso y Celulosa Reciclada. Bachelor's Thesis, UCE, Quito, Ecuador, 2019.
11. Thongnuanchan, B.; Suwanpetch, S.; Nakason, C. Utilization of Raw Gypsum as Hydrated Filler in Bagasse Particleboard Bonded with a Formaldehyde-free Epoxidized Natural Rubber Adhesive. In *Advanced Materials Research*; Trans Tech Publications Ltd.: Wollerau, Switzerland, 2013; Volume 626, pp. 44–49.
12. Espinoza-Herrera, R.; Cloutier, A. Physical and mechanical properties of gypsum particleboard reinforced with Portland cement. *Eur. J. Wood Wood Prod.* **2011**, *69*, 247–254. [CrossRef]
13. Tittelein, P.; Cloutier, A.; Bissonnette, B. Design of a low-density wood–cement particleboard for interior wall finish. *Cem. Concr. Compos.* **2012**, *34*, 218–222. [CrossRef]
14. Ahmad, Z.; Lum, W.C.; Lee, S.H.; Rameli, R. Preliminary study on properties evaluation of cement added gypsum board reinforced with kenaf (*Hibiscus cannabinus*) bast fibres. *J. Indian Acad. Wood Sci.* **2017**, *14*, 46–48. [CrossRef]
15. Rangavar, H.; Khosro, S.K.; Payan, M.H.; Soltani, A. Study on the possibility of using vine stalk waste (*Vitis Vinifera*) for producing gypsum particleboards. *Mech. Compos. Mater.* **2014**, *50*, 501–508. [CrossRef]
16. Arruda, L.M.; Del Menezzi, C.H.; Teixeira, D.E.; De Araújo, P.C. Lignocellulosic composites from Brazilian giant bamboo (*Guadua magna*) Part 1: Properties of resin bonded particleboards. *Maderas. Cienc. Tecnol.* **2011**, *13*, 49–58. [CrossRef]
17. Deltoro Torró, V.; Jiménez Ruiz, J.; Vilán Fragueiro, X.M. *Bases para el Manejo y Control de Arundo donax L. (Caña común). Colección Manuales Técnicos de Biodiversidad*; Conselleria d'Infraestructures, Territori i Medi Ambient, Generalitat Valenciana: Valencia, Spain, 2012; ISBN 978-84-482-5777-4.
18. Ferrandez-García, A.A.; Ortuño, T.G.; Ferrandez-Villena, M.; Ferrandez-Garcia, A.; Ferrandez-García, M.T. Evaluation of Particleboards Made from Giant Reed (*Arundo donax* L.) Bonded with Cement and Potato Starch. *Polymers* **2021**, *14*, 111. [CrossRef] [PubMed]
19. Ferrandez-García, M.T.; Ferrandez-Garcia, A.; Garcia-Ortuño, T.; Ferrandez-Garcia, C.E.; Ferrandez-Villena, M. Assessment of the physical, mechanical and acoustic properties of *Arundo donax* L. biomass in low pressure and temperature particleboards. *Polymers* **2020**, *12*, 1361. [CrossRef]
20. Ferrandez-Garcia, M.T.; Ferrandez-Garcia, C.E.; Garcia-Ortuño, T.; Ferrandez-Garcia, A.; Ferrandez-Villena, M. Experimental evaluation of a new giant reed (*Arundo Donax* L.) composite using citric acid as a natural binder. *Agronomy* **2019**, *9*, 882. [CrossRef]
21. Ferrández-García, C.E.; Andreu-Rodríguez, J.; Ferrández-García, M.T.; Ferrández-Villena, M.; García-Ortuño, T. Panels made from giant reed bonded with non-modified starches. *BioResources* **2012**, *7*, 5904–5916. [CrossRef]
22. Copeland, L.; Blazek, J.; Saman, H.; Tang, M.C. Form and functionality of starch. *Food Hydrocoll.* **2009**, *23*, 1527–1534. [CrossRef]
23. Ferrer García, M.; Marfisi Valladares, S.; Danglad Flores, J.Á.; Cecconello, L.; Rojas de Gáscue, B. Producción de espumas sólidas de celulosa y almidón de yuca. *Saber* **2013**, *25*, 439–444.
24. Whistler, R.L.; BeMiller, J.N.; Paschall, E.F. *Starch: Chemistry and Technology*; Academic Press: Cambridge, MA, USA, 2012. [CrossRef]
25. *EN 326*; Wood-Based Panels, Cutting and Inspection. Part 1: Sampling and Cutting of Test Pieces and Expression of Test. European Committee for Standardization: Brussels, Belgium, 1994.
26. *EN 309*; Particleboards. In Definitions and Classification. European Committee for Standardization: Brussels, Belgium, 2005.
27. *EN 323*; Wood-based panels. In Determination of Density. European Committee for Standardization: Brussels, Belgium, 1993.
28. *EN 317*; Particleboards and Fiberboards. In Determination of Swelling in Thickness after Immersion in Water. European Committee for Standardization: Brussels, Belgium, 1993.
29. *EN 319*; Particleboards and Fiberboards. In Determination of Tensile Strength Perpendicular to the Plane of de Board. European Committee for Standardization: Brussels, Belgium, 1993.
30. *EN 310*; Wood-Based Panels. In Determination of Modulus of Elasticity in Bending and of Bending Strength. European Committee for Standardization: Brussels, Belgium, 1993.
31. *EN 12667*; Thermal performance of building materials and products. In Determination of Thermal Resistance by Means of Guarded Hot Plate and Heat Flow Meter Methods: Products of High and Medium Thermal Resistance. European Committee for Standardization: Brussels, Belgium, 2001.
32. *EN 312*; In Particleboards—Specifications. European Committee for Standardization: Brussels, Belgium, 2010.
33. Hegazy, S.S.; Ahmed, K. Effect of date palm cultivar, particle size, panel density and hot water extraction on particleboards manufactured from date palm fronds. *Agriculture* **2015**, *5*, 267–285. [CrossRef]

34. Kim, S.; Kim, J.A.; An, J.Y.; Kim, H.S.; Kim, H.J.; Deng, Y.; Feng, Q.; Luo, J. Physico-Mechanical Properties and the TVOC Emission Factor of Gypsum Particleboards Manufactured with *Pinus Massoniana* and *Eucalyptus* sp. *Macromol. Mater. Eng.* **2007**, *292*, 1256–1262. [CrossRef]
35. Kim, S. Incombustibility, physico-mechanical properties and TVOC emission behavior of the gypsum–rice husk boards for wall and ceiling materials for construction. *Ind. Crops Prod.* **2009**, *29*, 381–387. [CrossRef]
36. Anglès, M.N.; Reguant, J.; Montané, D.; Ferrando, F.; Farriol, X.; Salvadó, J. Binderless composites from pretreated residual softwood. *J. Appl. Polym. Sci.* **1999**, *73*, 2485–2491. [CrossRef]
37. Pintiaux, T.; Viet, D.; Vandenbossche, V.; Rigal, L.; Rouilly, A. Binderless materials obtained by thermo-compressive processing of lignocellulosic fibers: A comprehensive review. *BioResources* **2015**, *10*, 1915–1963.
38. Peleteiro, S.; Rivas, S.; Alonso, J.L.; Santos, V.; Parajó, J.C. Furfural production using ionic liquids: A review. *Bioresour. Technol.* **2016**, *202*, 181–191. [CrossRef] [PubMed]
39. Suzuki, S.; Shintani, H.; Park, S.Y.; Saito, K.; Laemsak, N.; Okuma, M.; Iiyama, K. Preparation of binderless boards from steam exploded pulps of oil palm (*Elaeis guneensis Jaxq.*) fronds and structural characteristics of lignin and wall polysaccharides in steam exploded pulps to be discussed for self-bindings. *Holzforschung* **1998**, *52*, 417–426. [CrossRef]
40. Boon, J.G.; Hashim, R.; Sulaiman, O.; Hiziroglu, S.; Sugimoto, T.; Sato, M. Influence of processing parameters on some properties of oil palm trunk binderless particleboard. *Eur. J. Wood Prod.* **2013**, *71*, 583–589. [CrossRef]

MDPI
St. Alban-Anlage 66
4052 Basel
Switzerland
Tel. +41 61 683 77 34
Fax +41 61 302 89 18
www.mdpi.com

Molecules Editorial Office
E-mail: molecules@mdpi.com
www.mdpi.com/journal/molecules

www.ingramcontent.com/pod-product-compliance
Lightning Source LLC
LaVergne TN
LVHW070457100526
838202LV00014B/1743